Noise in Spintronics

Noise in Spintronics
From Understanding to Manipulation

Farkhad G. Aliev
Juan Pedro Cascales

Pan Stanford Publishing

Published by

Pan Stanford Publishing Pte. Ltd.
Penthouse Level, Suntec Tower 3
8 Temasek Boulevard
Singapore 038988

Email: editorial@panstanford.com
Web: www.panstanford.com

British Library Cataloguing-in-Publication Data
A catalogue record for this book is available from the British Library.

Noise in Spintronics: From Understanding to Manipulation

ISBN 978-981-4774-34-5 (Hardcover)
ISBN 978-1-315-11088-2 (eBook)

Contents

Preface

Spintronics is a rapidly expanding area of solid-state physics which aims to exploit the spin degree of freedom of the electron in addition to its charge—the basics of current electronics. The history of spintronics can be traced back to the 19th century when William Thomson (Lord Kelvin) found that the resistance of ferromagnets depended on the relative orientation between the electron (spin) current and the magnetization, which he controlled by an external field. The anisotropic magnetoresistance (AMR) effect has found application in a wide variety of devices, such as wheel speed control in cars. The basis of AMR is the so-called spin–orbit interaction, which gave way to the rise of spin orbitronics, aiming the control of electron spin direction and propagation through the interface engineering.

The area of spin-orbitronics is only the tip of the iceberg of all current and potential future applications of spin physics in technology. Practically all known textbooks on spintronics and related areas are dedicated to the analysis of time-averaged physical quantities. It could be in part, because it was easier experimentally to measure those averaged quantities and more challenging to observe and analyze their fluctuations. It took almost a decade after the giant magnetoresistance was discovered by Nobel Prize winners Albert Fert and Peter Grunberg in 1988 to demonstrate the importance of the spin current fluctuations as a new information source. Pioneering experiments carried out by Hardner and co-authors in 1993 on magnetic trilayers multilayers and by Ingvarsson and co-authors in 2000 on magnetic tunnel junctions opened a research field related with the investigation of noise in spintronics.

Noise is defined as the random fluctuations of a physical quantity. In electronic devices, this quantity is the voltage and/or the current at any port of the device. In general, the random nature of this inherent noise in electron transport comes from the thermal motion and intrinsic properties of the building blocks of solids. Measuring noise can be an incredibly rich source of information about the microscopic properties of spintronic devices that are often not accessible by other techniques. In this book, we cover the main physical mechanisms and the different contributions ($1/f$ noise, shot noise, etc.) behind electronic fluctuations in spintronic devices of varied nature.

Our book provides a general introduction of noise in spintronic devices. It is aimed not only for experts in spintronics but also for students, with many examples and case studies to illustrate points. The book covers a rich variety of electronic fluctuations throughout the chapters, in different sub-branches of spintronics: epitaxial, molecular spintronics, spintronics with quantum dots, with 2D materials, magnetic field sensors applications etc.

It is expected that the importance of the noise factor in electronics and particularly of spin and charge-related noise in spintronics will grow with time. This is mainly because of the natural shrinking trend in devices' dimensions as we enter the nanotechnology era. We expect, therefore, the readership to be mainly among advanced undergraduate- and graduate-level students in nanotechnology, spintronics, and nanomagnetism. Also, the book could be of interest for graduate students of physics, electrical and electronic engineering, and materials science and engineering; researchers in nanoscience; and engineers in nanotechnology and medicine, especially those with an interest in biosensing.

Acknowledgments

The authors would like to thank many colleagues and friends for their support. Among those who formed the Condensed Matter Physics Department of Universidad Autónoma de Madrid and our MAGNETRANS research group and contributed in one or other way to the opening of the new research line related with noise in spintronics: Raul Villar, Arkadi Levanyuk, Sebastian Vieira, Rodolfo Miranda, Ruben Guerrero, David Herranz, Isidoro Martinez, and Pablo Andres. Special acknowledgments and thanks go to our long-standing international collaborators and friends Jagadeesh Moodera, Józef Barnaś, Vitali Dugaev, Coriolan Tiusan, Michel Hehn, Stephane Andrieu, Fanny Greullet, Fred Bonell, Yaroslav Tserkovnyak, Stephen Russek, Minn-Tsong Lin, Jhen Yong Hong, Günter Reiss, Gernot Güntherodt, Victor Moshchalkov, Rainer Schad, Chris van Haesendonck, Rembert Duine, Ivo Vavra, Ursula Ebels, Alek Dediu, Mairbek Chshiev, Felix Casanova, and Luis Hueso. FGA thanks Stephane Mangin for his generous help with the establishment of collaboration with Cori Tiusan. We also thank the members of the other branch of the Magnetrans group for providing their expertise over the last decade (Juan F. Sierra, Ahmad Awad, Antonio Lara, and Vladimir Pryadun).

We thank Juan Carlos Cuevas and Thomas Frideriksen for their stimulating discussions of the origin of spin-dependent noise through molecular barriers; Ed Nowak for providing a simple deduction of the joint thermal and shot noise formula; and Michael Coey, Paolo Freitas, and Claude Fermon for their interest in the discussion of different aspects of noise in spintronics. The authors wish to acknowledge the Spanish Science Ministry (MEC, MINECO), Comunidad de Madrid, Fundacion Seneca (Murcia), and European

Science Foundation for financial support of the research on noise in spintronics.

FGA dedicates this book to his family, especially to his mother, Margarita; wife, Tatiana; children, Timour and Maria; and grandchildren, Artur, Kira, and Emma. JPC dedicates this book to his parents, Bernardo and Cecilia, and his wife, Ana. Thank you for unconditional help and support.

Chapter 1

Introduction

1.1 Magnetoresistance

Magnetoresistance is referred to as the change of the electrical resistance of a material or device when an external magnetic field is applied to it. There exist several magnetoresistance effects, including anisotropic magnetoresistance (AMR), giant magnetoresistance (GMR) and tunneling magnetoresistance (TMR). Recent studies show that GMR-based devices could be mainly implemented in the high current densities spintronic, while TMR structures are more useful in the situations when low power dissipation becomes the main priority.

Anisotropic magnetoresistance (AMR) was discovered by William Thomson in 1857 (Lord Kelvin) (Thomson, 1857) in iron and nickel, which presented a dependence on the angle between the electric current and the magnetization direction. The magnetic anisotropy in magnetic conductors is characterized by the resistivity of the material and depends on the angle between the external applied magnetic field and current running through the material. The physical origin of the AMR effect lies in spin orbit coupling. The electron cloud about each nucleus deforms slightly as the direction of the magnetization rotates, and this deformation changes the

Noise in Spintronics: From Understanding to Manipulation
Farkhad G. Aliev and Juan Pedro Cascales

ISBN 978-981-4774-34-5 (Hardcover), 978-1-315-11088-2 (eBook)
www.panstanford.com

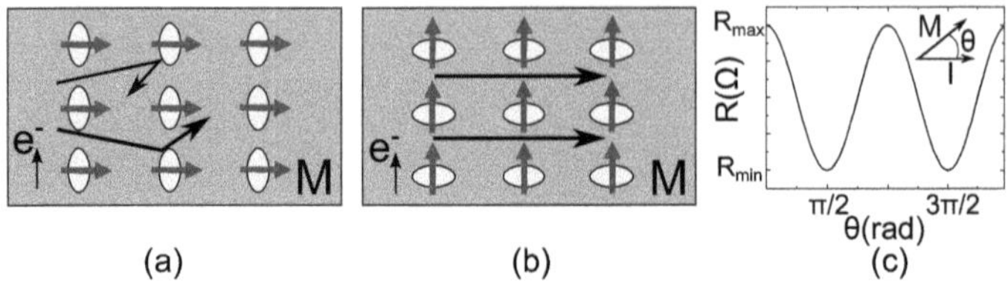

Figure 1.1 AMR effect. The distorted electron clouds of each atom scatter more electrons when the field is applied parallel (a) to the direction of the current, while the scattering is minimal when they are perpendicular (b). (c) Variation of resistance with the angle between current and magnetization.

amount of scattering undergone by the conduction electrons when traversing the lattice. A simplified picture of the effect is shown in Fig. 1.1. Also showing rather low (few percent) magnetoresistance, the AMR effect has broad application in automotive, bike and many other industries because of its simplicity and robustness of the AMR devices.

The discovery of **giant magnetoresistance** (GMR) (Baibich et al., 1988; Binasch et al., 1989) was done in thin-film structures composed of alternating ferromagnetic (FM) and non-magnetic (NM) conductive layers which show antiferromagnetic coupling (Cebollada et al., 1989; Grünberg et al., 1986). This discovery earned Albert Fert and Peter Grünberg the Nobel prize in Physics in 2007.

The simplest multilayer structure is called a *spin valve*, which is composed of two ferromagnetic layers separated by a non-magnetic, metallic spacer, which provide a technologically more robust GMR. The application of GMR by the use of spin valves in 1997 by IBM changed the landscape of magnetic data storage by dramatically increasing storage capacity. This helped pave the way for some of today's most popular devices, and it was the first application of spintronic devices to everyday technology.

The origin of the GMR effect is the spin-dependent scattering of electrons in the magnetic layers which constitute the structure (Camley and Barnas, 1989). Fig. 1.2 presents the simplest model, referred to as the *two-current model*, where the current running through the device is presented as composed of two parallel currents, one due to spin-up and the other due to spin-down electrons. If the magnetization directions of the FM layers are

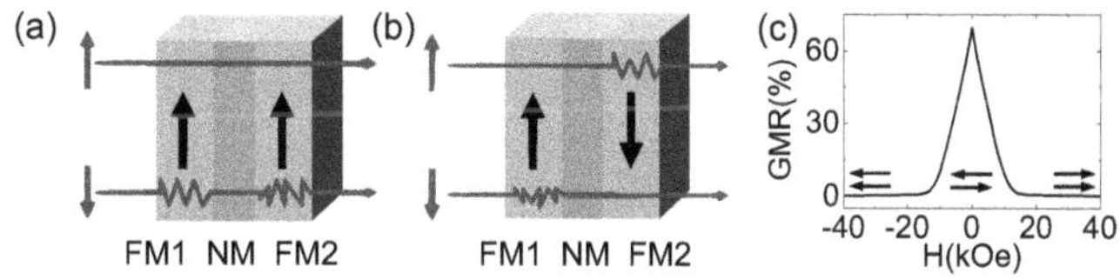

Figure 1.2 GMR effect. (a) Multilayers with alternating magnetization (antiferromagnetic system) present high scattering probabilities. (b) Minimal scattering when the magnetization directions of the layers are parallel. (c) Sketched dependence of the in-plain magnetoresistance typically observed for the exchange coupled ferromagnetic layers.

parallel (P state), the majority of electrons will have a spin parallel to this direction and they will encounter a minimum scattering (low resistance). If, on the other hand, the magnetization vectors of the layers are aligned in an antiparallel orientation (AP state), the majority of electrons coming from the first FM layer will scatter with high probability in the second FM layer (high resistance). These two resistance states may be used to read data bits in magnetic hard disk drives.

1.1.1 Tunnel Magnetoresistance

If the non-magnetic metallic spacer layer of the spin valve is replaced by a sufficiently thin non-magnetic insulating layer, the electrons can tunnel from one ferromagnetic layer to the other. If tunneling conserves the spin, the device is called a *magnetic tunnel junction* or MTJ (see Fig. 1.3).

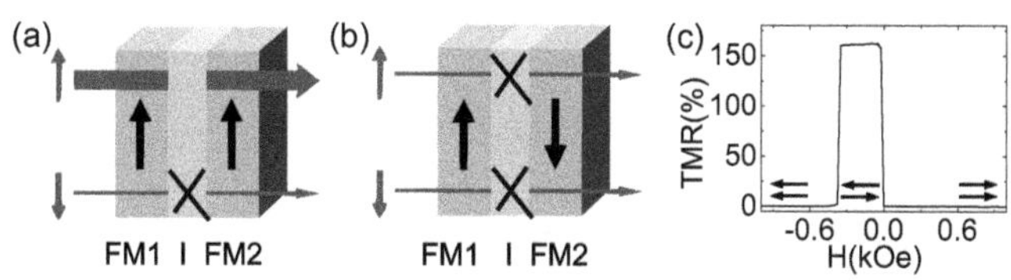

Figure 1.3 Simplified picture of the TMR effect for 100% polarized ferromagnets. (a) In the P state the spin majority electrons from the top electrode easily find available majority states in the bottom layer to tunnel into. (b) In the AP state the spin majority electrons from the top layer may only tunnel into the scarcely available minority states of the bottom layer. (c) Typical TMR vs. external magnetic field.

Jullier (Julliere, 1975) was the first to grow and study MTJ-based devices at low (liquid helium) temperatures, while Moodera with co-authors (Moodera et al., 1995) and Miyazaki and Tezuka (Miyazaki and Tezuka, 1995) observed much larger (about 10%) changes in the tunneling resistance with magnetic state at room temperature.

Classically, these structures usually have FM electrodes with different coercive fields $H_c^1 < H_c^2$. If an external magnetic field is applied so $H_{ext} > H_c^1 > H_c^2$, the magnetization of both layers is oriented along the same direction and this is referred to as the P state. In the range $H_c^2 > H_{ext} > H_c^1$ there is an antiparallel alignment of the layers' magnetization, the AP state. The difference of coercive field is realized either by choosing a hard and a soft FM material for each layer or by exchange-coupling one of the layers to an antiferromagnet (exchange biased).

Similarly as to what happens in spin valves, the parallel or antiparallel alignment of the layers' magnetization presents a low or high resistance state. This effect is called the tunneling magnetoresistance (TMR) effect. TMR is usually given as a percentage, corresponding to the difference in resistance between the P and AP states normalized by the P state resistance (referred to as the "optimistic" TMR):

$$\text{TMR} = \frac{R_{AP} - R_P}{R_P} \tag{1.1}$$

The following section will describe the details of tunneling between magnetic materials that are needed to understand the TMR effect.

1.2 Tunneling Involving Magnetic Materials

The main ingredient needed to understand the TMR effect is spin-dependent tunneling. After discussing the phenomenon in general, specific details of tunneling between magnetic materials will be introduced.

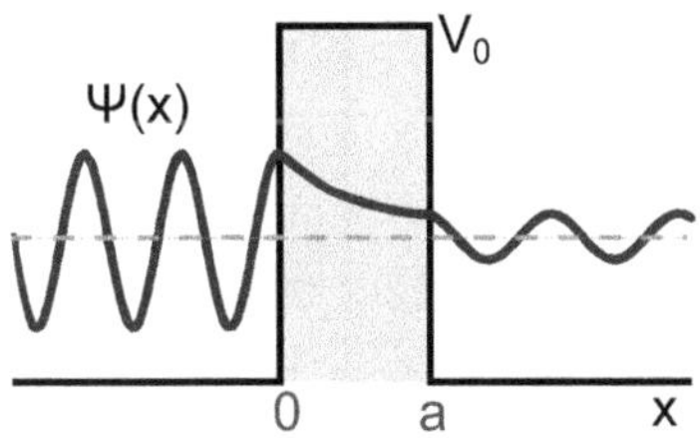

Figure 1.4 Transmitted wave through a rectangular potential barrier, known as quantum tunneling.

1.2.1 The Rectangular Potential Barrier

In classical physics, a particle cannot penetrate into or across a potential barrier if its energy is smaller than the potential of the barrier. In quantum mechanics, a particle may tunnel through a barrier it could not overcome classically with a probability that decays exponentially with the barrier width (Fig. 1.4). This is the well-known problem of the rectangular potential barrier of width a and height V_0. Considering an incoming wave from the left, the solutions to the wavefunction in each region are

$$\begin{aligned} \psi_1(x) &= e^{ik_1x} + re^{-ik_1x} && x \leq 0 \\ \psi_2(x) &= Ae^{ik_2x} + Be^{-ik_2x} && 0 \leq x \leq a \\ \psi_3(x) &= te^{ik_1x} && a \leq x, \end{aligned}$$

where $\kappa_2 = \sqrt{\frac{2m(E-V_0)}{\hbar^2}}$, r and t denote the reflected and transmitted amplitude, respectively. Applying the continuity conditions on the wavefunction and its derivative at $x = 0$ and $x = a$, one obtains the value of the coefficients. The case which we are interested in is when $0 < E < V_0$, which in the case of weak tunneling ($T \ll 1$) yields a transmission probability:

$$T \propto e^{-2\kappa_2 a} \tag{1.2}$$

where $T = \|t\|^2$. One may consider the three-dimensional problem where a wave traveling mainly in the Z direction encounters a rectangular potential barrier. The problem is solved by separation of variables, yielding a plane wave in the X and Y directions and the solution obtained above (let us call it $\Phi(z)$) for the Z direction, i.e.,

$$\Psi(x, y, z) = \Phi(z)\, e^{i\left(k_x x + k_y y\right)}$$

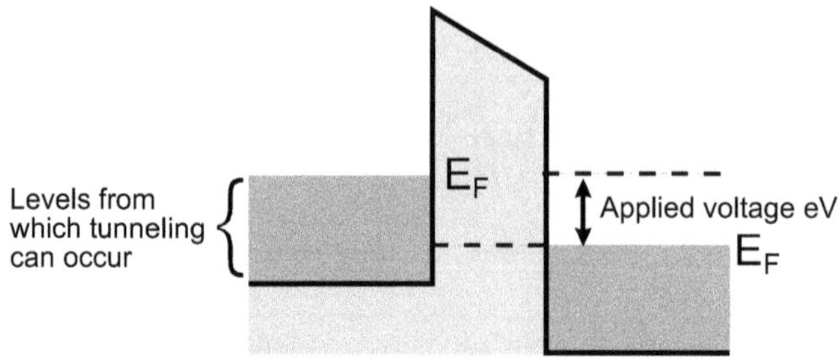

Figure 1.5 Sketch which following (Gasiorowicz, 2003) explains the energy diagram for tunneling between two metals separated by vacuum with an applied potential difference V.

In $\Phi(z)$, k is replaced by k_z and, κ and k_z present the following dependence with $\mathbf{k}_{\|} = (k_x, k_y)$:

$$k_z = \sqrt{2m\frac{E}{\hbar^2} - k_{\|}^2} \quad \text{and} \quad \kappa = \sqrt{2m\frac{(V_0 - E)}{\hbar^2} - k_{\|}^2}$$

This effect has numerous physical applications: the inversion of the ammonia molecule, the tunnel diode, the Josephson effect, the α-decay of certain nuclei, etc.

Now consider the tunneling occurs between two metals. Observing Fig. 1.5, without a potential difference V and at zero temperature, the tunneling is not possible because the levels on both sides of the barrier are filled. The electric field changes the shape of the barrier, bringing some empty levels in correspondence with the filled ones on the other side of the barrier. Now the tunneling can occur, with the transmission probability calculated above.

The current through a tunnel junction with an applied voltage V may be expressed as

$$I(V) = \int_{-\infty}^{+\infty} \rho_L(E)\rho_R(E+eV)|M|^2 f(E)(1-f(E+eV))dE, \quad (1.3)$$

where ρ is the density of states of the left (ρ_L) or right (ρ_R) electrode, $|M|^2$ is the transmission probability, $f(E)$ the occupied states of the left electrode and $(1 - f(E))$ the unoccupied states of the right electrode.

Simmons' model (Simmons, 1963) gives a very useful and simple approximation to the tunneling current by using a rectangular barrier. The barrier's thickness a and mean energy height V_0 of a

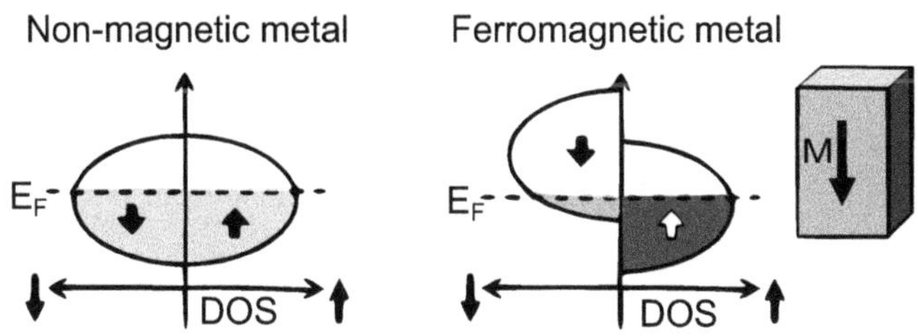

Figure 1.6 Left: sketch of the electronic bands of a normal metal. Right: sketch of the exchange split electronic bands of a ferromagnet.

junction may be obtained by fitting the following equation of the current density to an $I-V$ curve:

$$J(V) = J_0 \left[\left(V_0 - \frac{eV}{2} \right) e^{-A\left(V_0 - \frac{eV}{2}\right)^{\frac{1}{2}}} - \left(V_0 + \frac{eV}{2} \right) e^{-A\left(V_0 + \frac{eV}{2}\right)^{\frac{1}{2}}} \right], \tag{1.4}$$

where $J_0 = \frac{e}{(2\pi)^2 \hbar a^2}$ and $A = \frac{2a}{\hbar}\sqrt{2m}$.

1.2.2 Jullière Model for Spin-Dependent Tunneling

The model (Julliere, 1975) that Jullière proposed in 1975 (based on the free-electron model) has been extensively used to explain the TMR effect in magnetic tunnel junctions with amorphous barriers. It attributes the difference in resistance between the magnetic states to the spin polarization of the bands of the ferromagnets and consequently, to spin-dependent electron tunneling. The band structure $E(k)$ of a ferromagnet can be decomposed into two sets of bands, one for majority spin ($\uparrow$ or "spin up"), where spins are parallel to the external magnetic field, and the other for minority spin ($\downarrow$ or "spin down") where the spins are antiparallel. These bands are separated in energy by the magnetic exchange splitting. As can be seen in Fig. 1.6, for a non-magnetic metal, there is the same number of spin up and spin down conduction electrons at the Fermi level. For a ferromagnetic material, due to the magnetic exchange, there is a higher number of conduction electrons with a spin up than a spin down. Then, an electronic extracted from such a

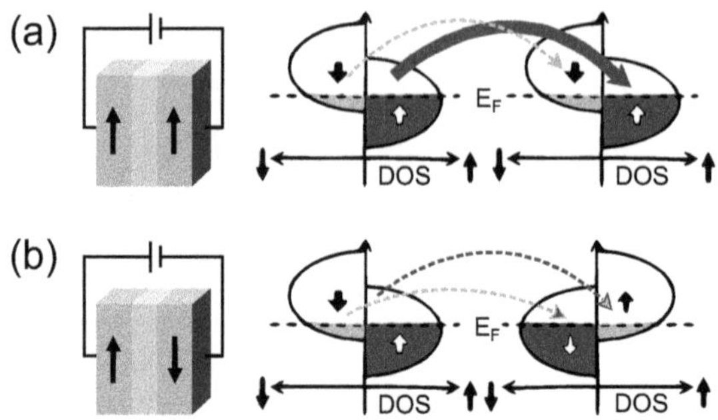

Figure 1.7 Sketch explaining Jullière model (Julliere, 1975). The majority electrons easily find available states in the P state (a) while much fewer states are available in the AP state (b).

material would be partially spin polarized. The **spin polarization** $\mathcal{P}$ of a ferromagnetic material is defined by

$$\mathcal{P} = \frac{n_i^{\uparrow} - n_i^{\downarrow}}{n_i^{\uparrow} + n_i^{\downarrow}}, \tag{1.5}$$

where $n_{i\sigma}$ is obtained from the spin-dependent density of states. A nonmagnetic material yields $\mathcal{P} = 0$ while a fully spin-polarized at E_F gives $\mathcal{P} = 1$. If $\mathcal{P}_i$ denotes the polarization of the i-th electrode ($i = 1, 2$), then the TMR may expressed as a function of the polarization:

$$\text{TMR} = \frac{G_P - G_{AP}}{G_{AP}} = \frac{\left(n_1^{\uparrow}n_2^{\uparrow} + n_1^{\downarrow}n_2^{\downarrow}\right) - \left(n_1^{\uparrow}n_2^{\downarrow} + n_1^{\downarrow}n_2^{\uparrow}\right)}{n_1^{\uparrow}n_2^{\downarrow} + n_1^{\downarrow}n_2^{\uparrow}}$$
$$= \frac{2\mathcal{P}_1\mathcal{P}_2}{1 - \mathcal{P}_1\mathcal{P}_2} \tag{1.6}$$

In order for this model to be valid, the tunnel barrier should be thick enough so the wavefunctions experience only a very small overlap. The Jullière model implies that the TMR depends exclusively on the density of states at the Fermi level of the electrodes, and thus is only valid when a small voltage is applied. It also does not take into account the filtering properties of the barrier or the influence of interface states (De Teresa et al., 1999).

All these arguments make the Jullière model mainly valid for MTJs with amorphous barriers such as AlO_x, where the various symmetries of the electron wavefunction can couple with evanescent states in AlO_x and have finite tunneling probabilities (Yuasa and Djayaprawira, 2007). The Julliere model is based on the assumption that the tunneling probabilities are equal for all Block states corresponds to symmetry independent or **incoherent tunneling**, for which the tunneling electrons do not have a preferred symmetry in momentum space (discussed in detail in the next paragraph).

1.2.3 Coherent Magnetic Tunneling

Fe, Ni and Co (and some of their alloys) owe their magnetic properties to unfilled 3*d* orbitals, and Bloch states with different symmetries of wave functions exist in these materials. Bloch states with Δ_1 symmetry (*spd* hybridized states) usually have a large positive spin polarization at the Fermi energy. Bloch states with Δ_5 and Δ_2 symmetry (*d* states) usually have a much smaller (or even negative) spin polarization at E_F.

Coherent tunneling (Yuasa and Djayaprawira, 2007) in this context indicates that a state of a given symmetry tunnels into a state of the same symmetry, that is, $\Delta_1 \rightarrow \Delta_1$, etc. If only Δ_1 states (which are highly spin-polarized) were to tunnel coherently through a barrier, a highly spin-polarized tunneling current would be obtained. Incidentally, since this spin-polarized state would be forbidden in the AP state, this would result in a high TMR ratio. For this effect to occur, the filtering of the other Bloch states (Δ_2, Δ_5) by the barrier is essential. A schematic view of coherent tunneling is shown in Fig. 1.9.

Crystalline MgO(001) barriers can be epitaxially grown over bcc Fe(001) with a 3% lattice mismatch (see the matching lattices in Fig. 1.8), which is compensated by lattice distortions in the layers and/or by dislocations formed at the interface. Such a high-quality interface allows coherent tunneling to take place. Coherent spin-dependent tunneling transport was theoretically predicted for epitaxial Fe(001)/ MgO(001)/ Fe(001) tunnel junctions with crystalline MgO barriers (Butler et al., 2001; Mathon and Umerski, 2001), and later confirmed experimentally (Parkin et al., 2004; Yuasa et al., 2004). The tunneling probability is highest for $k_\parallel = 0$, for which three kinds of tunneling (or evanescent) states exist in the

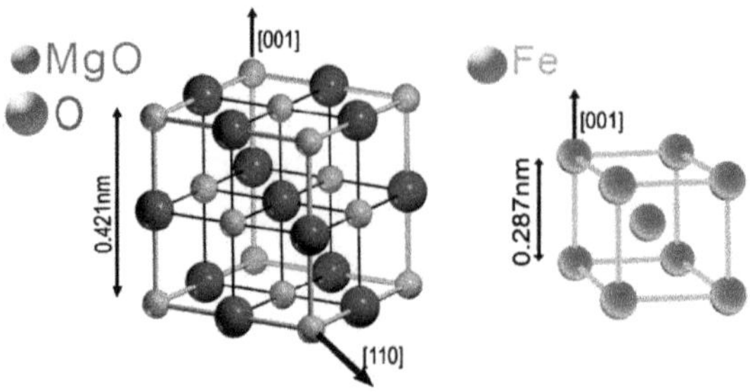

Figure 1.8 Sketch explaining epitaxial growth of Fe(100) on MgO(100) with an almost perfect matching. Note that Fe lattice is rotated by 45° with respect to the MgO lattice.

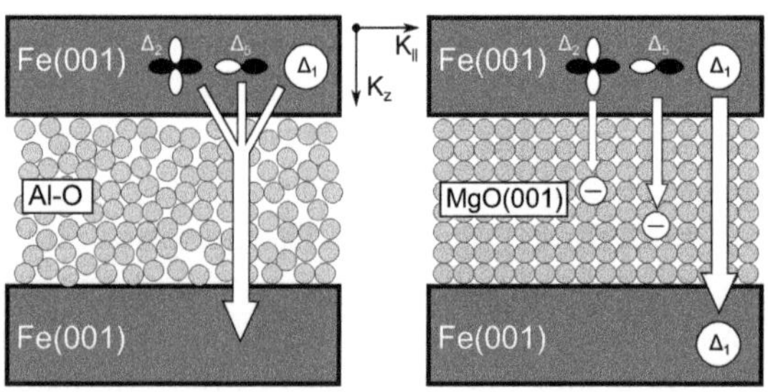

Figure 1.9 Diagram of electron tunneling through (a) amorphous Al-O barrier and (b) crystalline MgO(001) barrier (adapted with permission of IOP Publishing Ltd. from Yuasa and Djayaprawira (2007)).

band gap of MgO(001), Δ_1, Δ_2, and Δ_5. These states have specific orbital symmetries, and when the symmetry of the tunneling wave function is conserved, so each Fe Δ_i Bloch state couples with its corresponding MgO Δ_i evanescent state (as illustrated in Fig. 1.9). Due to its symmetry, the MgO Δ_1 evanescent states have the longest decay length, so in the parallel state, the tunneling is dominated by the Fe $\Delta_1 \Leftrightarrow \Delta_1$ MgO $\Leftrightarrow \Delta_1$ Fe electron channel.

The Δ_1 Fe band is fully spin-polarized at the Fermi energy (Fig. 1.10), which makes the conductance in the P state much larger than in the AP state. Therefore, a very large TMR effect is expected, as well as for other FM metals and alloys based on Fe and Co (bcc FeCo, bcc CoFeB, and some Heusler alloys) (Tsymbal and Zutic, 2011).

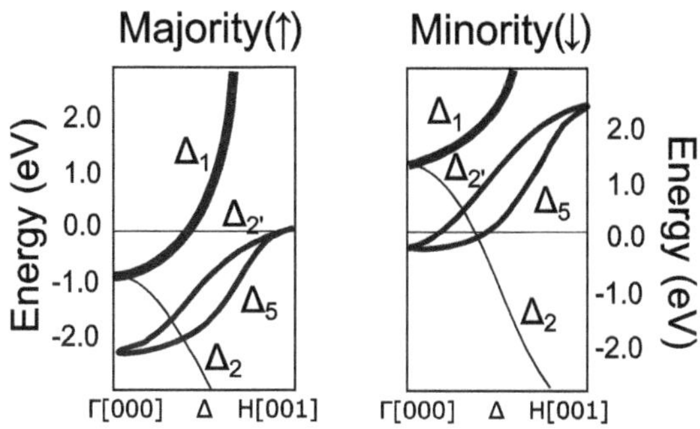

Figure 1.10 Sketch which following redrawing by (Yuasa and Djayaprawira, 2007) describes the bulk band structure for the majority and minority spin of bcc Fe.

1.3 General Motivation behind Noise Measurements in Spintronics

The introduction part paves the way to the explanation of physical mechanisms behind *time averaged* spin-dependent electron transport. As we shall see in the following chapters, by precise quantitative analysis of *fluctuations* in those parameters (spectrum of the magnetization fluctuations of the electrodes or of the number of electrons transmitted during a short period of time and related voltage fluctuations of the biased spintronic devices) could provide a valuable new information important both for fundamentals aspects of the electron and spin transport as well as for applications of those devices. For example, as we shall see in the chapter dedicated to discuss noise in single barrier magnetic tunnel junctions, shot noise and its variation with alignment of ferromagnetic electrodes contains unique information on the electron-electron and electron-localized spin interactions during the tunneling process. Along these lines, shot noise measurements may provide important information on spatial distribution of the defects inside barrier contributing to sequential tunneling. Moreover, double barrier magnetic tunnel junctions containing three ferromagnetic electrodes allow unprecedented magnetic state control over shot noise. Knowledge and control of other contributions to the low frequency noise such as $1/f$ or random telegraph noise are crucial for optimization of magnetic field sensors and spin torque oscillators. Additionally, low frequency noise spectroscopy of epitaxial magnetic tunnel junctions could be used for the band edge spectroscopy of buried interfaces. Moving towards the microwave range, both low and high frequency noise response in nanoscale magnetic tunnel junctions allow to discriminate between different regimes where spin torque effects switch lead to steady magnetization precession. Last but not least, although noise studies in low dimensional materials including graphene, 2D materials in general and molecular spintronic are in their infancy, initial steps in this direction will also be resumed in this book. Outlook and perspectives of noise studies in spintronic will be provided in the last chapter.

Chapter 2

Noise in Spintronic Systems

2.1 A Formal Description of Noise

Noise is defined as the random fluctuations of a physical quantity. In electronic devices, this quantity is the voltage and/or the current at any port of the device. In general, the random nature of this inherent noise comes from the thermal motion and intrinsic properties of the building blocks of solids, such as the generation and recombination of carriers, the discreteness of the current, the existence of deep traps or the influence of magnetic fluctuations on electron transport. Measuring noise gives information on the properties of the system that are often not accessible by other techniques. Besides, noise is in general a figure of merit of a device, as it determines the noise floor and thus the *signal to noise ratio* (SNR) of a system.

2.1.1 Stochastic Processes

Let us define the fluctuations of a variable $x(t)$ as the expected value of the squared deviation from the mean $\langle \delta x^2 \rangle$, that is, the variance. Fluctuations appear naturally in all physical systems in contact with a thermal bath or where quantum effects (like tunneling) become important. The first of these processes ever studied was the random

Noise in Spintronics: From Understanding to Manipulation
Farkhad G. Aliev and Juan Pedro Cascales

ISBN 978-981-4774-34-5 (Hardcover), 978-1-315-11088-2 (eBook)
www.panstanford.com

motion of microscopic particles submerged in a fluid, discovered by Brown and later known as Brownian motion (Brown and Mazey, 1828).

The particles that Brown observed move because of the random collisions of the liquid's molecules with the microscopic particles. This random force presents a variance that is directly proportional to the damping of the liquid. Relations of this kind between the energy dissipated by a system and the fluctuations of a generalized force are known as *fluctuation-dissipation relations* (Callen and Welton, 1951). Thermal noise in a resistor is a manifestation of this theorem, where the voltage is the generalized force and the role of dissipation is played by the electrical resistance and the temperature (Nyquist, 1928). Another example of the fluctuation dissipation theorem is the magnetic noise observed in soft ferromagnets, which is related with frequency independent imaginary permeability at low temperature (Durin et al., 1993).

Let us consider a random or stochastic variable $X(t)$, characterized by a distribution function $F_X(x)$ which is defined as the probability P of the variable X being equal or smaller than a certain value x:

$$F_X(x) = P(X \leq x)$$

The derivative of the distribution function with respect to x is referred to as the probability density:

$$f_X(x) = \frac{dF_X(x)}{dx}$$

The different statistic moments m_n associated to the random variable may be defined by using the probability density by

$$m_n = \int_{-\infty}^{+\infty} x^n f_X(x)dx.$$

The first moment corresponds to the mean value of our random variable. The central moments μ_n of the variable are given by

$$\mu_n = \int_{-\infty}^{+\infty} (x - m_1)^n f_X(x)dx.$$

Clearly, $(x - m_1)$ corresponds to the fluctuation δx of x with respect to the mean value, so μ_1 corresponds to the average value of

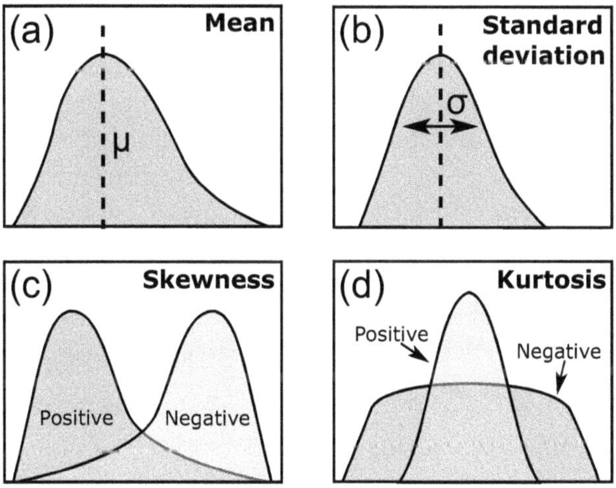

Figure 2.1 Diagram representing the first four statistic moments of random variable.

δx, which for a random process is zero. The second central moment μ_2 corresponds to the average value of δx^2, i.e., the variance σ^2. The third moment or skewness gives an idea of how asymmetric the distribution is, the fourth moment or Kurtosis gives a measure of the distribution's peakedness, etc. As an example, a graphical interpretation of the first four moments is shown in Figure 2.1.

If the process under study can be modeled as a sum of N random variables, then independently of the distribution of each variable, the distribution of the sum will approach a Gaussian distribution if N is large enough. This is known in Statistics as the Central Limit Theorem. Only the first two cumulants, which for the Gaussian distribution are given by the mean μ and the variance σ^2, are non-zero for the Gaussian distribution. *Therefore, if we can guarantee that our process has a Gaussian probability density, determined by a mean m_1 and a variance μ_2, determining these two quantities experimentally will be enough to characterize the random process.* In the context of electronic transport through spintronic devices, the first and second moments may be obtained by measuring the voltage (mean) and its variance, which we extract from the Fourier transform of the voltage fluctuations (see below).

Case study 2.1: Central limit theorem

Simulate the Central Limit Theorem in **Octave**. The MatLab code below will plot the probability distribution of one of the n random variables $x_i(t)$ and the probability distribution of the sum of all these variables. By increasing n, it can be seen that the distribution of the sum tends to a Gaussian when the n is large enough. Comment and uncomment the different probability distributions for the random variables to see that the distribution of the sum tends to a Gaussian independently of the distribution of the random variables.

```
n=2; %number of random variables x_i(t)
t=1000; %maximum time
x=randn(t,n); % Gaussian distribution
%x=betarnd(10,4,t,n); % beta distribution
%x=chi2rnd(2,t,n); % chi-squared distribution
%histogram of x_1(t)
[N,X]=hist(x(:,1),20);
figure %we plot it in a new figure
bar(X,N)
%histogram of the sum
[N,X]=hist(sum(x,2),20);
figure %we plot it in a new figure
bar(X,N)
```

2.1.2 Time Series Analysis

The noise of an electronic device may be understood within the framework of the fluctuation-dissipation theorem, if, for example, the voltage is taken as the generalized force. Then, if we design an experiment which measures the variance of our generalized force, ($\langle \Delta V^2 \rangle$), at thermal equilibrium and at some fixed range of frequency Δf, the **spectral density** *function S_V may be defined as the variance of the voltage per unit frequency*:

$$S_V = \frac{\langle \Delta V^2 \rangle}{\Delta f} \tag{2.1}$$

The frequency-dependent spectral density is obtained by differentiating the variance with respect to the frequency:

$$S_V(f) = \frac{d\langle \Delta V^2 \rangle}{df} \tag{2.2}$$

Thus, *the spectral density tells us how a signal or a time series is distributed with frequency.*

Let us now discuss an alternative way of defining the spectral density of a process through the correlation function.

The correlation function (here defined for real signals) is a measure of similarity between two signals $X(t)$, $Y(t)$, when one of them is delayed by a lag τ. The definition slides the $Y(t)$ function along the x-axis and calculates the integral of their product at each position. When the functions match, the correlation is maximized. It is given by

$$R_{XY}(\tau) = \lim_{T\to\infty} \frac{1}{2T} \int_{-T}^{T} X(t)Y(t+\tau)dt. \tag{2.3}$$

The autocorrelation function of a variable $X(t)$ is defined as the correlation of the signal with itself:

$$R_{XX}(\tau) = \lim_{T\to\infty} \frac{1}{2T} \int_{-T}^{T} X(t)X(t+\tau)dt \tag{2.4}$$

Finally, the Wiener–Khinchin theorem relates the autocorrelation function to the power spectral density via the Fourier transform:

$$S(f) = \int_{-\infty}^{\infty} R_{XX}(\tau)e^{-i2\pi f\tau}d\tau \tag{2.5}$$

$$R_{XX}(\tau) = \int_{-\infty}^{\infty} S(f)(\tau)e^{i2\pi f\tau}df \tag{2.6}$$

From now on we will refer the fluctuations as a spectral density instead of a variance.

Case study 2.2: Wiener–Khinchin theorem

Below is a demonstration in Octave of the Wiener–Khinchin theorem. It shows that the same result is obtained by calculating the power spectral density of a random variable from the modulus squared of the Fourier transform of the fluctuations or from the Fourier transform of the autocorrelation of $x(t)$.

```
t=1000; %maximum time
Fs=1;
L=t;
x=randn(t,1); % x(t)
[y,lag] = xcorr(x,x);
% FT(Rxx)
NFFT = 2^nextpow2(L);
Y = fft(y,NFFT)/L;
f = Fs/2*linspace(0,1,NFFT/2+1);
% FT(x(t))
NFFT = 2^nextpow2(L);
Y2 = fft(x,NFFT)/L;% FT(x(t))
% Plot both functions
loglog(f,2*abs(Y(1:NFFT/2+1)))
xlabel('Frequency (Hz)')
ylabel('S(f)')
hold all
loglog(f,2*Y(1:NFFT/2+1).*conj(Y(1:NFFT/2+1)))
legend('|FT(Rxx(\tau)|^2','|FT(x(t)|^2')
```

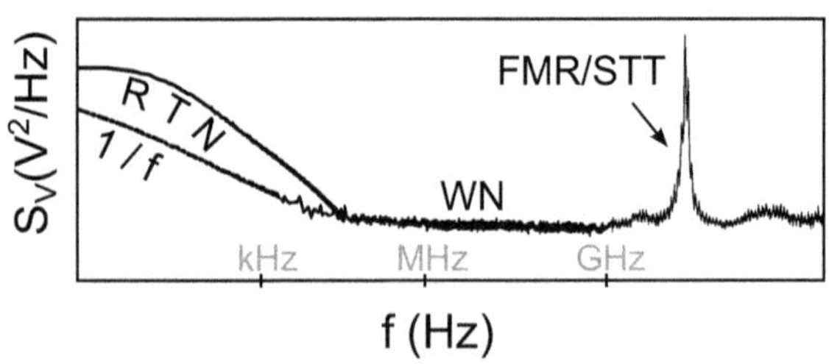

Figure 2.2 Diagram of the main noise contributions in spintronic devices. RTN is random telegraph noise, WN is white noise, and FMR/STT refers to resonant exctiation of the moment by spin transfer torque.

2.2 White Noise

White noise is a random signal (acoustical, electrical, etc.) with a constant power spectral density up to quantim limit usually in the GHz range. It receives its name from an analogy of white light, which contains light of all visible frequencies. There are two components of white noise in spintronic systems: thermal and shot noise.

2.2.1 Thermal Noise

Thermal or Johnson–Nyquist noise was discovered in 1928 by John B. Johnson at Bell Labs (Johnson, 1928), who described his results to his co-worker Harry Nyquist, who produced a theory explaining the observation (Nyquist, 1928). The thermal noise of a conductor depends on its resistance and temperature. It is one of the contributions to "white noise" (independent of frequency), and it is due to the random motion of electrical charge carriers caused by thermal agitation. This noise is the manifestation of the fluctuation-dissipation theorem (Callen and Welton, 1951) which relates thermodynamically the dissipation in a dynamic system (resistance), determined through a non-equilibrium property, with the fluctuations in that system. The spectral density S_V of the voltage due to thermal noise is given by

$$S_V = 4k_B T R. \tag{2.7}$$

where k_B is Boltzmann's constant, T is the temperature and R is the electrical resistance of the device under study. The above expression is one of the best known examples of the fluctuation-dissipation theorem, which relates the equilibrium noise (S_V) with out of equilibrium measureable in the linear regime physical parameter (electric resistance here). Measuring thermal noise can be used to calibrate the gain of an experimental setup or it can provide a nice experiment for a laboratory class, as is explained below.

2.2.2 Shot Noise

In statistics, the Poisson distribution expresses the probability of a number of events occurring in a fixed interval of time (or space) if the events occur with a known average rate (or mean), and are independent of the time elapsed since the last event. This distribution was first introduced by Siméon Denis Poisson (1781–1840) and its first practical application was carried out by Ladislaus Bortkiewicz in 1898 when he investigated the number of Prussian soldiers accidentally killed by horse kicks. A varied number of processes obey Poisson statistics, from the emission of electrons in a vacuum tube to the photons gathered per pixel in a CCD camera, or the number of molecules found in a small volume at some time.

Case study 2.3: Thermal noise

A simple way to test this law is by measuring the voltage noise of a resistor with no applied current by using a circuit like the one shown below. Typically, thermal noise can be used to calibrate the unknown gain of an amplifier. As can also be seen in the figure below, thermal noise may be also used to estimate the value of Boltzmann's constant, e.g., in a laboratory class for students. The figure shows that a good estimate of k_B can be obtained from the slope of S_V vs. RT with just a few resistors.

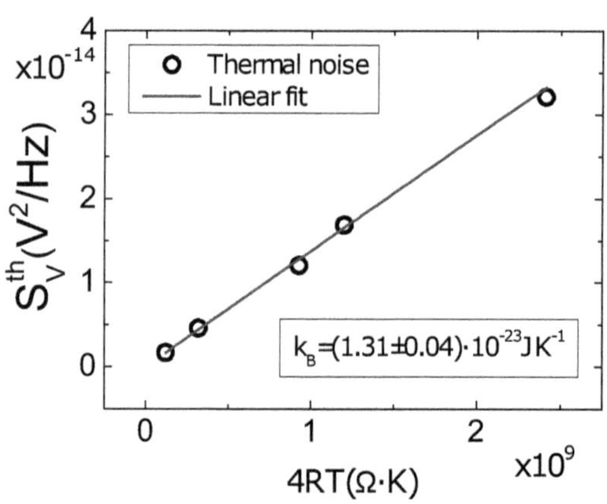

A discrete random variable X is said to have a Poisson distribution with a mean $\lambda > 0$, if the probability of having k events in a given interval $P(X = k)$ for $k = 0, 1, 2, \ldots$ is given by

$$f(k; \lambda) = P(X = k) = \frac{\lambda^k e^{-\lambda}}{k!}. \tag{2.8}$$

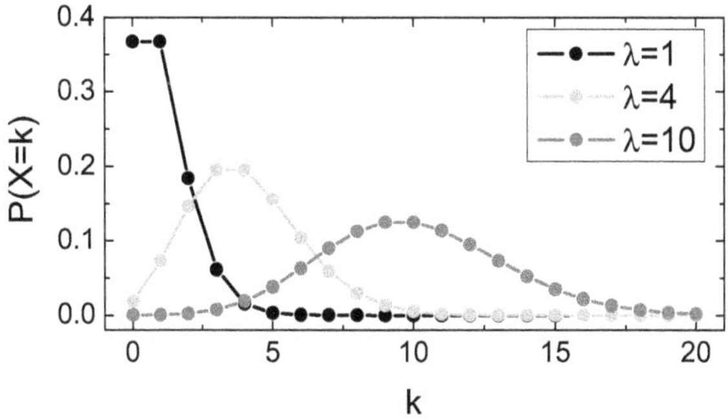

Figure 2.3 Poisson distribution for different values of λ.

When the average number of events occurring in the fixed interval of time or space becomes very large, i.e., $\lambda \to \infty$, the Poisson distribution is equivalent to the Gaussian distribution, as seen in Fig. 2.3.

In electronic devices, shot noise is due to the discreteness of the electrical current. It was first detected by Schottky in 1918 in a vacuum tube, where electrons are emitted by the cathode following a Poisson probability distribution, and received by the anode after a time of flight. In this kind of random process, the mean squared fluctuation of the number of emission events equals the average count of emission events. Thus, the spectral density of the current fluctuations in a vacuum tube is given by Schottky's formula $S = 2eI$, where e is the charge of the electron and I the average current. The value $2eI$ is referred to as Poissonian of full shot noise. Shot noise presents a frequency independent spectrum (white noise) up to a certain cut-off frequency, typically in the GHz range. This cut-off frequency is related to the time taken for an electron to travel through the conductor (Schoelkopf et al., 1997).

In mesoscopic devices like point contacts or tunnel junctions, the shot noise comes from the fact that the transmission probability T of an electron tunneling from one lead to another is very small ($T \ll 1$), which yields an expression for the shot noise such as (Blanter and Büttiker, 2000a)

$$S_I = \frac{e^3|V|}{\pi\hbar}\sum_{\mathrm{n}} T_{\mathrm{n}}(1 - T_{\mathrm{n}}),$$

where V is the bias voltage, T_n is the transmission probability in the n channel and $\hbar$ is Planck's constant. This equation yields the Poissonian value only in the limit of a low transparency system as, for example, a tunnel junction. Variations in the transparency of the system may lead to sub-Poissonian or, even, super-Poissonian shot noise (Blanter and Büttiker, 2000a).

The equation corresponding to Poissonian voltage shot noise is

$$S_V = 2eI R_d^2, \tag{2.9}$$

where R_d is the differential resistance of the device. This equation describes white noise only at $T = 0$ K, since for $T > 0$ and $V = 0$ the system would exhibit zero fluctuations at zero bias, which violates the fluctuation-dissipation theorem. The problem is solved by adding thermal noise.

Shot noise and thermal noise are often, and incorrectly, viewed as additive and independent noise sources. This independence has been questioned for a long time and it can be shown that both types of noise have the same explanation if the electrical conductance is treated as a quantum-mechanical transmission phenomenon (Landauer, 1993).

The joint expression for voltage white noise (Landauer, 1993) (shot and thermal) for a tunnel junction is

$$S_V = 2eI R_d^2 \coth\left(\frac{eV}{2k_B T}\right). \tag{2.10}$$

We will derive this equation in the following subsection in an intuitive way.

2.2.3 Unified Equation for Shot Noise in Tunnel Junctions

A unified expression for the thermal and shot noise can be *informally* (for a more mathematical derivation see Ref. (Blanter and Büttiker, 2000a)) derived from the same physical model by using several assumptions from various branches of physics and signal analysis (Callegaro, 2006).

Consider an ideal tunnel junction J with an insulating gap g (vacuum or dielectric) separating two metal contacts A and B. The junction is connected to an ideal voltage source G which applies a constant bias voltage V.

Let us denote $I(t)$ as the current flowing through the circuit, which is kept at a constant temperature T (isothermal). The current is made up of discrete carriers which have an elementary charge q, referred to as electrons from now on.

Considering the electrons as quantum-mechanical particles, there is a finite probability per unit time P_{AB} (P_{BA} respectively) that an electron (instantaneously) tunnels from A to B (B to A respectively). Quantum mechanics gives us the values of P_{AB} and P_{BA} knowing the nature of the junction.

Suppose the detector of our thought experiment allows us to watch individual electrons tunneling through the barrier, because its sampling time τ is small enough. The bandwidth of the measurements would be

$$\Delta f = \frac{1}{2\tau}. \tag{2.11}$$

Then, we can observe three different events:

(1) An electron crosses $A \to B$. Then $I(1) = +q/\tau$, occurring with probability $P_{I(1)} = P_{AB}\tau$.
(2) No electrons tunnel, so $I(0) = 0$, occurring with probability $P_{I(0)} = 1 - (P_{AB} + P_{BA})\tau$.
(3) An electron crosses $B \to A$. Then $I(-1) = -q/\tau$, occurring with probability $P_{I(-1)} = P_{BA}\tau$.

Since these are all the possible outcomes, the average and mean square values of the current are

$$\langle I \rangle = \sum_{i=-1}^{1} I(i) p_{I(i)} = q(P_{AB} - P_{BA}) \tag{2.12}$$

$$\langle I^2 \rangle = \sum_{i=-1}^{1} I(i)^2 p_{I(i)} = \frac{q^2}{\tau}(P_{AB} + P_{BA}). \tag{2.13}$$

Let us think of the tunneling event as a transition between two states, having the electron in A or B. So n_A, n_B represent the occupation of the states A and B. If the system is in a steady state, n_A and n_B are constant and the following is true:

$$P_{AB} n_A = P_{BA} n_B \tag{2.14}$$

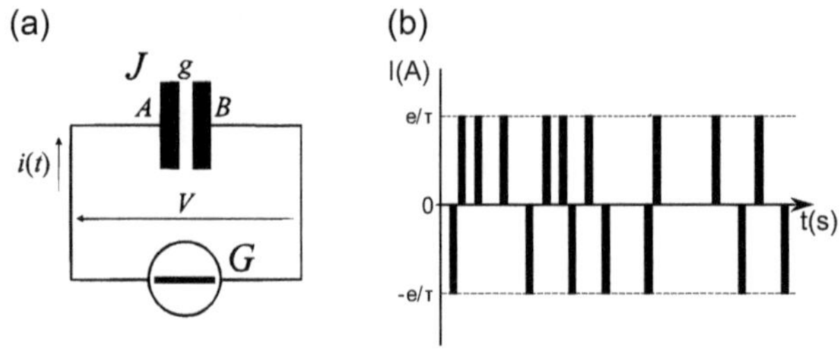

Figure 2.4 (a) The tunnel junction J has contacts A and B separated by a gap g, connected to a source G that supplies a constant voltage V. The current $I(t)$ flows through the junction at temperature T. (b) Three realizations of $\overline{I}$ are observed.

This comes from the *detailed balance* principle (MacEachern, 1997), which informally states that for kinetic systems which are decomposed into elementary processes, then "*At equilibrium, each elementary process should be equilibrated by its reverse process*".

The states A and B have energies E_{A} and E_{B}, so $E_{\mathrm{A}} - E_{\mathrm{B}} = qV$. Since the occupation numbers satisfy the Boltzmann distribution (Carroll, 1971):

$$\frac{n_{\mathrm{A}}}{n_{\mathrm{B}}} = e^{-\frac{qV}{k_{\mathrm{B}}T}} \tag{2.15}$$

Combining equations 2.14 and 2.15, we arrive at

$$\frac{P_{\mathrm{BA}}}{P_{\mathrm{AB}}} = e^{-\frac{qV}{k_{\mathrm{B}}T}},$$

so

$$\langle I \rangle = qP_{\mathrm{AB}}(1 - e^{-\frac{qV}{k_{\mathrm{B}}T}})$$

$$\langle I^2 \rangle = \frac{q^2}{\tau}P_{\mathrm{AB}}(1 + e^{-\frac{qV}{k_{\mathrm{B}}T}}),$$

which combined, along with equation 2.11, yields

$$\langle I^2 \rangle = 2q\langle I \rangle \frac{1 + e^{-\frac{qV}{k_{\mathrm{B}}T}}}{1 - e^{-\frac{qV}{k_{\mathrm{B}}T}}} \Delta f = 2q\langle I \rangle \coth\left(\frac{qV}{2k_{\mathrm{B}}T}\right) \Delta f.$$

Finally, if one is interested in obtaining the voltage spectral density, the resistance R of the junction must be included:

$$S_{\mathrm{V}} = S_I R^2 = \frac{\langle I^2 \rangle}{\Delta f} = 2q\langle I \rangle R^2 \coth\left(\frac{qV}{2k_{\mathrm{B}}T}\right)$$

If the resistance of the sample presents a non-linear IV, the differential conductance R_d must be used in place of R, as stated by Gupta's theorem (Gupta, 1978, 1982). The previous equation reduces to the expression for shot noise (equation 2.9) or thermal noise (equation 2.7), in the limits $T \to 0$ or $V \to 0$.

Case study 2.4: Shot noise thermometry

Shot noise may be used to fabricate very sensitive and accurate thermometers and the Shot Noise Thermometer reported in Ref. Spietz et al. (2003) is based on the noise from a tunnel junction. In this thermometer, temperature is related to the voltage across the junction by a relative noise measurement with only the use of the electron charge, Boltzmann's constant, and Fermi-Dirac statistics. If the noise power is normalized to the zero bias (thermal) noise, and the bias voltage is scaled relative to temperature, labeled by x, then the noise follows a universal function $x \cdot \coth(x)$.
This is because the measured noise power should scale with applied bias (V) as

$$G\left[T_\mathrm{n} + \frac{eV}{2k_\mathrm{B}} \coth\left(\frac{eV}{2K_\mathrm{B}T}\right)\right]$$

with G being average gain-bandwidth product, T_n- average noise temperature, and T the fit parameter to obtain the temperature (Spietz et al., 2003).

2.3 Random Telegraph Noise

Random telegraph noise (RTN) is a type of electronic noise characterized by a time-dependent signal where step-like transitions between two or more discrete voltage (or current) levels take place at random times, as shown in Fig. 2.5. These fluctuations are non-Gaussian; so they do not satisfy the central limit theorem.

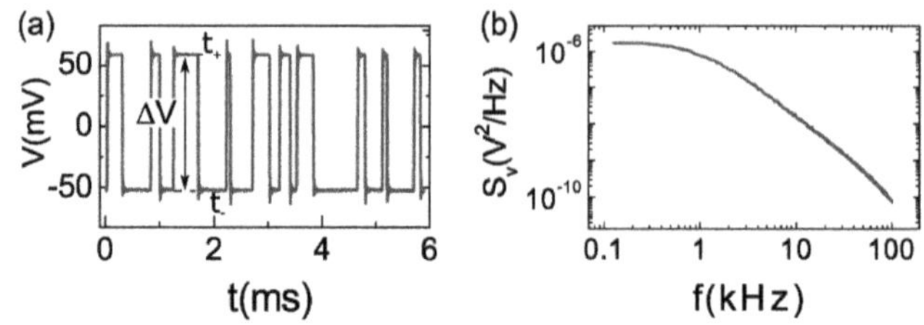

Figure 2.5 (a) Diagram of random telegraph noise between two charged states. (b) corresponding noise power spectrum.

RTN can be originated by different mechanisms depending on the structure being studied, like random trapping and releasing of charge carriers at defect sites in bulk semiconductor crystals or local changes in the magnetization of a ferromagnetic electrode.

Let us denote τ_+ and τ_- as the average times spent in the upper and lower states. It is found that the random times spent in each state, t_+ and t_-, follow an exponential probability distribution (Wolf and Holler, 1967):

$$p(t_\pm) = \tau_\pm^{-1} \exp\left(-\frac{t_\pm}{\tau_\pm}\right) \tag{2.16}$$

The lifetime of each level can be found by an exponential fit to a histogram of the times spent in them. For a two-level RTN, the spectral density is Lorentzian (Kirschenbaum et al., 1995; Machlup, 1954) and is given by

$$S_V^{RTN} = \frac{4\Delta V^2}{(\tau_+ + \tau_-)} \frac{T^2}{(1 + (2\pi T f)^2)}, \tag{2.17}$$

with $T = (1/\tau_+ + 1/\tau_-)^{-1}$, where τ_+ and τ_- are the characteristic dwell times of each level and ΔV is the voltage difference between the states. This equation can be generalized for multilevel RTN.

Analyzing this type of noise may be quite useful for determining different properties, depending on the nature of the fluctuations. If the RTN is due to charge trapping/detrapping, for example in oxide traps in MOSFETS (Hung et al., 1990), then properties like the trap depth may be extracted from the dependence of $\ln(\tau_+/\tau_-)$ with the voltage. An Arrhenius-type dependence is supposed for each characteristic time, and so one obtains

$$\ln(\tau_+/\tau_-) = K - \frac{q}{k_B T}\left(\frac{x_T}{t}V + \ldots\right), \tag{2.18}$$

where K is a constant, x_T the trap depth, t the oxide thickness and V the applied voltage. Other terms which depend on the temperature may be added to estimate surface potentials, etc.

The resistance fluctuations may also have a magnetic origin, for example, due to the occurrence of magnetization fluctuations in the electrodes of a magnetic tunnel junction. In this case, the fluctuating magnetic moment may be calculated from the dependence of $\ln(\tau_+/\tau_-)$ with the applied magnetic field. Since these two-level systems have been shown to have thermally activated rates, an Arrhenius law is postulated for each characteristic time (Kirschenbaum et al., 1995), so one arrives at

$$\ln(\tau_+/\tau_-) = K + \frac{2\Delta m}{k_B T}H, \tag{2.19}$$

where K is a constant, Δm is the fluctuating magnetic moment and H the applied magnetic field.

Case study 2.5: Random telegraph noise

Simulate random telegraph noise in **Octave**. The MatLab code below generates random telegraph fluctuations where the times spent in each level obey an exponential probability distribution. By increasing the number of iterations in the first for loop, it can be seen that the averaged spectral density improves. The theoretical expression for the spectral density of RTN is also plotted using the parameters ΔV and $\tau_{\text{up/down}}$ which are estimated from the generated time series. The reader is encouraged to vary the lifetime of the states and the amplitude of the fluctuations in the code and observe the changes in the time trace and spectrum.

```
 for j=1:50 %# of averages
 % generate RTN
n=1000; dv=0.1; dt=0.001; Fs=1/dt;
```

```
tauu=0.25; taud=0.18; x=[];
for i=1:n
if mod(i,2) == 1 %up state
  tu=tauu*abs(exprnd(tauu));
  pts=round(tu/dt);
  x=[x;dv*(1+0.1*rand(pts,1))];
else %down state
  td=taud*abs(exprnd(taud));
  pts=round(td/dt);
  x=[x;dv*0.1*rand(pts,1)];
end
end
t=dt*(1:length(x));

NFFT = 2^nextpow2(length(t));
f = Fs/2*linspace(0,1,NFFT/2+1);
df=f(2)-f(1);
Y = fft(x,NFFT)/length(t);% FT(x(t))

if j == 1
    figure1 = figure;
    psd=2*Y(1:NFFT/2+1).*conj(Y(1:NFFT/2+1))/df;
else
    psd=psd+2*Y(1:NFFT/2+1).*conj(Y(1:NFFT/2+1))/df;
end
% RTN PSD from FFT
clf; loglog(f,psd/j); hold all
xlabel('Frequency (Hz)')
ylabel('S_\textrm{V} (V^2/Hz)')
xlim([1 300]); ylim([1e-7 2*psd(65)/j])
% RTN theoretical PSD
x2=x-0.05;%estimate tau_up/down
tauu2=dt*numel(find(x2>0))/(n*0.5);
taud2=dt*numel(find(x2<0))/(n*0.5);
tau=1/(1/tauu2+1/taud2);
```

```
s=dv^2*4*tau^2/(tauu2+taud2)./(1+(2*pi*f*tau).^2);
plot(f,s,'LineWidth',2,'Color','red')
legend('PSD[x(t)]','S_{RTN}^{theor}')
axes2 = axes('Parent',figure1,...
'Position',[0.2 0.17 0.35 0.35]);
xlim(axes2,[0 1]); ylim(axes2,[-0.01 1.2*dv]);
box(axes2,'on'); hold(axes2,'all');
plot(t,x,'Parent',axes2); pause(0.1)
end
```

2.4 1/*f* Noise

The low-frequency $1/f$ noise, also referred to as pink noise or flicker noise is a ubiquitous type of noise found in many physical, biological and even economic systems: fluctuations of the sea level, intensity in a music recording, human heart rates or electrical currents in semiconductor devices.

$1/f$ noise was discovered in 1925 by Johnson, in an experiment designed to test Schottky's theory of shot noise in vacuum tubes. After almost a century of research, the origin of $1/f$ noise in solid state remains unclear and only phenomenological models are used to characterize it.

The noise spectra in spintronic devices (as well as in electronic devices in general or other systems in nature) present a dependence, in some range of frequency, of the form $1/\mathrm{f}^{\beta}$, where the exponent β is usually close to 1. This type of noise is referred to as $1/f$ noise, flicker or pink noise. These fluctuations are present in a wide range of different electronic devices, and they are attributed to fluctuations in the conductance, which is reflected in the fact that $S_{1/f} \propto V^2$. For example, in Si MOSFETs, the resistance of the channel fluctuates due to electron capture into and emission from traps that lie in the oxide layer. In a MOSFET with a large area (above 1 $\mu\mathrm{m}^2$), the $1/f$ noise appears from a superposition of the effects of a large number of traps with various activation energies and relaxation times (Dutta and Horn, 1981; Kogan, 1996).

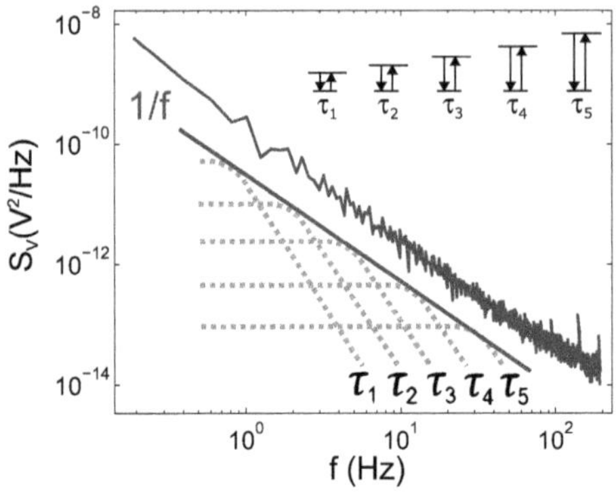

Figure 2.6 Sketch of $1/f$ noise as a sum of defects with different relaxation times.

A fluctuation which is characterized by a single relaxation time τ is defined by a spectral density which is a Lorentzian function of frequency:

$$S_X(f) = \langle \Delta X^2 \rangle \frac{4\tau}{1+\omega^2\tau^2} \tag{2.20}$$

Let us now consider that the kinetics of the fluctuating quantity $X(t)$ may be described as a superposition of several relaxation processes with different relaxation times (or even a continuous distribution of relaxation times), schematically represented in Fig. 2.6. Then the spectral density of such a process is given by averaging the Lorentzian spectral density of a this distribution with a weight function $p_X(\tau)$ which contains the number of subsystems with relaxation time τ and the corresponding variance of fluctuations. The spectral density is then given by

$$S_X(f) = \int_0^\infty d\tau p_X(\tau) \frac{4\tau}{1+\omega^2\tau^2}. \tag{2.21}$$

The spectral density has the form $S_X(f) \propto 1/f$ at some interval of frequency if $p_X(\tau) \propto 1/\tau$. Du Pré (Du Pré, 1950) and Van der Ziel (Ziel, 1950) proposed that the fluctuations are a result of the superposition of activation processes with different relaxation times, with $\tau = \tau_0 e^{-(E/k_B T)}$, where E is the activation energy. Then, if the distribution in activation energies $F_X(E)$ is chosen to be constant, the distribution of relaxation times $p_X(\tau)$ has the required form, since

$$p_X(\tau)d\tau = F_X(E)dE \Rightarrow p_X(\tau) = F_X(E)/(d\tau/dE) = k_B T\, F_X(E)/\tau$$

and if $F_X(E) = \text{const.}$, $p_X(\tau) \propto 1/\tau$.

Case study 2.6: 1/*f* noise

Simulate $1/f$ noise in **Octave**. The MatLab code below generates Lorentzians whose amplitude is weighted appropriately by their characteristic time. As can be seen by varying the number of curves n from 1 to 100, an overlap of random processes with a Lorentzian spectrum may yield a $1/f$ spectrum.

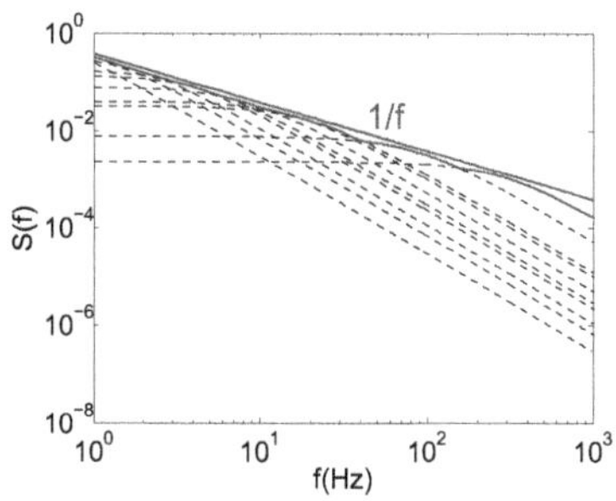


```
n=1; %# of lorentzians
df=1; span=1e3; f=1:df:span;
tau1=1/(10*span);tau2=1/df;
p0=0; figure1=figure;
psdm=[];

for j=1:n
tau=10^(-unifrnd(0,3.5));
psd=4*tau./(1+(2*pi*f*tau).^2);
if psd(1) >= p0
    p0=psd(1);
end
psdm=[psdm,psd'];
loglog(f,psd,'--','Color','black')
```

```
xlabel('f(Hz)');ylabel('S(f)');
hold all; pause(0.2)
end

plot(f,max(psdm,[],2),'LineWidth',2,'Color','red')
pause(0.2);
plot(f,1.2*p0/f(1)./f,'LineWidth',2,'Color','green')
annotation(figure1,'textbox',...
[0.56 0.69 0.041 0.073],...
'EdgeColor',[1 1 1],'String',...
{'1/f'},'FontSize',16,...
'Color','green');
```

The most widely used phenomenological description of $1/f$ noise is the one proposed by Hooge (Hooge et al., 1981a). This model describes the $1/f$ noise in terms of the variation of the density of charge carriers, which is proportional to the dimensionless Hooge parameter α. The Hooge parameter allows comparing the $1/f$ noise power in different devices, and is given by

$$S_V = \frac{\alpha V^2}{\Omega f^{\beta}}, \tag{2.22}$$

where V is the applied DC voltage and Ω the volume or lateral size of the conductive region. In the case of tunnel junctions, the noise power is normalized by the area A of the junction, so the Hooge parameter typically has dimensions of μm^2.

$$S_V = \frac{\alpha V^2}{A f^{\beta}} \tag{2.23}$$

The exponent β varies between 0.9 and 1.4, and variations from this range are usually due to additional noise contributions, such as random telegraph noise (Ralls and Buhrman, 1988) or generation-recombination processes (Kogan, 1996).

$1/f$ fluctuations appear in practically all electronic devices and are associated with low frequency fluctuations in resistance. Spintronic devices, like GMR or TMR systems are influenced by defects in their structure which give rise to $1/f$ type fluctuations.

Besides, due to the link between magnetic order and conductivity in spintronic devices, these may exhibit an additional $1/f$ noise source. This term is directly related to the magnetization noise that can be described using the fluctuation-dissipation theorem.

2.4.1 Thermal Magnetic Fluctuations

As was described in the introduction, there exist a relation between energy dissipated by a system and the fluctuations of a certain generalized force.

In ferromagnetic systems, the energy losses can be described by the complex susceptibility $\chi(\omega) = \chi'(\omega) + i\chi''(\omega)$. This quantity is dependent on frequency and its real and imaginary parts are described as the in-phase and out-phase, respectively, response of the magnetization to a small AC magnetic field. By using this quantity the thermal fluctuations of the magnetization ($S_M(\omega)$) can be expressed (Prodi et al., 1989) as

$$S_M(\omega) = \Omega \frac{4k_B T}{\omega} \chi''(\omega). \tag{2.24}$$

where Ω is the volume of the ferromagnetic body. The observation of this noise in a ferromagnet was reported at low frequencies in Refs. (Prodi et al., 1989; Vitale et al., 1992). The above expression is another well-known example of the application of fluctuation-dissipation theorem in nanomagnetism. The obtained power spectrum is consistent with ferromagnetic losses independent of frequency, thus the fluctuations exhibit $1/f$ noise.

2.4.2 Magnetic Noise in Spintronics

We have already mentioned that in magnetoresistive devices, due to the link between magnetization and resistivity, a magnetic noise from the electrodes can be detected in resistance fluctuations. This relationship was first reported in a multilayer structures which exhibited GMR, in Ref. (Hardner et al., 1993b). In this work, the detected $1/f$ noise is accurately predicted from a fluctuation-dissipation relation using the imaginary (or out-of phase) susceptibility of the system, and it was found that the fluctuations in

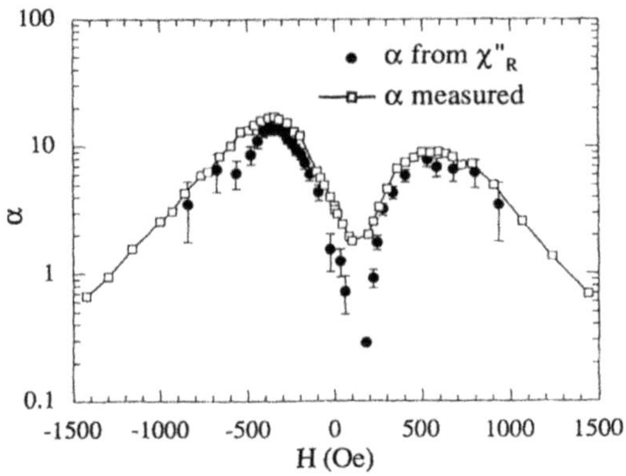

Figure 2.7 Comparison of Hooge parameters obtained from the resistance noise measurements and the AC susceptibility. Adapted with permission from Hardner et al. (1993b).

resistance where proportional to the magnetization noise and the derivative $dR(H)/dH$, see Fig. 2.7. It can also be seen in Fig. 2.7 that at $H = 0$ Oe the $1/f$ noise predicted by the fluctuation-dissipation theorem underestimates the resistance noise actually observed. This failure of the fluctuation-dissipation theorem is, in fact, expected as in this field region the main noise contribution is the domain wall dynamics, as long as domain walls are out of thermodynamic equilibrium. This out of equilibrium noise is called Barkhausen noise and has been observed in other spintronic devices as magnetic tunnel junctions in Ref. (Guerrero et al., 2002). Indeed this kind of noise is not $1/f$ noise but rather $1/f^2$ noise, that is consistent with random walk dynamics and avalanches, as is expected in domain wall dynamics which may contribute to noise in the small field range.

The equilibrium noise is exhibited by every spintronic system, e.g., magnetic tunnel junctions, limits the use of GMR or TMR systems in magnetic field sensing and high frequency applications. In general, the design of sensors using crossed anisotropy (Grigorenko and Mapps, 2003) or easy axis biasing (Paperno and Kaplan, 1995) drives the Barkhausen noise negligible. However the presence of this kind of noise and its proportionality to the sensitivity (dR/dH) is a big drawback for using them in general applications.

2.5 High Frequency Noise in Spintronics

If the lateral size of MTJs is reduced below 1 μm several new effects may be observed. First of all, due to the "inverse magnetoresistance" effect, large currents can be used to change the magnetic state of these small MTJs. Moreover, the influence of the magnetization dynamics on high frequency noise becomes an important factor (Fuchs et al., 2004). Therefore, it is worth dedicating a few lines to high frequency noise in spintronic devices. In equilibrium, the magnetization of a ferromagnet aligns with an effective field H_{eff}. The effective field corresponds to the combination of the externally applied magnetic field and the demagnetizing field of the ferromagnetic material under study. When a magnetic configuration is away from equilibrium, the magnetization precesses around the instantaneous local effective field. The Landau–Lifshitz–Gilbert (LLG) equation (equation 2.25) describes the time evolution of the magnetization as due to: the precession around H_{eff} (first term), dissipation (second term) and the influence of spin torque (third term), which opposes the dissipation. The spin torque term in the LLG equation was proposed by Slonzewski in Ref. (Slonczewski, 1996), which describes the torque exerted on the magnetization by a spin-polarized current.

$$\dot{\mathbf{M}} = -\gamma \mathbf{M} \times \mathbf{H}_{\text{eff}} + \frac{\alpha}{M_s} \mathbf{M} \times \dot{\mathbf{M}} + \gamma \frac{a_j(\theta)}{M_s} \mathbf{M} \times (\mathbf{M} \times \mathbf{P}), \qquad (2.25)$$

where γ is the gyromagnetic ratio, α the Gilbert damping parameter, M_s the saturation magnetization, $\mathbf{P}$ a unit vector in the direction of the current polarization and $a_j(\theta)$ is proportional to the current density J (Ralph and Stiles, 2008). The different terms are schematically shown in Fig. 2.8.

Then, we see that we can alter the magnetization of a ferromagnetic component of a spintronic device of small size either by applying an external field or by running a spin-polarized current through it. With a high enough spin-polarized current density, it is even possible to invert the magnetization of the material. So if we carry out noise measurements at high frequency (in the range of several GHz), thermal ferromagnetic resonances and spin torque effects could be observed.

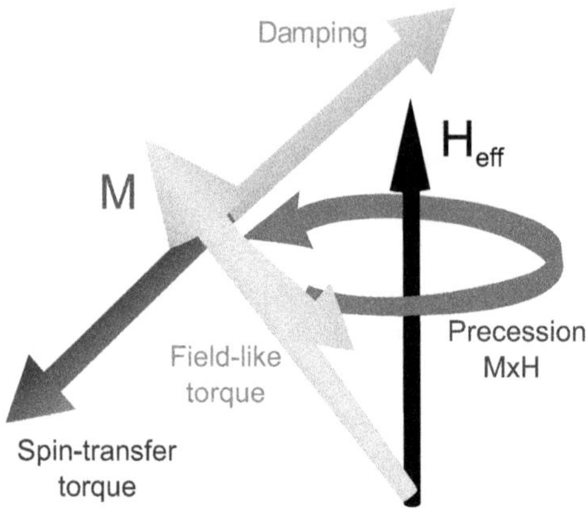

Figure 2.8 Sketch of the different contributions to the LLG equation.

2.5.1 Thermal Ferromagnetic Resonance (T-FMR)

A ferromagnetic resonance (FMR) is the precession of the magnetic moment of a ferromagnetic material about an externally applied magnetic field (Kittel, 1951). The precession of the magnetization may be excited by applying a transverse rf field. The microwave energy is strongly absorbed from the source when its frequency is equal to the magnetization's precessional frequency. FMR techniques are extensively used to extract the magnetic properties of materials such as the effective magnetization M_{eff}, the damping coefficient α, the gyromagnetic ratio γ in the metal, etc.

Thermal excitation may cause the spins in a ferromagnetic layer to deviate from equilibrium, and initiate a damped gyromagnetic motion (Zhu, 2002). As a result, a ferromagnetic resonance in the material is excited. It is worth noting that this random thermal field only performs the role of exciting the material's resonating spin waves. If a nanoscale magnetic device, such as a magnetic tunnel junction, is biased by a small dc current (under the value for spin-torque-induced dynamics), the resulting high frequency, voltage noise spectrum presents an FMR peak due to the thermal excitation (Guan et al., 2009).

Determining the dependence of the thermally excited FMR (T-FMR) with field and bias may give information on such mechanisms as spin-torque driven magnetic switching in nanoscale magnetic tunnel junctions. Also, the noise contribution originating from T-FMR is referred to as mag-noise, and it could present the ultimate limitation on the signal-to-noise ratio in hard drive read heads. This threatens the miniaturization of read heads for ultrahigh area recording density applications (Zhou et al., 2002).

2.5.2 Excess Microwave Noise due to Spin-Transfer Torque

As mentioned earlier, a high density, spin-polarized current can exert a torque on the magnetization of a magnetic layer. Typically, the critical current density in magnetic tunnel junctions is around 10^7 A/cm^2 (Ralph and Stiles, 2008), although values as low as 10^6 A/cm^2 have been achieved (Diao et al., 2007). This technique could be used as an efficient, fast, and reliable way of writing data bits in magnetic random access memories (MRAM) (Gallagher and Parkin, 2006). It would also make the devices simpler (there would be no moving parts), smaller, and more energy efficient.

These effects have been extensively studied in magnetic tunnel junctions (Ralph and Stiles, 2008). A transition from T-FMR to a spin-torque induced steady state, or even a reversal of the magnetization of the free layer may occur for some critical value of current I_c, which can be inferred from the bias dependence of the resonance frequency, obtained from high-frequency noise measurements (Houssameddine et al., 2008). It is typically observed that the frequency and width of the resonance peaks changes little at small currents, but decreases dramatically when $|I| \geq |I_c|$. The decrease is accompanied by a sudden increase in output power, which makes the spin torque effect in MTJs interesting for creating a new generation of microwave oscillators, as seen in Fig. 2.9.

2.5.3 Phase Noise

Phase noise is also another important factor limiting the applications of nanosized devices as microwave oscillators. As we have argued above, a sinusoidal voltage (microwave emission) may be

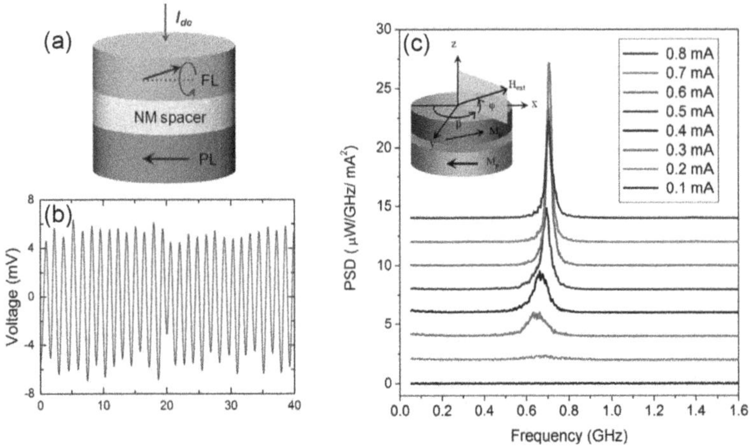

Figure 2.9 MTJs used as high-frequency oscillators (adapted with permission from Ref. Zeng et al. (2013)). (a) The fixed layer, which acts as a spin polarizer (PL) and excites a steady state precession of the magnetization in the free layer (FL) (b) Voltage oscillations produced by the oscillations of the free layer in the nanopillar sample, in response to the spin transfer torque from a 0.8 mA current. (c) Microwave emission spectra measured in a STNO magnetic tunnel junction.

obtained from spin-torque oscillators (Zeng et al., 2013). In this signal, both the amplitude and the total phase have small fluctuating contributions. The high-frequency spectrum of the voltage signal presents a peak with some linewidth. The phase fluctuations (which indicate how well the oscillator maintains the oscillations without phase-breaking) and the non-linear coupling between the phase and amplitude both contribute to the linewidth. Both these contributions are of great importance to understand the linewidth broadening and to optimize the quality of the devices. A useful alternative to measuring the spectra of the voltage fluctuations is to obtain the time trace of the oscillations (e.g., using a one shot oscilloscope). With this time trace, the Hilbert transform of the signal can be calculated, separating the contributions of amplitude and phase noise (Kim et al., 2008b). The fluctuations of each contribution may now be studied using the Fourier transform. The work in Ref. (Quinsat et al., 2010) applies this method to characterize the phase and amplitude noise of MgO based magnetic tunnel junctions which are used as high frequency oscillators.

Having introduced the readers into different general physical aspects of the noise in spintronic, we proceed to overview the experimental methods to measure electronic noise, mainly concentrating of the low frequency part of the spectra.

Chapter 3

Experimental Methods

With the previous tools in hand, the problem of studying noise in electronic devices becomes only a matter of constructing an appropriate experimental design. We hope to give a general overview the main experimental techniques that are used to measure electronic noise. The systems described here in no way constitute a definite guide to the world of noise measurements, although the reader will be able to form a good idea of the basic tools used and the important differences between various techniques. The discussion will center on determining the voltage fluctuations of a spintronic (or electronic) device.

In general, there exist two main ways of obtaining the spectral density of a signal: (i) by the use of the fast Fourier transform (FFT) or (ii) a filtering technique. The FFT method is used for low frequency experiments, where a resolution from a fraction of Hz up to tens of kHz is needed. On the other hand, the filtering technique is used for the detection of high frequency dynamics (in the range of GHz), where having a resolution of kHz is sufficient.

In the FFT method, the signal of interest is sampled by $2N$ points during a time T with a resolution in time of Δt. The FFT of this discrete time trace is calculated, giving a spectrum with N data points (or bins) up to a maximum frequency $f_{\max} = 1/2\Delta t$ and a frequency step of $\Delta f = 1/T$.

Noise in Spintronics: From Understanding to Manipulation
Farkhad G. Aliev and Juan Pedro Cascales

ISBN 978-981-4774-34-5 (Hardcover), 978-1-315-11088-2 (eBook)
www.panstanford.com

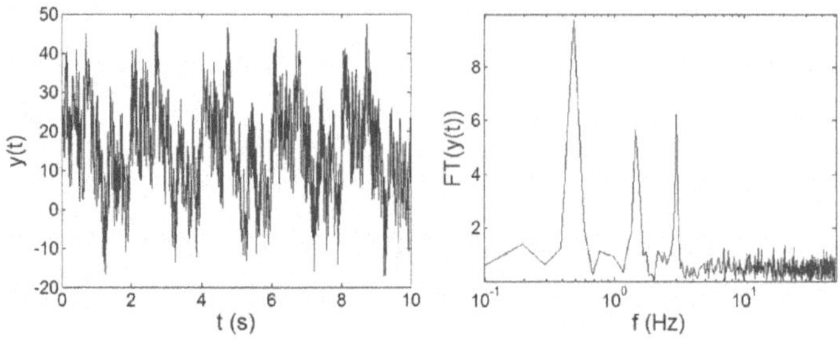

Figure 3.1 Example of a time-dependent signal and its Fourier transform (FT).

The FFT of a signal may be calculated with the use of equation 2.5 from the time trace of the fluctuations obtained by an oscilloscope or a digitizer card, as shown in Fig. 3.1. A spectrum analyzer is a device which is designed specifically for this purpose, and which allows obtaining directly the FFT of a signal with the possibility of having different frequency resolution and span settings.

In the filtering technique, the spectral density for a range of frequencies (or span) is obtained by sweeping the central frequency of the filter. Since the bandwidth is inversely proportional to the time of taking one measurement, having a high resolution in frequency (a small frequency step) would take a very long time. This is how spectrum or network analyzers which have bandwidths in the GHz range obtain the spectral density of high frequency signals.

In general, good spectra are obtained after averaging many cross-correlation spectra at each voltage. This procedure reduces the variance of our measurement by $1/N$ (the deviation as $1/\sqrt{N}$), with N the number of averages. In the case of the FFT technique, the number of averages chosen is determined by the time T for taking one spectrum, which is given by the Δf and $f_{\max}$ settings of the spectrum analyzer.

3.1 Amplifying the Signal

The signals we are interested in measuring have a very small amplitude, and they need to be treated (amplified, filtered, etc.) appropriately through some amplifier or other linear system. One

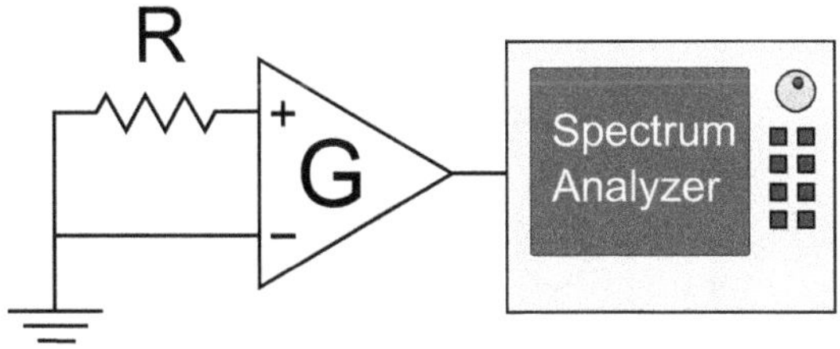

Figure 3.2 A basic circuit for measuring thermal noise in a resistor. G is amplifier gain and R is input resistance.

could think of the amplifier as a "black box" with an input signal $V_{\text{in}}(t)$ and an output signal $V_{\text{out}}(t)$. A linear system can be characterized by a transfer function defined by

$$H(f) = V^{\text{out}}(f)/V^{\text{in}}(f). \tag{3.1}$$

Consequently, the transfer function can be expressed as a function of the power spectral densities $S_{\text{V}}^{\text{in}}(f)$, $S_{\text{V}}^{\text{out}}(f)$ through the relation:

$$S_{\text{V}}^{\text{out}}(f) = S_{\text{V}}^{\text{in}}(f)|H(f)|^2 \tag{3.2}$$

valid only if the setup does not add any noise, which is general not true. The level of noise inherent to the system is called the noise floor. Therefore, the noise of the signal of interest should be significantly higher than the noise floor. The signal to noise ratio is defined as the ratio between the signal power and the noise floor power. Ideally, the SNR should be as high as possible so the fluctuations added by the electronics do not obscure the voltage fluctuations from the sample.

The simplest circuit which can be set up for noise measurements is shown in Fig. 3.2 and can be used to carry out the Case Study discussed in Section 2.2.1. This simple scheme is valid if the noise added to the signal by the amplifying circuit has a lower amplitude than the fluctuations of interest, that is, if there is a low SNR. In the case of spintronic devices, one often deals with fluctuations of low amplitudes comparable to the noise added by most amplifying equipment.

Let us now consider the noise added by our amplifying circuit to our signal, which we consider to be made up of a DC contribution plus some fluctuations $V(t) = V_{\text{DC}} + \delta V(t)$. The fluctuations have

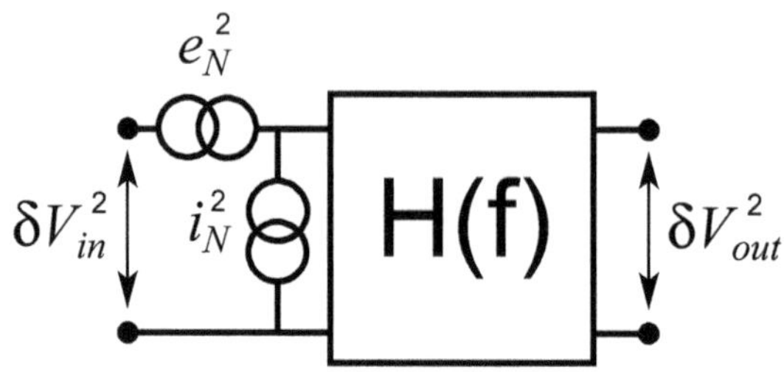

Figure 3.3 Representation of a linear system with intrinsic voltage (e_N^2) and current (i_N^2) noise.

a power spectral density of δV_{in}^2 and we define the SNR of the input signal by $SNR = V_{\text{DC}}^2/\delta V_{\text{in}}^2$. We can model the noise introduced by the linear system as two independent sources of fluctuations each represented by their respective spectral densities e_N^2 (with voltage units) and i_N^2 (with current units). These sources are placed as shown in Fig. 3.3. The current noise i_N^2 becomes voltage fluctuations by acting on the resistance R_{S} or device connected to the input of our linear system, so that $\delta V_{i_N}^2 = i_N^2 R_{\text{S}}^2$.

By connecting an arbitrary input resistance R_{S}, the contributions to the total noise in the system are: the voltage noise generated by our resistance (signal of interest), and the voltage and current noise from the linear system. Therefore, the resulting spectral density which is measured, normalized by the gain (transfer function $H(f)$), is given by

$$S(f) = \delta V_{\text{in}}^2 + e_N^2 + i_N^2 R_{\text{S}}^2. \tag{3.3}$$

The *noise figure* (NF) of an amplifier is defined as

$$NF = \frac{SNR_{\text{in}}}{SNR_{\text{out}}}. \tag{3.4}$$

This quantity gives an estimate of how our signal of interest δV_{in}^2 degrades when it is amplified. An ideal amplifier will have $NF = 1$, i.e., it will not modify the input signal. In our case, we have that $SNR_{\text{out}} = V_{\text{DC}}^2/S(f)$ and for our input resistance R_{S}

$$NF(f, R_{\text{S}}) = \frac{S(f)}{\delta V_{\text{in}}^2} \tag{3.5}$$

Amplifiers usually include their voltage and current noise specifications, and the appropriate one can be chosen depending on the resistance and the voltage noise of the device under study.

Case study 3.1: Choosing the right amplifier

Consider the INA217 and the INA111 amplifier chips. These devices have the following intrinsic noise: $e_N = 1.3\ \text{nV}/\sqrt{\text{Hz}}$, $i_N = 0.8\ \text{pA}/\sqrt{\text{Hz}}$ for the INA217 and $e_N = 10\ \text{nV}/\sqrt{\text{Hz}}$, $i_N = 0.8 \cdot 10^{-3}\ \text{pA}/\sqrt{\text{Hz}}$ for the INA111. It can be easily calculated (and shown in the Figure below) that if the resistance of the sample $R < 12.4\ \text{k}\Omega$, then the INA217 amplifier adds less noise to the measurement, while for higher values of resistance the INA111 chip is a better option because of its lower current noise.

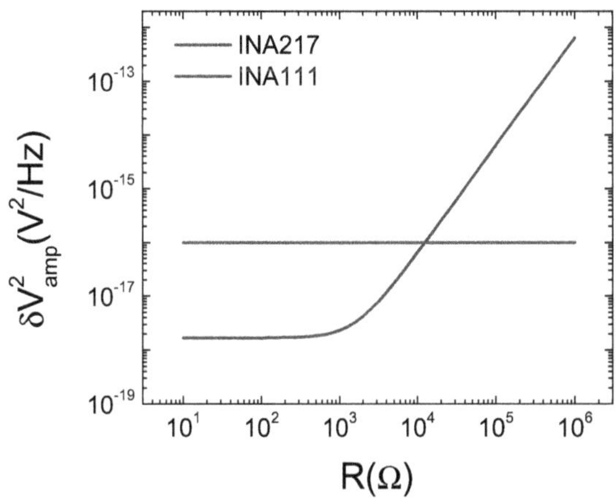

Usually, two amplifying stages are used for noise measurements. Typically, the first pre-amplifying stage adds very little noise (high SNR) but has a relatively low gain. The preamplified signal is then input into a second amplifier, which can have a higher intrinsic noise (because its input signal has already been pre-amplified) and ideally a gain yielding a final signal which falls into the range of the spectrum analyzer.

3.1.1 The Cross-Correlation Technique

If the noise of the amplifiers available to the researcher is higher than the fluctuations under study, the cross-correlation technique may be used to reduce the influence of the electronics on the measurement. This method can be implemented by having two channels which amplify the signal of interest using amplifiers of the same model. Both channels, $X(t)$, $Y(t)$, have therefore the signal of interest plus noise from the electronics, which is of the same amplitude but independent between channels. The cross-spectrum (or cross-correlation spectrum) may be obtained by the use of a spectrum analyzer.

If we are dealing with complex-valued functions, the correlation function changes to

$$R_{\mathrm{XY}}(\tau) = \lim_{T\to\infty} \frac{1}{2T} \int_{-T}^{T} X^*(t)Y(t+\tau)dt. \tag{3.6}$$

This is known as the cross-correlation function of two complex-valued, time-dependent variables. Taking the conjugate of $X(t)$ ensures that aligned peaks (or aligned troughs) with imaginary components will contribute positively to the integral.

Each of the two channels mentioned above contains the signal of interest $V_{\mathrm{R}}(t)$ and the voltage noise of its preamplifier, i.e., $X(t) = V_{\mathrm{R}}(t) + X_{\mathrm{pre}}(t)$ and $Y(t) = V_{\mathrm{R}}(t) + Y_{\mathrm{pre}}(t)$. $X_{\mathrm{pre}}(t)$ is independent (or uncorrelated) from $Y_{\mathrm{pre}}(t)$, so using equation 3.6:

$$\begin{aligned} R_{\mathrm{XY}}(\tau) &= \lim_{T\to\infty} \frac{1}{2T} \int_{-T}^{T} (V_{\mathrm{R}}(t) + X_{\mathrm{pre}}(t))^*(V_{\mathrm{R}}(t+\tau) + Y_{\mathrm{pre}}(t+\tau))dt \\ &= \lim_{T\to\infty} \frac{1}{2T} \left(\int_{-T}^{T} V_{\mathrm{R}}^*(t) V_{\mathrm{R}}(t+\tau) + \cancelto{0}{\int_{-T}^{T} V_{\mathrm{R}}^*(t) X_{\mathrm{pre}}(t+\tau)} \right. \\ &\quad \left. + \cancelto{0}{\int_{-T}^{T} X_{\mathrm{pre}}^*(t) V_{\mathrm{R}}(t+\tau)} + \cancelto{0}{\int_{-T}^{T} X_{\mathrm{pre}}^*(t) Y_{\mathrm{pre}}(t+\tau)} \right) \end{aligned}$$

we see that the only term that survives is the term involving $V_{\mathrm{R}}^*(t)V_{\mathrm{R}}(t+\tau)$.

Spectrum analyzers which include this feature obtain the cross-spectrum, which is given by

$$\mathcal{F}\{R_{\mathrm{XY}}\} = (\mathcal{F}\{X(t)\})^* \cdot \mathcal{F}\{Y(t)\}. \tag{3.7}$$

A diagram of the experimental circuit is shown in 3.4. The signal (DC + fluctuations) should be first duplicated into two channels, denoted by X and Y, and then amplified by two amplifiers of the same model. The DC part of the signal should be filtered out, and

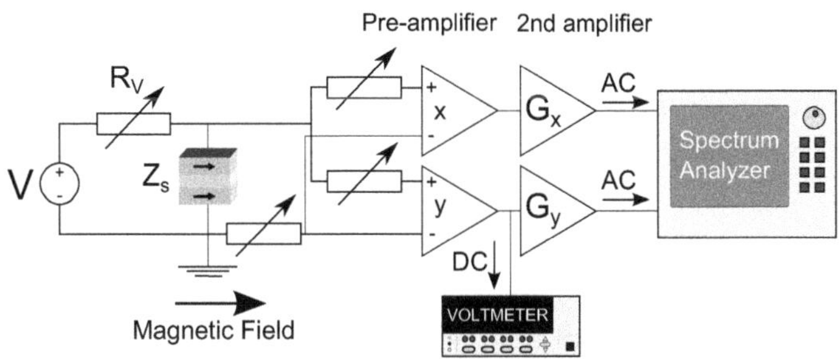

Figure 3.4 Diagram of the low frequency measurement circuit.

some low-noise amplifiers include this option. As we discussed earlier, depending on the signal of interest, it may be necessary to amplify the signal in two stages. Both channels containing the amplified and filtered signal are then input into a spectrum analyzer, which calculates the cross-spectrum.

Since both channels of the signal are amplified nearly identically but independently of one another, both channels contain the sample noise signal δV_{R_S} amplified by the same gain, plus a random noise δV_{preamp} due to the detection circuit. The noise due to the electronics of the system is of the same magnitude for both channels, but the noise of one channel is uncorrelated (independent) from the noise in the other channel.

If we consider the preamplifiers from both X and Y channels to have the same i_{N}^2, the total fluctuations reaching the spectrum analyzer will be

$$S_{\text{V}} = e_{N_X}^2 + e_{N_Y}^2 + \delta V_{R_S} + 2i_{\text{N}}^2 R_{\text{S}}^2. \tag{3.8}$$

The details of how this result is derived may be found in Ref. (Guerrero, 2007). By using a cross-correlation technique (explained in the next subsection) we may neglect the voltage noise contributions from the preamplifiers. Then, we can see that the only unwanted fluctuations correspond to the current noise of the preamplifiers. So we should choose a preamplifier with the lowest current noise possible. The homemade preamplifiers considered in the previous section employ the INA111 chip, which presents the following characteristics: $e_{\text{N}} = 10$ nV/$\sqrt{\text{Hz}}$ and $i_{\text{N}} = 0.8 \times 10^{-3}$ pA/$\sqrt{\text{Hz}}$. The current noise of this operational amplifier is 3 orders of magnitude lower than for other widely used chips, such as the INA217 and the INA114.

The spectrum analyzer used in the experiments obtains the cross-spectrum, which is given by

$$\mathcal{F}\{R_{\mathrm{XY}}\} = (\mathcal{F}\{X(t)\})^{*} \cdot \mathcal{F}\{Y(t)\}. \tag{3.9}$$

3.2 Data Analysis

In relation to low frequency noise, one has to carry out an extensive work regarding the analysis of the data. The main goal is to improve the relative error of the estimated parameters extracted from the spectra or time series, as well as converting the analysis of the data to a semi-automatic and reliable process.

For a good analysis, it is very convenient to deal with large sets of data to write scripts (for example in Octave) which incorporate fixed criteria, and only require a minimum input by the researcher. Regardless of the good functioning of the programs, the analysis, curve fittings, etc., are always supervised should strange results arise. The improvements in data analysis have focused on estimating the parameters of $1/f$ noise, shot noise and random telegraph noise.

3.2.1 1/*f* Noise Analysis

For spectra presenting $1/f$ noise (equation 2.22), we have

$$S_{\mathrm{V}}(f) = \frac{\alpha V^2}{A f^{\beta}} \Rightarrow \log\left(S_{\mathrm{V}}(f)\right) = \log\left(\frac{\alpha V^2}{A}\right) - \beta \log(f)$$

so if we perform the linear fit $y(x) = mx + n$ to the logarithm of the spectrum (with $x = f$), $\alpha = Ae^{n}/V^2$ and $\beta = -m$. Each parameter is obtained with its standard deviation.

The range of frequencies $[f_{\mathrm{min}}, f_{\mathrm{max}}]$ for which the fit is carried out is determined before running the program. Also, the band-pass filter of the SR560 amplifiers is undone by dividing the spectra by the filter's transfer function. For this, the high-pass and low-pass cutoff frequencies must be indicated as well. Unwanted peaks sometimes appear in the spectra which come from the power grid (50 Hz and its multiples), and they are also automatically removed.

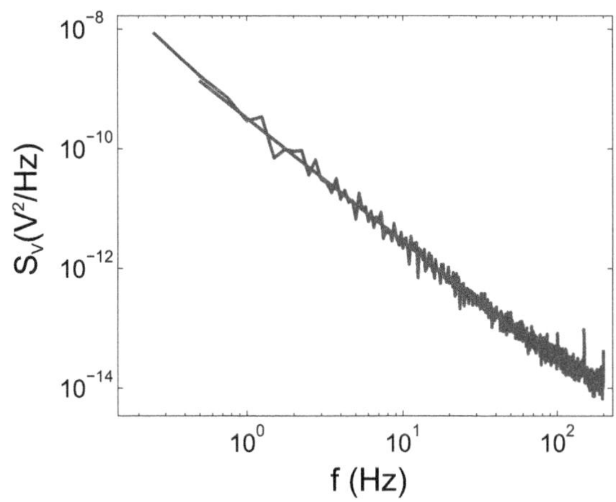

Figure 3.5 Fit of equation 2.22 to a noise power spectrum presenting a typical $1/f$ dependence.

An example of a successful fitting is shown in Fig. 3.5. The analyzed spectra usually contain a dominant $1/f$ contribution from the sample for a bias above 10 mV, while the measurement lines and current source contribute negligibly.

3.2.2 Shot Noise Analysis

For spectra presenting a frequency independent region (i.e., flat), we have two different options for extracting the value of shot noise.

If the spectrum is $1/f$-like at first, and is frequency independent at higher frequencies, we can consider describing it by

$$S_V(f) = \frac{\alpha V^2}{A f^\beta} + S_V^{\text{white}}(f), \tag{3.10}$$

where S_V^{white} corresponds to the white noise (shot noise, thermal noise and noise from the amplifiers). If we fix $\beta = 1$, a fit of the type $y(x) = a/x + b$ yields $\alpha = aA/V^2$ and b is equal to the white noise value. The minimum and maximum frequency between which the fit is carried out is defined at the start. The result of such a fitting is shown in Fig. 3.6(a).

Another alternative is to calculate the histogram of the spectrum. Since the flat part of the spectrum oscillates around the value

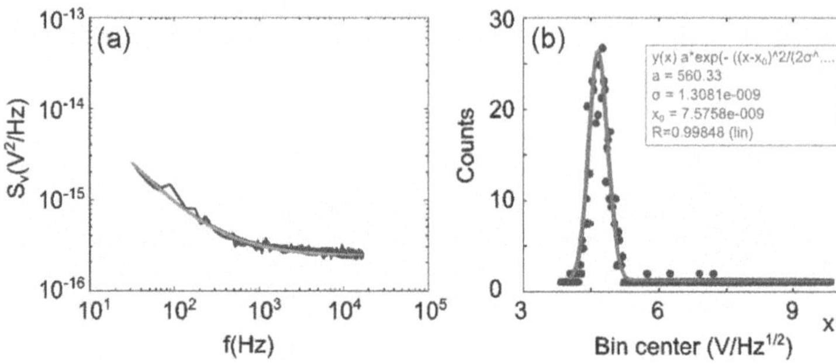

Figure 3.6 Fit of (a) equation 3.10 to a spectrum presenting $1/f$ and shot noise or (b) a Gaussian to the histogram of a spectrum presenting shot noise.

corresponding to the white noise amplitude, the result of the histogram is a Gaussian curve centered at S_V^{white}. By fitting the histogram with $y(x) = Ae^{-\frac{(x-x_0)^2}{\sigma^2}}$, the white noise amplitude is given by x_0 and its standard deviation by σ. A typical fit of this sort is presented in Fig. 3.6(b).

When the temperature is low enough ($T = 0.3$ K) so $eV \ll k_B T$, then the white noise obtained corresponds to the shot noise from the sample, plus some electronic noise from the experimental setup. Supposing this noise remains constant with the applied bias, we extrapolate the value of noise from the electronics from shot noise measurements. The shot noise must scale linearly with the voltage (equation 2.9), and must be zero at zero bias. So the noise from the electronics is estimated as the offset which must be subtracted from the measurements so the shot noise is zero at zero bias. For higher temperatures, also the thermal noise is subtracted from the measurements to extract the shot noise using equations 2.9 and 2.10.

Previously described analysis suggested shot noise measurements done at very low (typically well below 1 K were condition $k_B \ll eV$ is fulfilled) temperatures. Somewhat different approach to obtain Fano factor has been suggested by Arakawa et al. (2011). This analysis method could be better suited for the shot noise measurements carried out in the temperature range close to liquid Helium temperatures (1.5–4.2 K) were transition from $k_B < eV$ to $k_B \ll eV$ regimes could take place. In the low bias range, when the random telegraph and $1/f$ noise contributions are yet smaller than the frequency independent white noise (with both thermal and shot noise) contribution, the current noise power spectrum S_I as a function of bias V_{sd} may be fitted to the relation 3.11 with Fano factor F being the only fitting parameter:

$$S_I(f) = 4k_{\mathrm{B}}T/\frac{dV}{dI} + 2F\left[eI\cot\left(\frac{eV_{\mathrm{sd}}}{2k_{\mathrm{B}}T}\right) - 2k_{\mathrm{B}}T/\frac{dV}{dI}\right] \quad (3.11)$$

Evidently, differential resistance $\frac{dV}{dI}$ should be obtained from the separate experiment.

3.2.3 Random Telegraph Noise Analysis

Two different methods of analyzing random telegraph noise have been developed.

3.2.3.1 Using the time series and spectrum

This is the most usual method found in the literature to obtain the time constants of the process. Two data files are needed for the time series and the spectrum, at each value of the parameter that is being varied (current, voltage or magnetic field).

If one studies two-state RTN, the histogram of a time series will have two Gaussians, centered around the voltage values of each state. One calculates the value of each time-series, and fit two Gaussian peaks to it. The difference in voltage ΔV between the levels is estimated from the difference between the center of the peaks $\Delta V = x_0^{\mathrm{up}} - x_0^{\mathrm{down}}$. Then, using equation 2.17, we can fit the spectrum by $y(x) = \frac{AT^2}{1+(2\pi Tx)^2}$ where $A = \frac{4\Delta V^2}{(\tau_{\mathrm{up}}+\tau_{\mathrm{down}})}$ and $T^{-1} = 1/\tau_{\mathrm{up}} + 1/\tau_{\mathrm{down}}$. Thus, we have two equations, and since we know ΔV from the time series, only two unknowns, τ_{up} and τ_{down}. The program should discard bad fits by checking the χ^2 value of the fit. An example of a good fit is shown in Fig. 3.7.

3.2.3.2 Using only the time series

The time constants of the RTN process may be obtained only using the time series if a sufficiently large number of up-down, and down-up transitions are recorded. This way, fitting the spectrum with the

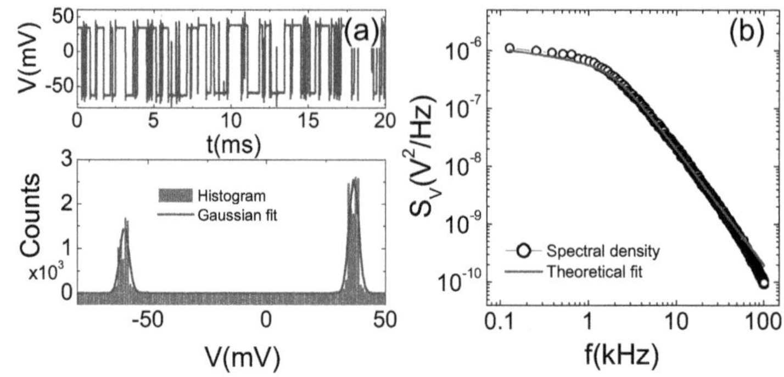

Figure 3.7 Fit of (a) two Gaussian peaks to the histogram of the time series and (b) the theoretical RTN equation to its respective equation.

theoretical curve is not necessary. The time constant τ_{up} may be estimated by the total time the time trace is in the *up* state, divided by the number of *up* terraces in the trace, which is equivalent to calculating the average of the duration of the *up* terraces. The same applies for τ_{down}. We have

$$\langle \tau_{\text{up}} \rangle = \frac{\sum_{i=1}^{N_{\text{up}}} t_i^{\text{up}}}{N_{\text{up}}}$$

$$\langle \tau_{\text{down}} \rangle = \frac{\sum_{i=1}^{N_{\text{down}}} t_i^{\text{down}}}{N_{\text{down}}}.$$

An appropriate program first calculates the histogram of the time series, makes sure there are two Gaussian peaks (two-state RTN), calculates the number of transitions, and finally gives an estimate for τ_{up} and τ_{down}.

Chapter 4

Noise in Metallic Spin Valves

This chapter discusses noise in metallic spintronic devices, paying main attention on spin valves (SVs) which are common for high current and relatively high dissipation spintronic applications. Noise in some other metallic structures such as exchange-coupled multilayers will also be discussed. The main advantage of metallic SV is their robustness respect the possible external damage (radiation, current peaks, etc.)

4.1 Resistance Low-Frequency Noise in Spin Valves and Its Relation with Fluctuation-Dissipation Theorem

Soon after GMR effect was discovered in the antiferromagnetically coupled multilayers, Hardner et al. considered the applicability of FDT to these early stage spintronic devices (Hardner et al., 1993b). The goal was to investigate link between fluctuations of the magnetic moment of the electrodes (linked through FDT to the out of equilibrium characteristic known as imaginary contribution to the

Noise in Spintronics: From Understanding to Manipulation
Farkhad G. Aliev and Juan Pedro Cascales

ISBN 978-981-4774-34-5 (Hardcover), 978-1-315-11088-2 (eBook)
www.panstanford.com

magnetic susceptibility) and fluctuations of the device resistance. The simplest form of such relation could be written as (Hardner et al., 1993b)

$$S_R(f) = \left(\frac{\partial R}{\partial m}\right)^2 S_{\mathrm{m}}(f), \tag{4.1}$$

where m is the fluctuating (free) magnetic moment and R is the resistance of spintronic device. For a sample in thermal equilibrium and exhibiting the linear response, the magnetization noise $S_{\mathrm{m}}(f)$ is given by the following FDT relation:

$$S_{\mathrm{m}}(f) = \Omega \frac{2k_{\mathrm{B}}T}{\pi\mu_0 f}\chi''_{\mathrm{m}}(f), \tag{4.2}$$

where χ'' is the imaginary part of the susceptibility, Ω represent the volume and μ_0 is the vacuum permeability. In the ferromagnetic system the χ'' is a constant. Substituting 4.1 in 4.1, we have

$$S_R(f) = \left(\frac{\partial R}{\partial m}\right)^2 S_{\mathrm{m}}(f) \Rightarrow \left.\frac{\partial R}{\partial H}\right|_{f=0} = \frac{2m\mu_0}{k_{\mathrm{B}}T\,\Delta R}\int_0^{\infty} S_R(f)df, \tag{4.3}$$

where $2m$ is the changes in magnetic moment and ΔR is the resistance under inversion of the magnetization of the fixed layer. In the case when the low-frequency noise is close to $1/f$ type, the integral of the equation 4.3 is proportional to Hooge factor α, therefore

$$\frac{\partial R}{\partial H} = \frac{2m\mu_0}{k_{\mathrm{B}}T\,\Delta R}\frac{R^2}{A}\ln\left(\frac{f_{\mathrm{max}}}{f_{\mathrm{min}}}\right)\alpha. \tag{4.4}$$

If close to the transition between both magnetic states, the $1/f$ noise is dominated by the magnetic fluctuations, the Hooge factor scales as dR/dH:

$$\alpha \propto \frac{dR}{dH} \tag{4.5}$$

This relation has been experimentally confirmed (Hardner et al., 1993b) in the case of resistance noise in Co/Cu multilayers.

An extensive study of the relation between normalized equilibrium low-frequency noise and the resistance susceptibility has been verified separately for the free and pinned magnetic layers in

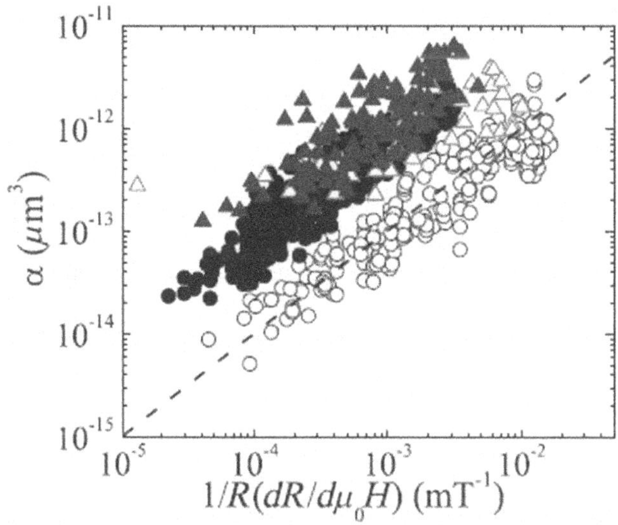

Figure 4.1 Hooge parameter during magnetization switching of different samples with free (open symbols) and fixed (closed symbols) magnetic electrodes of spin valve. The evident linear dependence in this compilation plot confirms that LFN near the switching fields is equilibrium noise of magnetic origin. Figure adapted with permission from Ref. Diao et al. (2011).

yoke-shaped, giant magnetoresistance spin valve sensors (Diao et al., 2011). Figure 4.1 shows predicted linear dependences both for free and fixed layers.

The normalized noise power was found to increase linearly with the inverse of the voltage contacts distance along the arm of the yoke, indicating a local noise mechanism (Diao et al., 2011). By analyzing dependence of the $1/f$ noise on the fluctuation volume, in the conduction of having observed $1/f$ type noise even for the smallest volume (where existed at least 10 fluctuators), the upper limits of 9×10^4 and 4.5×10^4 nm^3 for the fluctuating magnetic volume of the free and pinned layers, have been estimated respectively. Local fluctuators are suggested to originate from the micromagnetic disorder formed during the deposition and processing of the SVs.

4.2 Is There Excess Low-Frequency Noise in the AP State?

Besides equilibrium (described by FDT) part, LFN could contain also non-equilibrium contribution which could result in substantial difference in the normalized LFN under change of the magnetic state of the SV device. We note that Hardner et al. (1993b) considered the applicability of FDT mainly to the magnetic field regions where FM layers varied their relative orientation LFN and where LFN may be strongly enhanced with respect to the well-defined P and AP configurations. In the case when AP and P states are well defined, a separate question could arise whether the normalized by V^2 or I^2 noise in these states are the same. The answer is generally no, because the excess noise could originate from non-equilibrium spin shot noise related with spin accumulation or some other factors to be discussed in the following paragraphs.

Among the first experimental indications of the excess normalized noise in the well-defined AP state was report by Guerrero and co-authors (Guerrero et al., 2005) on electron transport and noise in epitaxial Fe(110)/MgO(111)/Fe(110) MTJs with suppressed coherent tunneling. Figure 4.2 shows typical noise power spectra for different bias currents and dependence of Hooge (obtained by the noise normalization by V^2) on the magnetic state. It was observed that the large lateral size of MTJs suppressed magnetic coupling between the ferromagnetic electrodes. Excess of the Hooge factor in the AP state is suppressed in smallest size junctions with dipolar coupling.

As long as TMR remains rather flat in the AP state providing negligible absolute values of $dR/dH(AP)$, the excess normalized noise (except in the transition regions between AP to P state) could not be directly attributed to the magnetization fluctuations. Among possible explanations could be influence of the non equilibrium spin shot noise (see separate discussion on spin shot noise in MTJs) or other (for example, magnetostriction related) factors discussed further below.

It is worth to mention that there have been also suggestions on the possible extrinsic character of the observed experimentally

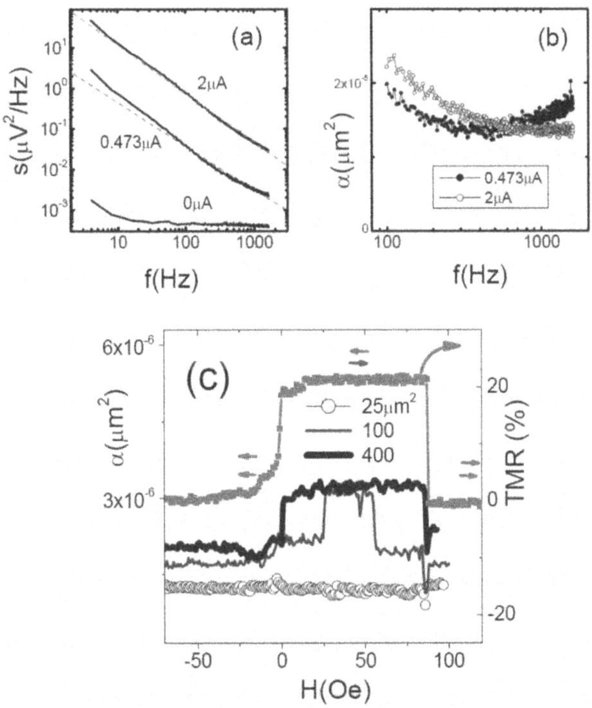

Figure 4.2 (a) typical noise power spectrum in Fe(110)/MgO(111)/Fe(110) MTJs for different bias currents. (b) Weak dependence of Hooge factor as a function of frequency for two bias currents corresponding to close to $1/f$ noise. (c) Dependence of the normalized noise (Hooge factor) on the increasing magnetic field for three MTJs with different lateral sizes indicated by labels. For the comparison a typical TMR curve (red solid square dots) is also shown (see Ref. Guerrero et al. (2005) for more details).

excess noise in the AP state in the SVs. For example Ozbay et al., (Ozbay et al., 2006) reported that the $1/f$ noise normalized by the resistance square was the largest in the AP state and decreased monotonically as the free and fixed layers become parallel in the P state. Later a strong upward deviation in the normalized resistance noise in the AP state in the above micron size SVs has been confirmed (Ozbay et al., 2009) by extensive comparison of the results with expected from FDT. It has been suggested that FDT model underestimates the noise in the close to AP state because of some finite fluctuation of both FM layers in the AP state. Later report by Guo et al. (Guo et al., 2013) did not observe any measureable differences in the noise spectra between the P state and AP states nor any noticeable noise excess in the magnetic transition region where dR/dH is largest. It was claimed that the previous results on LFN vs. magnetic state were influenced by the magnetic aftereffect.

In the following and throughout the book we shall discuss different possible reasons for non-trivial variation of the LFN with magnetic state. Particularly, the possible influence of spin flips at interfaces, sample size affecting coupling between electrodes and presence of spin torque effects in the smaller junctions. Indeed, current and magnetization noise coupling through spin torque is one of multiple reasons for the normalized noise excess in the AP state.

4.3 Coupled Current and Magnetization Noise

Indeed, one possible reason for the excess normalized noise in the AP state could be the coupling between current and magnetization noise (Foros et al., 2009). We have previously seen that relation between the equilibrium voltage noise and the out-of-equilibrium energy dissipation (proportional to electric resistance) as a standard example of the fluctuation dissipation theorem. In the case of static (non-fluctuation) magnetization in spintronic devices, the relative variation of equilibrium noise between P and AP states could be expected to be proportional to GMR, just because current/voltage noise would follow resistance change. This would lead to the magnetic state independent Hooge factor - in line with shown in Figure 4.2c for the smallest in size MTJs with strongly coupled (i.e., weakly fluctuating) ferromagnetic electrodes.

Situation, however, could be qualitatively different if in addition to the current, the total magnetization of the device fluctuates as well (Foros et al., 2009). Generally, the fluctuation in magnetization of insulated single domain nanomagnet could be described by another manifestation of the FDT. It has been shown that white noise in the magnetization fluctuations due to thermal agitation is proportional to magnetization damping (Safonov and Bertram, 2005). Viscous (Gilbert) damping that relaxes the magnetization towards the ground state configuration was introduced as phenomenological parameter in Landau–Lifshitz–Gilbert (LLG) equation. In the SVs, when the ferromagnetic layer is in the contact with metallic environment (non-magnetic spacer- NM), an additional loss of the angular momentum due to spin pumping at FM/NM interphase results in enhanced Gilbert damping (Tserkovnyak et al., 2005). Adsorbed at the interface spin-current components lead to fluctuation in spin transfer torque which produces spin waves (Berger, 1996). Corresponding extra noise, in accordance with the FDT, is directly related with the above discussed excess in Gilbert damping. Foros and co-authors (Foros et al., 2009) investigated the interplay between zero frequency current and magnetization noise in multilayers composed of monodomain ferromagnetic and non-magnetic films. Magnetoelectronic circuit theory (Brataas et al., 2006) was used to calculate the charge and spin-current fluctuations in symmetric (both layers fluctuate) and asymmetric (one of the FM is fixed) configurations.

It has been suggested therefore the existence of coupled current and magnetization fluctuations associated with dissipation for the SVs in out-of-equilibrium conditions (Foros et al., 2009). Resulting interaction between current and magnetization dynamics (via the reciprocal effects of spin-transfer torque and spin pumping) affect not only the time averaged characteristics (GMR) but also the noise of the device. Resulting cross-talk of current and magnetization dissipations is predicted to give rise to a considerable excess in the resistance noise in AP configuration due to spin-current noise (Foros et al., 2009).

On the experimental side, as we have mentioned in brief, there has been controversy on the excess (normalized by voltage square) low-frequency noise in the AP state in comparison with P state in spin valves. Early works on LFN carried out for macroscopic (above micron size SVs) found anomalies in the normalized $1/f$ noise (Hooge factor) in the field regions where FM electrodes reorient (Hardner et al., 1993b). Later reports examined Hooge factor in P and AP orientations in SVs and MTJs with relatively small TMR (not exceeding correspondingly 10% and 20%) found normalized by V^2 noise power to be similar if the noise is measured in the true equilibrium conditions (Guo et al., 2013). As we shall see in the following chapters, the situation with Hooge factor dependence on

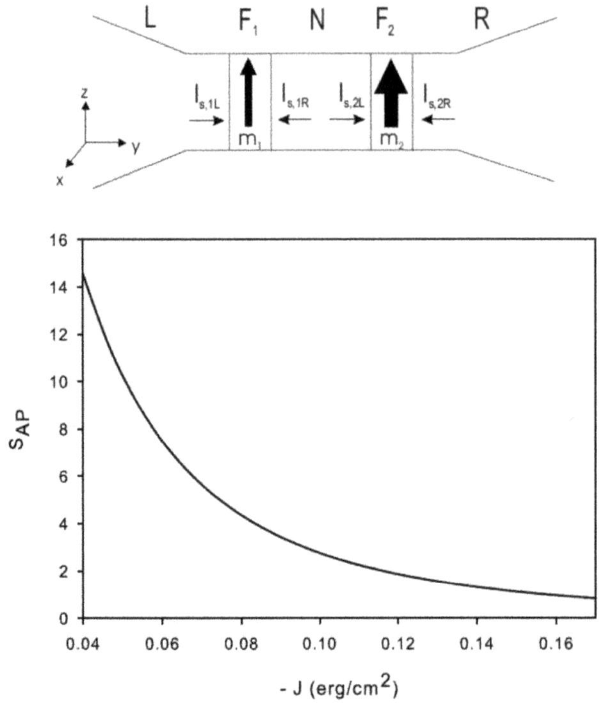

Figure 4.3 Coupled current and magnetization noise. Upper part shows sketch of a spin valve with two ferromagnets F1 and F2 and currents. Lower part shows the resistance noise in the AP configuration as a function of the dipolar and exchange couplings between the ferromagnets. Adapted with permission from Ref. Foros et al. (2009).

the magnetic state in MTJs with MgO barriers (with TMR exceeding 200%) could be different. We are not aware on extensive studies of the $1/f$ noise in nanometer size metallic SVs for which theoretical predictions (Foros et al., 2009) have been made. It seems however, that the predicted coupled magnetization and current fluctuations could be mainly relevant for the nm size MTJs with ultrathin (below 1 nm) semi-transparent MgO barriers. Indeed, as we shall see further below in the chapter analyzing LFN in single barrier MTJs, the predicted by Foros et al. (Foros et al., 2009) strongly enhanced Hooge factor in the AP respect P state could probably be observed in sub-100 nm elliptic CoFeB/MgO/CoFeB MTJs with 0.9 nm-thick

MgO barrier (Cascales et al., 2013a), indicating possible coupled magnetization and current fluctuations

Interestingly, the current-induced spin-transfer torques have been suggested to act also in the opposite direction, i.e., suppress the effects of thermal fluctuations (Bandopadhyay et al., 2011). The idea is to increase (with small extra power) the fidelity of the SV device by incorporating feedback loop which controls the spin-transfer torque depending on the instantaneous magnetization configuration. Magnetic configuration is continuously measured through electric resistance and correcting current pulses are generated by state electronics.

Alternative route to suppress the magnetic noise through the spin torque suggests using periodic spin torques to enhance the stability of the system (Rebei, 2011). In the real devices with relatively low energy barriers, it is shown that the spin torque effects are capable to inhibit interwell transitions suppressing in this way the RTN from a spontaneous switching between neighboring states.

4.4 Link between Domain Wall Related Noise and Spin Torque in Spin Valves and Synthetic Antiferromagnets

Ferromagnetic (or antiferromagnetic) line structures with domain walls present a peculiar example of the natural spin valve. High enough current densities could displace domain wall due to spin torque or self-field. The study of domain wall motion under spin transfer torque (STT) is currently one of the hottest topics due to potential development of the novel types of magnetic state based memories (such as race-track memories (Parkin et al., 2008)).

Investigation of low-frequency noise due to DWs or other topological anomalies such as magnetic inhomogenieties in the SVs, antiferromagnetically coupled magnetic multilayers or MTJs (for noise in single barrier MTJs, see Chapter 6) could provide a valuable information on the presence of STT through the analysis of the LFN as a function of both density and sign of the current. Here we discuss such a possibility on two different examples: one includes Co/Pt and of Ni/Pt SVs etched into nanopillars with an elongated hexagonal shape of 100 by 200 nm^2 and perpendicular anisotropy (Cucchiara et al., 2009). Other example considers STT related noise in $[Fe/Cr]_{10}$ multilayers patterned into few micron size bridges (Herranz et al., 2009).

Detailed measurements of RTN and the dwell times corresponding to DW pinning and nucleation (Cucchiara et al., 2009) showed that the experimentally observed noise is consistent with thermal activation over an effective potential barrier. The different metastable magnetization states originate from the DW wall pinning induced by the presence of defects. It has been found that the presence of spin transfer torque helps to reach magnetic states not accessible using applied fields only through an additionally reduced energy barrier (Cucchiara et al., 2009).

One of the major problems of the proposed race-track memory devices are high current densities needed to move DWs exceeding 10^8 A/cm^2. Those currents heat the devices very close to the Curie temperature (Yamaguchi et al., 2005). Spin torque in antiferromagnetic system has been suggested as an alternative way to solve the problem of heating due to high critical currents (Nuñez et al., 2006).

Experimentally, such possibility has been investigated by comparison the noise from low-frequency DW dynamics in Co line structures as reference ferromagnetic structure and in $[Fe/Cr]_{10}$ synthetic antiferromagnetic (SAF) systems (Herranz et al., 2009). A detailed study of DW magnetoresistance and low-frequency noise in SAFs revealed an unexpected dependence of the DW-induced LFN on the current density (Herranz et al., 2009). This observation has been attributed to current-induced magnetization dynamics that occurs at a critical current density which is smaller than the typical current densities observed in the ferromagnetic materials, typically being above 10^7 A/cm^2. Fig. 4.4(a) shows the low-temperature in-plane magnetoresistance in a sample with composition [Fe (2.4 nm)/ Cr (1.3 nm)]$_{10}$. The normalized magnetoresistance, measured at a temperature of 77 K in this type of MML, for the low magnetic field regions where DWs are nucleated at $H = H_n$ and annihilated at $H = H_a$.

When the current density exceeds 2×10^5 A/cm^2 at $T = 77$ K, the excess noise observed for low magnetic fields, where DWs

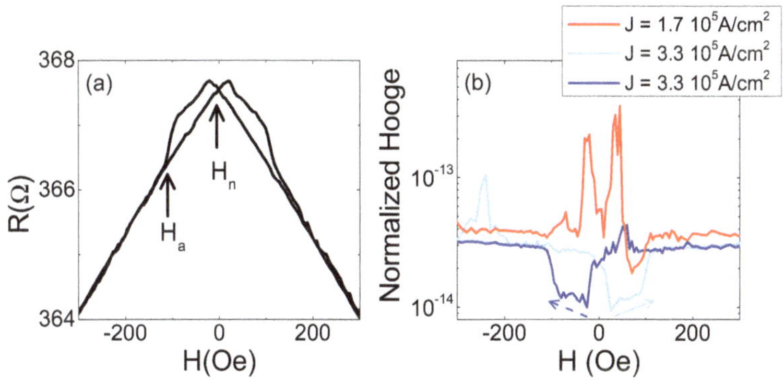

Figure 4.4 Spin torque in synthetic antiferromagnet detected via LFN. (a) Low-field magnetoresistance measured at $T = 77$ K with very small current density and with the current parallel to the external magnetic field. (b) Hooge factor (without area normalization) plotted on a logarithmic scale vs. magnetic field for [Fe(2.4 nm)/Cr(1.3 nm)]$_{10}$ at 77 K with two different applied current densities. The normalized noise vs. field measurements are done with a current density of $J = 3.3 \times 10^5$ A/cm^2 and the results are shown both for increasing and decreasing the magnetic field. The experiments have been carried out with the *dc* current parallel to the external magnetic field and to the magnetic easy (001) axis.

are created and propagated, is suppressed in the field interval for stronger DW pinning. The excess noise and the deviation from the $1/f$-like frequency dependence are attributed to non-equilibrium noise present at low fields to a RTN-type contribution to the noise at low biases. Despite this deviation, the Hooge factor has been used as a simplified parameter for the analysis of the normalized noise. A possible scenario for the substantial suppression of the magnetic noise in the vicinity of the depinning field at current densities $J_c > 2 \times 10^5$ A/cm^2 (see Fig. 4.4b) could be the reduction in the local magnetic disorder induced by the large scale DW motion. In other words, it has been proposed that the downward and upward variations in the magnetic noise at low fields appear due to the competition between local magnetic disorder (which provides the background magnetic noise) and its reduction by the moving DWs at larger current densities. This experimental study indicates on the perspectives of the future uncompensated SAF based race-track spintronics and magnetic memories.

Closing this paragraph, we mention that the following chapters will also discuss qualitatively similar effects observed for DW dynamics due to STT effect created by perpendicular currents in MTJs. As long as MgO tunnel barrier provides symmetry filtering effect (Cascales et al., 2013a; Herranz, 2012; Herranz et al., 2011), MTJs of such kind dramatically enhance effectiveness of perpendicular current through current focusing into topological anomalies (or DWs) so that the STT related DW dynamics could be observed at rather small total current densities.

4.5 Shot Noise in Spin Valves

Shot noise in non-magnetic phase-coherent diffusive conductor (which is substantially longer than the mean free path but shorter than an inelastic scattering length) is suppressed down to 1/3 of its full value (Beenakker and Büttiker, 1992a). This is the result of the bimodal distribution of the transmission eigenvalues (Beenakker and Büttiker, 1992a). Magnetic degree of freedom in SVs potentially allows control over the shot noise through combined influence of spin flips and relative alignment of the ferromagnetic electrodes. Despite numerous theoretical predictions considered below, up to now this possibility remained unexplored experimentally. The main difficulty is probably related with too small bias signals possible to apply to SVs. Most of the theoretical considerations on the shot noise in spin valves look at possible influence of spin orbital coupling on shot noise.

Influence of SO effects on shot noise of the spin-polarized electrons has been first studied in a beam splitter geometry where cross-correlation between two entangled spin-polarized electrons (with phase affected by Rashba SO coupling in one of the leads) (Egues et al., 2002). Spin orbit interaction was found to determine the Fano factor, implying that the Rashba coupling could be measured via shot noise (Egues et al., 2002).

Influence of SO interaction-induced spin flips on the SN of spin-polarized electrons injected into diffusive metal was considered at $T = 0$ by Lamacraft (Lamacraft, 2004). The main parameter to vary was relation between length of the wire (L) and spin diffusion length

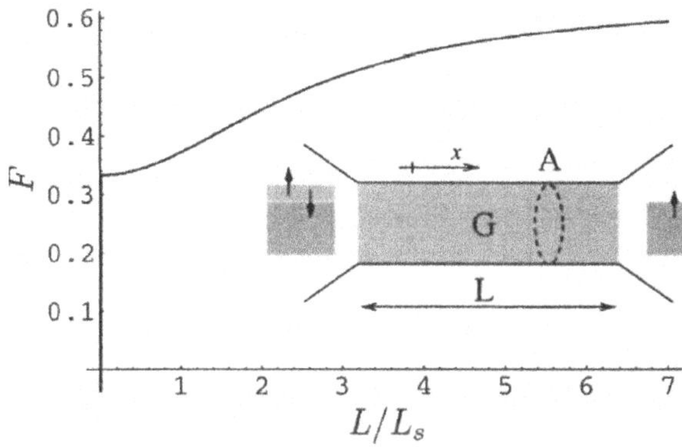

Figure 4.5 Fano factor for spin-polarized current a diffusive wire as a function of inverse normalized by wire length L spin-relaxation length L_S. Inset shows experimental geometry. Adapted with permission from Lamacraft (2004).

L_S. As long as $L/L_S \ll 1$ at $T = 0$, the Fano has value remains equal to $F = 1/3$ expected for the unpolarized electrons. This result indicates that, as expected in the absence of spin relaxation, spin polarization of the electrons cannot affect the shot noise. In the limit $L_S/L \gtrsim 1$, the Fano factor, however, gradually increased from 1/3 to 2/3 (see Fig. 4.5).

In open mesoscopic system the Fano factor could therefore be a sensitive tool to measure the spin relaxation length (Lamacraft, 2004). As we shall see further in Chapter 6, such possibility has been proven experimentally through investigation of the magnetic state dependence of shot noise in double barrier magnetic tunnel junctions (Cascales et al., 2012).

Mishchenko (Mishchenko, 2003) considered shot noise of spin-polarized electrons propagating through diffusive conductor of the length L and connected through two barriers to the different ferromagnetic leads. This configuration is close to one realized in lateral SVs (see Fig. 4.6). Spin-flip events were supposed to be independent, energy conserving and obeying Poissonian statistics. It is therefore suggested that the typical energy change during a spin flip is below $k_B T$ or applied bias eV.

Parameter $s = L/2L_S$ used in plots presented in Figure 4.6 is the dimensionless measure of the amount of spin-flip scattering in the

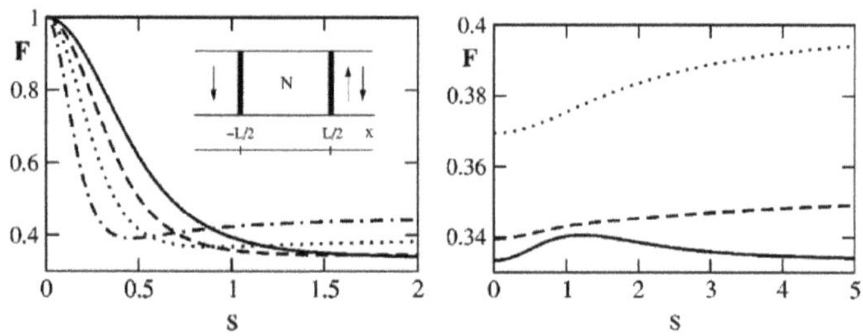

Figure 4.6 Left: Influence of spin flips on shot noise for AP alignment for short diffusive conductor connected to two ferromagnetic electrodes. Contact resistance r reflects the barriers strength with solid line showing results for $r = 0$, dashed for $r = 1$, dotted for $r = 3$, dot dashed for $r = 10$. Inset sketches the device under study. Right part shows influence of spin flips on the shot noise in P configuration. Adapted with permission from Ref. Mishchenko (2003).

system (with L_S being spin diffusion length and L- thickness of the normal metal spacer). When the polarizations of the half metallic leads are antiparallel, electrons cannot be transferred across the valve without changing its spin direction and shot noise becomes Poissonian in the limit $s = 0$. For large spin-flip scattering $s \gg 1$, independent of the relative magnetization of the leads the shot noise approaches the universal value $F = 1/3$ present in a conventional diffusive conductor, since an injected electron likely loses its spin polarization before reaching the opposite electrode (see Fig. 4.6).

For intermediate spin flip strength situation is dramatically different. Depending on the relative alignment of the ideal FM electrodes (P vs. AP) and contact resistance, the dependence of the Fano on the spin flip rate inside diffusion central conductor could substantially change. For the AP configuration and without spin flips, the central conductor with its barriers represents just a single barrier, leading therefore to the Poissonian shot noise (see Fig. 4.6). Spin flips in the AP state, however, open new conductance channels (affected partially by spin diffusion in the central electrode) and result in suppression of the Fano to the values somewhat less than 0.5 for the large spin flip scattering rates. However, in the P state, the introduction of the spin flips trends to enhance the shot noise, similarly to the observed for the previously discussed case with photons propagating in the disordered media. While for the zero contact resistance, the shot noise is retained around 1/3 (disordered metals) and only slightly enhanced by the spin flip scattering (because electron transport is due to mainly spin down (up) states), for the larger contact resistance ($r = 1$), the Fano approaches 1/2 typical for the sequential tunneling regime.

Using the language of spin particles only, one could say that the number of particles in each of spin up (down) sub-systems could not be now conserved due to spin-flip scattering with total spin fluctuation equal to one. In some sense, such boson-like spin fluctuations resemble the well-known fluctuations of the radiation in random optical media. In the absence of photon conservation law, Poissonian photons emitted on the input by laser are converted to super-Poissonian after being propagated in the disordered media (Beenakker and Patra, 1999).

Surprisingly different predictions on the magnetic state dependent shot noise in quasi-equilibrium spin valves with FM electrodes connected through diffusive wire were obtained by Heikkila and Nagaev (Heikkilä and Nagaev, 2013) once they also considered the heat transfer between spin-polarized currents. In the conditions when energy relaxation couples the energies in the two spin systems of the electrons, Wiedemann–Franz law could be violated leading to non-universal shot noise (Heikkilä and Nagaev, 2013).

Figure 4.7 shows that in the P configuration, introduction of non-magnetic spacer may result in a strong suppression of the Fano factor below value $\sqrt{3}/4$ obtained for spin subsystems in local equilibrium and without heat exchange. In the AP configuration the electron-electron scattering balances otherwise thermally uncoupled spin-up and spin-down channels leading to a strongly suppressed Fano factor.

Spin flip shot noise in multi-terminal spin valve with two left (up-down magnetization) and two right (up-down magnetization) and connected by normal metal has been analyzed by Belzig and Zareyan (Belzig and Zareyan, 2004). In was suggested that cross-correlations between currents in right terminals with opposite magnetization directions may provide the spin-flip scattering rate. Although the work assumed fully polarized terminals, authors point out that their

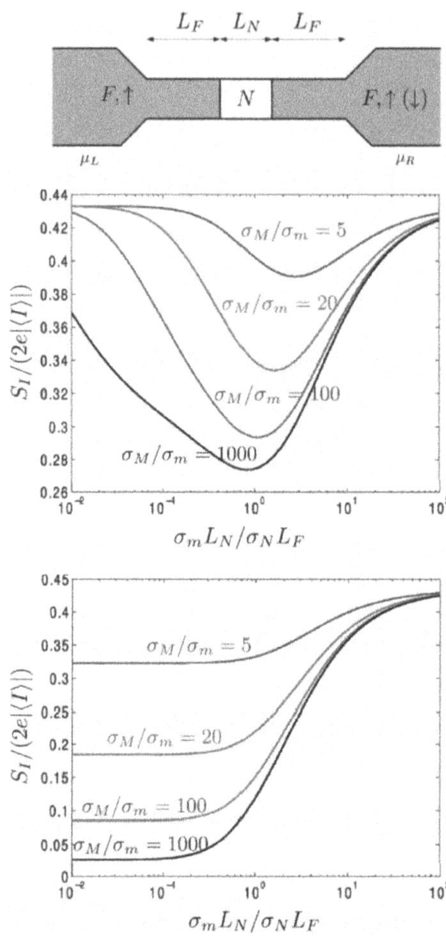

Figure 4.7 Shot noise in spin valve. Upper part: sketch of the SV under study. Middle figure shows for different normalized conductances Fano in the P configuration as a function of normalized spacer and conductance of the ferromagnet. The lower part shows same results for AP state. Adapted with permission from Ref. Heikkilä and Nagaev (2013).

method could also be generalized to the leads with arbitrary spin polarizations.

Summarizing this chapter, we annotate the main points. First of all, most reports seem to agree that LFN in SVs close to transition between different magnetic alignments could be described by FDT.

Secondly, there are some suggestions that LFN in the AP state could be affected by non equilibrium process such as spin accumulation and link between current and spin noises. Finally, while the shot noise in SVs remains practically unexplored experimentally, there are number of interesting theoretical predictions pointing out to rather informative character of SN measurements in SVs potentially providing information on spin flips times. Influence of spin-flip scattering on the shot noise in SVs is analogous to the influence of disorder on shot noise of photons propagating in disordered media, both increasing Fano factor. Last but not least, LFN measurements in ferromagnetic and antiferromagnetic metallic line structures have been shown to be an alternative and rather precise tool to detect current-induced depinning of domain walls.

Chapter 5

Shot Noise in Single-Barrier Magnetic Tunnel Junctions

This chapter presents a summary of some of the most relevant studies of shot noise in inorganic single-tunnel-barrier spintronic devices. Although we shall deal mainly with charge-related shot noise, we shall see that spin degree of freedom could influence the shot noise through spin flip scattering or spin accumulation. We first briefly consider shot noise in Al_2O_3-based MTJs which were historically the first type of MTJs discovered. The main deal of analysis will be dedicated to shot noise in MgO-based MTJs which revolutionized spintronics since 2004. We shall conclude with discussion of recent proposals and attempts to observe pure spin shot noise in MTJs.

5.1 Shot Noise MTJs with AlO_x Barriers

We have already mentioned in the introduction part that the large tunneling magnetoresistance at room temperature was first reported in MTJs with single AlO_x barriers (Miyazaki and Tezuka, 1995; Moodera et al., 1995, 1999a). Since AlO_x usually grows

Noise in Spintronics: From Understanding to Manipulation
Farkhad G. Aliev and Juan Pedro Cascales

ISBN 978-981-4774-34-5 (Hardcover), 978-1-315-11088-2 (eBook)
www.panstanford.com

amorphously, the electrode/barrier interfaces are far from ideal and have numerous structural defects.

The work of Nowak et al. (Nowak et al., 1999) reported earliest study of the noise in MJTs. Although the paper concentrated on $1/f$ noise, the authors also mentioned to observe a full shot noise ($F = 1$) in CoFe/Al_2O_3/CoFe MTJs in both the parallel (P) and antiparallel (AP) magnetic states.

The detailed shot noise measurements in spintronic devices was first reported for low-resistance MTJs (George et al., 2002). The study made on less than 100 Ω MTJs (possibly designed to be used in read heads applications) showed that the results are not consistent with the resistance of the samples examined. The measurements yielded shot noise that was lower than expected, and which pointed to the existence of parallel paths of conduction, pinholes. An electrical model of the sample is proposed where the tunnel junction has an ideal resistance R_T which produces thermal noise when unbiased and shot noise when biased and each pinhole is described as a parallel resistance R_P which exhibits thermal noise only. Through this simple model, the authors were able to account for the decrease of the shot noise as well as the tunneling magnetoresistance, correlating it with the low resistance-area products of the junctions. The model suggests that there exists a relation between the existence (or not) of pinholes and the TMR and shot noise values. Qualitatively similar measurements recently carried out for MTJs with ultrathin (below 1 nm) MgO barriers (Arakawa et al., 2015) will be discussed later in this Chapter as a case study.

Among other earliest studies regarding SN presented by Jiang et al. (Jiang et al., 2004a), reported the observation of full shot noise ($F \approx 1$) in the pinhole free MTJs in the AP state. Later, the same group (Jiang et al., 2004b) measured a strong suppression (down to $F \approx 0.45$) of the SN in magnetic tunnel junctions, which was not understood. The junctions with the smallest resistance present a suppressed Fano factor (below 0.5 for some MTJs), which along with the reduced resistance and the appearance of random telegraph noise, points to an inhomogeneous current flow and thus a defective barrier. Still, the suppression of shot noise under the Poissonian value was not well understood.

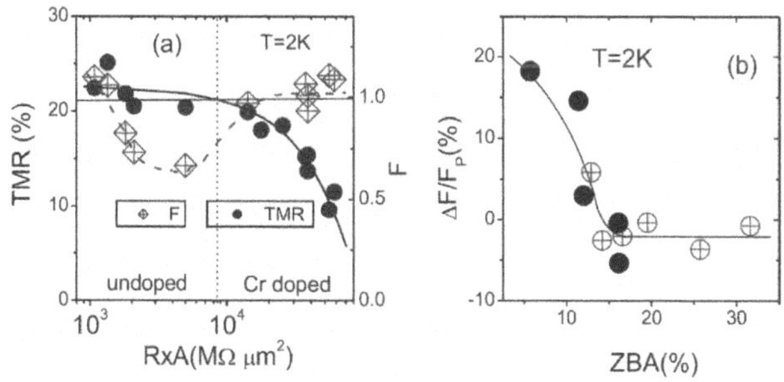

Figure 5.1 Shot noise in magnetic tunnel junctions with sequential tunneling. (a) Dependence of the TMR and F (for the P state) on the resistance area product (R × A). Solid horizontal line marks $F = 1$. Vertical dashed line separates the Cr-free and Cr-doped regions. (b) Dependence of the relative variation of F with alignment $\Delta F/F_{\mathrm{P}}(\%) = 100(F_{\mathrm{AP}} - F_{\mathrm{P}})/F_{\mathrm{P}}$ on the relative strength of the conductance zero bias anomaly (ZBA). Solid symbols point to the undoped samples. The lines are guides for the eye. Adapted with permission from Guerrero et al. (2006b).

Guerrero et al. (2006b) reported the first detailed study (experiment and modeling) of shot noise in single-barrier magnetic tunnel junctions, which varied with the barrier conditions and the magnetic state. This study contains an investigation of shot noise in Co(80Å)/ Al_2O_3(14Å)/ Py(100Å) MTJs. The experimental results showed that Cr impurities seem to be localized near one of electrodes. Therefore, a shot noise reduction down to $F = 0.66$ occurs for non-doped junction, as shown in Fig. 5.1a, indicating homogeneously distributed defects (according to theoretical predictions (Nazarov and Struben, 1996)). Also, they found that the Fano factor depended on the magnetic configuration, being enhanced in the AP state. Interestingly, inverse correlation between strength of the zero-bias conductance anomaly (ZBA) and dependence of the Fano on the magnetic state has been observed (see Fig. 5.1b). Once the ZBA depth reflects the inverse spin flip (spin relaxation) time, shot noise was found to be magnetic state dependent in the limit of excess spin accumulation on the paramagnetic impurities on the barrier defects. This points out possible presence of the spin accumulation-related shot noise, which will be discussed in more details below in this Chapter and in the part describing shot noise in the double-barrier magnetic tunnel junctions.

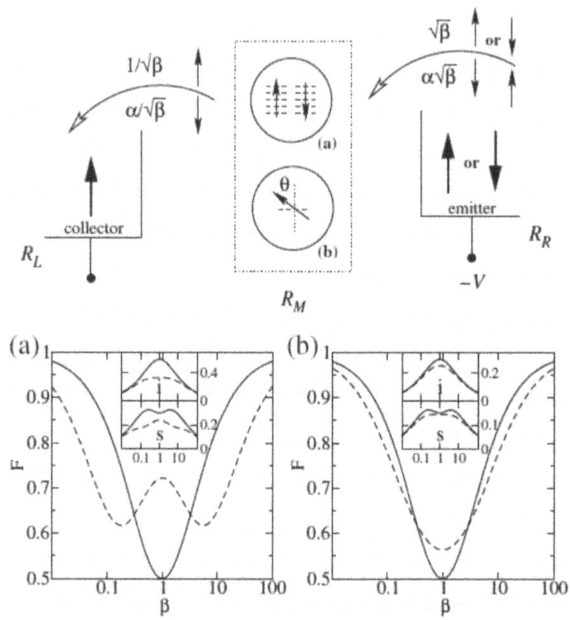

Figure 5.2 Modelling shot noise in magnetic tunnel junctions with sequential tunneling. Upper part sketches two models considered to describe magnetic state dependent shot noise in MTJs. Model (a) has large normal region inside the barrier so that electrons could be treated as non-interacting but coupled asymmetrically to emitter and collector representing FM electrodes. Model (b) has is a single spin-polarized impurity level that can hold only one electron with spin at a (random) angle with respect to the collector magnetization. Lower part shows calculated Fano factor for each of the models as a function of barrier asymmetry (β) and magnetic state which are solid lines for the P state and dashed for the AP state. Adapted with permission from Guerrero et al. (2006b).

To capture qualitatively the main experimental observations on the reduction of shot noise and its dependence of the magnetic state of the electrodes two simple models for the sequential tunneling through an island inside the barrier have been considered (Guerrero et al., 2006b) with main results presented in Fig. 5.2. Comparison of the experimentally observed AP-P asymmetry with the model require some structural preference towards tunneling through the middle of the barrier affected by some inelastic spin-flip mechanism in the barrier to provide reduction of the Fano in P state and enhancement of Fano in the AP configuration (Guerrero et al., 2006b).

Later Garzon et al. (2007) reported on the surprising observation of enhanced spin-dependent shot noise in magnetic tunnel barriers. suggesting transport through localized states within the barrier. One should, however, carefully analyze the form of experimentally measured frequency dependence of the noise power which might be affected by some additional RTN contribution with characteristic time well above the typical experimental bandwidth of few kHz. It these circumstances other (apparently white for the given bandwidth noise) contribution may emerge in addition to the intrinsic frequency independent to GHz range noise from the tunneling electrons.

To summarize the early shot noise studies on MTJs, the amorphous AlO_x present localized states due to impurities, which allow the electrons to tunnel sequentially. This yields a shot noise which is at or below the Poissonian level and which could be controlled magnetically for relatively long spin flip times.

5.2 Shot Noise with MgO Barriers

The TMR record at room temperature in Al_2O_3 (Wang et al., 2004a) was set at 70%. A TEM image of the cross section of a tunnel junction with an AlO_x (Wang et al., 2004b) reveals considerable interface roughness. Having the highest possible, room temperature, TMR is quite desirable for technological purposes. The scientific community was very excited when crystalline MgO barriers were suggested as an effective spin filter (Butler et al., 2001; Mathon and Umerski, 2001). Soon after, the ambient TMR record was improved up to 220% (Parkin et al., 2004; Yuasa et al., 2004), finally reaching an impressive 604% (Ikeda et al., 2008) (Fig. 5.3a). As could be seen in Fig. 5.3b, epitaxially grown Fe/MgO/Fe tunnel junctions present a very good interface between the different layers (Tiusan et al., 2006).

The first experimental study of shot noise in tunnel junctions with MgO(100) barriers was done by Guerrero et al. (Guerrero et al., 2007). The junctions showed 185% TMR at room tem-

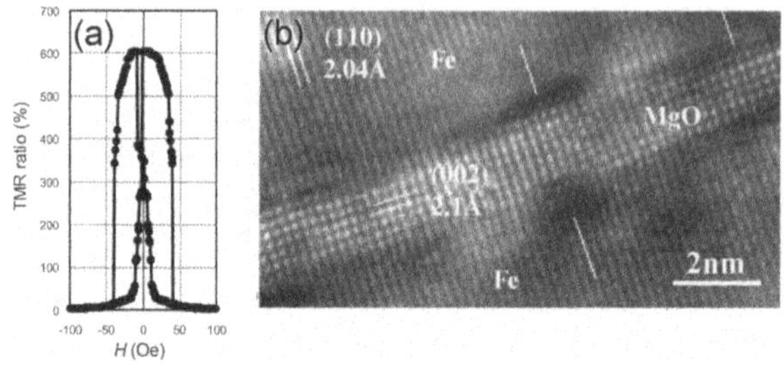

Figure 5.3 (a) Example of record high TMR reached at room temperature for MTJs with MgO barrier. (b) Transmission electron microscopy (TEM) image of the cross section of a Fe/MgO/Fe junction. Adapted with permission from Ref. Ikeda et al. (2008).

perature, quite interesting for applications, and 300% at 4 K (see Fig. 5.4). The layer structure of the epitaxial junctions was Fe(100)/Fe–C/MgO(100)/Fe(100). Shot noise as a function of the magnetic state was measured at 10 K and up to 1 V, and it presented

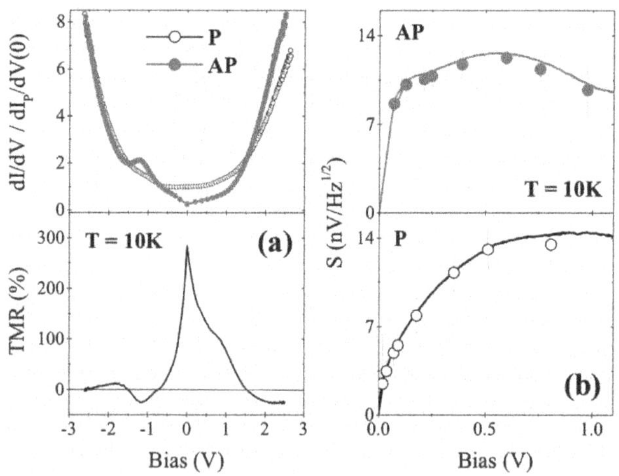

Figure 5.4 High bias conductance and shot noise in epitaxial MTJs with MgO barrier. (a) Differential conductivities at 10 K (top panel) and the related bias dependent TMR (V) (bottom panel). (b) Shot noise measurements in P and AP states measured at 10 K in the bias range when the electrons are injected from the top towards the bottom MTJ electrode (negative voltage). Adapted with permission from Ref. Guerrero et al. (2007).

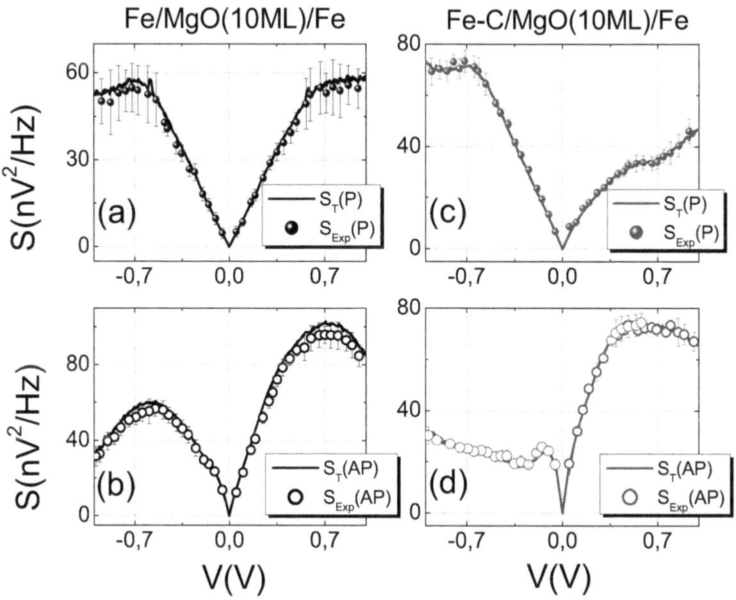

Figure 5.5 Shot noise in P and AP states measured at 4 K as a function of bias in MTJs with MgO barrier. The lines (black line for C free and red line for C doped) correspond to theoretical value of the Poissonian transport while points correspond to experimental results.

a Poissonian character (direct tunneling) in both P and AP states. This demonstrated that a pure spin-dependent direct tunneling mechanism was governing the transport, giving the final evidence of the high structural quality of MgO barriers.

More detailed shot noise measurements on fully epitaxial MgO-based MTJs with and without Carbon doping of the interface carried out later by Herranz et al. (see Ref. Herranz (2012)) showed that presence of Carbon at Fe/MgO interface (which diffuses from the MgO substrate) does not reduce the Fano factor below $F = 1$, indicating immunity of MgO to Carbon doping (see Fig. 5.5).

Figure 5.5 represents a typical shot noise measurements with frequency independent component in noise power obtained from the Gaussian distribution of noise with frequency. Independently of the $S_V(f)$, carried out at $T = 5$ K, for both kinds of MTJs samples (free and Carbon doped). The solid line shows the "theoretical" expectation for the shot noise, for electron tunneling of Poissonian type with $S_V = 2e\langle I\rangle/G^2$, where G is the differential conductance measured in Fig. 5.5. The points represent the mean value of the power spectral density in the range where it is independent of the frequency. The error bars show the dispersion of the shot noise (white spectrum) obtained as half-width the Gaussian distribution. The experimental data reveals a Poissonian shot noise, i.e., Fano factor equal to one, which clearly indicates the absence of electron correlations and/or sequential tunneling phenomena (Guerrero et al., 2007). This was found to be valid both for the P and AP states, confirming that the spin dependent conductance is due to a direct tunnel between electron bands, as it is expected for the coherent tunneling (Blanter and Büttiker, 2000b).

Follow up report on shot noise in MTJs (Sekiguchi et al., 2010) corroborated the initial finding (Guerrero et al., 2007) revealing full shot noise in CoFeB/MgO/CoFeB tunnel junctions with 1.5 nm-thick barriers and down to very low biases (± 5 mV). Similarly to (Guerrero et al., 2007), the shot noise was Poissonian for both magnetic states (P and AP states) - a sign of the absence of electron correlations in the low bias regime. The authors presented an interesting work, since they are able to fit the equation of white noise (thermal plus shot noise) to the data at several low temperatures (between 3 and 20 K).

Contrary to what happens with the amorphous Al_2O_3 barriers, crystalline MgO-based MTJs present interfaces presumable nearly defect free (in the plastic relaxation regime for MgO), but could give rise to coherent interband tunneling in the presence of spin filtering and localized surface states. Possible influence of symmetry filtering was suggested to be seen by Arakawa et al. (2011) who observed almost Poissonian shot noise (see Fig. 5.6) in MTJs with a much thinner MgO barriers (1.05 nm) and with high TMR ratios (over 200% at 3 K). The Fano factor in the antiparallel state was found to be quite close to 1, but different to the parallel state where Fano was $F = 0.91 \pm 0.01$. These values are found not to depend on either temperature or magnetic field. Since the junctions present a high TMR ratio and $F = 1$ in the AP state, then the barrier does

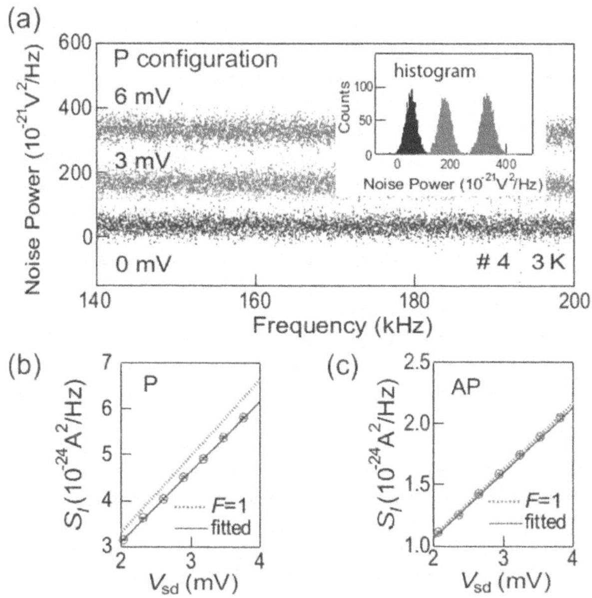

Figure 5.6 Shot noise in CoFeB/MgO/CoFeB MTJs. (a) Measured noise power spectra in MTJs with MgO barrier in the P state for voltages of 0, 3, and 6 mV at 3 K. The inset shows the histograms of each spectrum between 140 and 180 kHz. (b) Estimated current shot noise (open dots) for the P state. The solid line is the fitted curve, while the dashed line shows the curve corresponding to $F = 1$. (c) Same graph for the AP state. Adapted with permission from Ref. Arakawa et al. (2011).

not seem to have interfacial defects, even though it is quite thin. The authors (Arakawa et al., 2011) initially argued that the sub-Poissonian statistics detected in the P configuration was due to a antibunching during coherent transport of a Δ_1 state in the low bias regime, since this state is filtered out in the AP configuration.

Indeed, first-principles calculations were able to describe such slight suppression of the Fano factor at very thin MgO barriers (Liu et al., 2012) qualitatively reproducing the experimental results discussed above (Arakawa et al., 2011). Liu et al. (Liu et al., 2012) computed the shot noise in ballistic regime in Fe/MgO/Fe tunnel junctions with different barrier thicknesses, and accounted for the decrease in shot noise by adding some disorder at the Fe/MgO interface. The addition of oxygen vacancies as interface defects is the most favorable energetically. Their inclusion in the calculation of a tunnel junction with a barrier thickness of 5 atomic layers (1.05 nm, as in (Arakawa et al., 2011)) brings the calculated TMR ratio from 3580% (concordant with theoretical predictions) to around 250%, which agrees quite well with experimental results. The calculations for this barrier thickness yield a suppression of the Fano factor which depends on the magnetic state, $F_{\mathrm{P}} = 0.87$ and $F_{\mathrm{AP}} = 0.98$, which is quite close to the experimental values $F_{\mathrm{P}} = 0.91$ and $F_{\mathrm{AP}} = 0.98$. The authors (Liu et al., 2012) also predict an onset of resonant tunneling in the antiparallel configuration at some threshold bias.

Alternative and probably more realistic view on the origin of reduced Fano observed in ultrathin MgO barriers (Arakawa et al., 2011) could be presence of defects and/or pseudo-pinholes. Indeed, shot noise measurements in the single-barrier tunneling provide a tool to discriminate between direct and sequential tunneling processes. For the uniform distribution of defects, sequential tunneling is predicted (Blanter and Büttiker, 2000b) and observed (Guerrero et al., 2006b) reducing the Fano factor to about 2/3. This situation could be different in the MgO-based MTJs without defects where direct tunneling process may be expected.

In this contest, the suppressed $F < 1$ observed in a single-barrier MTJs should most probably be a consequence of the anomalously reduced barrier thickness ($d_{\mathrm{MgO}} = 1.05$ nm). It is very difficult to grow a perfect thin barrier, without defects and dislocations, and the contribution of the defects in the barrier has a strong impact on the sequential tunneling process. A possible scenario of this situation is shown in the sketch of Fig. 5.7, where for MTJs with barrier thickness well above the plastic relaxation (>5 ML) the Fano factor is close to 1. On the contrary, when barrier is below <5 ML, the density of defects is high, inside the barrier, and the sequential tunneling process appears to become important. Further experimental work is needed to investigate the shot noise in magnetic tunnel junctions with better control over the intermediate defect states, for example by growing double-barrier MTJs, which will be discussed in a separate chapter.

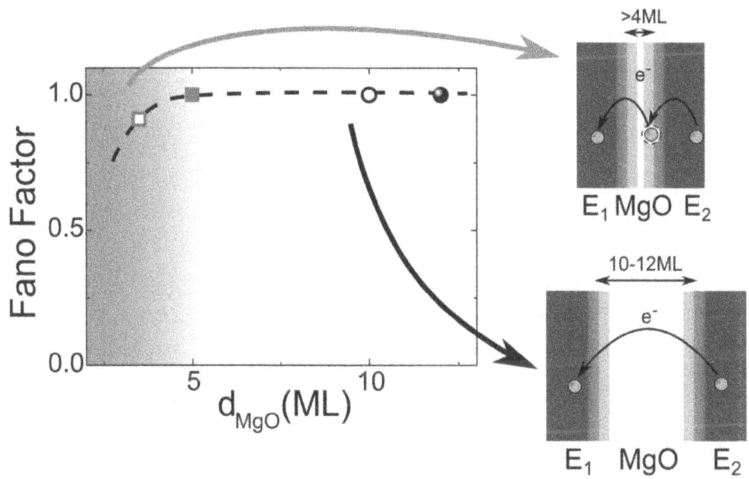

Figure 5.7 Sketch explaining increasing influence of defects on the electron tunneling in MTJs with reduction of the MgO barrier thickness. Main part summarizes the Fano factor presented in this Chapter as well as in the previous works (Arakawa et al., 2011; Guerrero et al., 2007; Sekiguchi et al., 2010).

More detailed later investigation of the shot noise on MTJs with ultrathin MgO barriers strongly supported the above discussed inhomogeneous barrier scenario and will be discussed below in details in form of case study.

Interestingly, the shot noise could be strongly reduced not only by the pinholes (George et al., 2002) or barrier inhomogeneinties (Tanaka et al., 2014), but also when multiple magnetic tunnel junctions are connected in series (Gokce et al., 2011). In such devices with shot noise in individual single MTJ (with the expected Poisson value) varies between $F = 1$ (single junction) to well below 0.5. Similar to the observed by Guerrero et al. (Guerrero et al., 2006b) deviations from $F = 1$ for single-barrier MTJs were described in terms of sequential tunneling through defects in the tunnel barrier. For series MTJs arrays the Fano scales as 1/N with N being the number of junctions in series. Such scaling points out the incoherent electron tunneling with each junction contributing to total nose as independent noise source.

Case study 5.1: Leakage conductance in MTJs with ultrathin MgO barriers determined from shot noise

Suppose that the tunnel barrier of MTJs is decreased below 1 nm so that in addition to tunneling current some leakage current (pseudo pinholes from barrier inhomogeneities) may appear. Influence of such noiseless ($F = 0$) leakage current I_l additional to the tunneling one I_t with Poissonian statistics so that corresponding shot noise power $S_I = 2eI_t$ has been considered by Tanaka et al. (Tanaka et al., 2014). Considering that total current $I = I_l + I_t$ and defining the observed Fano factor as $F = S_I/2eI$, by substituting S_I and I into expression for Fano, one obtains the following dependence of Fano on the magnetic state (α = P or AP) as a function of $G_{t,\alpha}$ and leakage conductance G_l:

$$F_\alpha = G_{t,\alpha}/(G_{t,\alpha} + G_l) \tag{5.1}$$

The below expression allows deducing the leakage conductance from the dependence of the measured Fano vs. magnetic state of MTJs.

$$G_l = (1 - F_{\mathrm{P}})G_{\mathrm{P}} \tag{5.2}$$

and the relation between the Fano in the P and AP configurations:

$$F_{\mathrm{AP}} = 1 - (1 - F_{\mathrm{P}})G_{\mathrm{P}}/G_{\mathrm{AP}} \tag{5.3}$$

This case study demonstrates that the leakage current contribution in the MTJs with ultrathin barriers may be determined from the shot noise.

5.3 Spin-Related Shot Noise

Pure spin current has recently emerged as an important new observable in different systems ranging from topological insulators (Hasan and Kane, 2010), triplet states at ferromagnet-superconductor interfaces (Eschrig, 2015), or Bose-Einstein condensates (BEC)

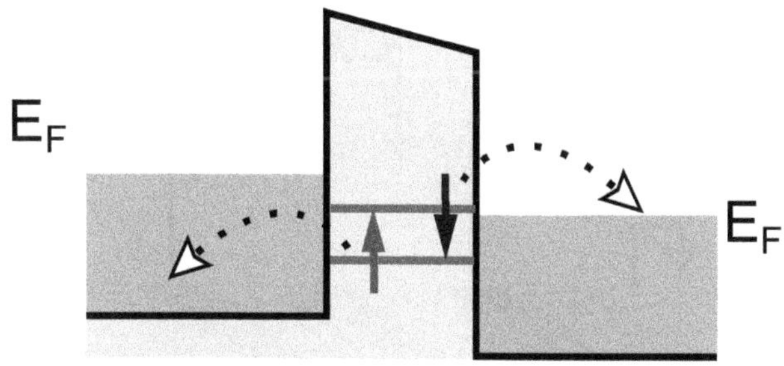

Figure 5.8 Sketch describing one of the possible sources of spin shot noise in quantum dot connected two leads.

(Oshima and Kawaguchi, 2016). Although most current experimental studies analyze mainly time averaged spin current, there has been important progress recently with understanding the spin current-related shot noise in spintronics, which we shall discuss below. We remark that spin current noise in spintronics yet remains largely untamed in comparison, for example, with semiconductor physics where it is currently a well established technique (Oestreich et al., 2005).

One simple example of how spin fluctuations could be activated is QD in the CB regime when transport of electron charge but not spin-related transport is forbidden. In the example sketched in Fig. 5.8, a spin shot noise in FM/QD/FM could be qualitatively understood as a consequence of simultaneous fluctuations of two electron spin forming singlet ground state in the opposite (toward electrodes) directions, so that charge-related shot noise is then canceled out, but spin shot noise which is finite. As we shall see further below, in some cases the spin fluctuations could be key ingredient providing even super-Poissonian current noise, for example in molecular spintronics.

By the time when the book is being written, to our best knowledge, there have been no reports on direct experimental detection of purely spin shot noise in FM/QD/FM or other spintronic devices. However, there have been number of experimental indications on the influence of spin accumulation and spin flips on the modification of the mainly charge-related shot noise. We have already mentioned previously in this Chapter that the first study of

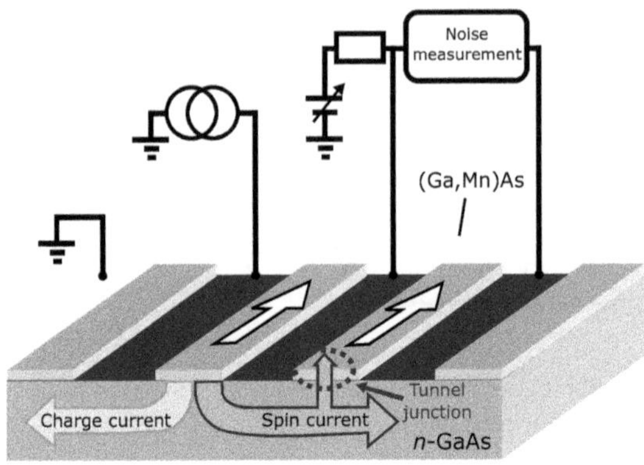

Figure 5.9 Sketch of the experiment realized in (Arakawa et al., 2015) to measure spin accumulation-related excess noise.

the shot noise in MTJs with single Al_2O_3 (Guerrero et al., 2006b) revealed dependence of the Fano on the magnetic state in the limit of slow spin flips, which could be considered as indirect proof for the presence of spin accumulation-related shot noise modification (see Fig. 5.1b). Further below we shall see that qualitatively similar effects could be observed in double-barrier magnetic tunnel junctions (Cascales et al., 2012) where spin dependent shot noise allows numerical evaluation of the spin relaxation time by fitting shot noise measured in four different possible magnetic states (existing due to unavoidable barrier asymmetry) to the model.

Recently, spin accumulation-related shot noise has been also investigated by Arakawa et al., but by using somewhat different (lateral) experimental configuration (Arakawa et al., 2015). It was shown that spin accumulation could be detected via shot noise of the tunneling current in the distant from the ferromagnetic electrode locations (Arakawa et al., 2015). The corresponding experiments are schematically sketched in Fig. 5.9.

The experimental idea is based on purely electric protocol proposed to measure spin accumulation between contacts attached to mesoscopic cavity (Meair et al., 2011). It has been proposed that for non-equilibrium spin accumulation in the limit $T \rightarrow 0$ and bias $\rightarrow 0$ shot noise saturates at some finite value reflecting spin diffusion and accumulation along the lateral direction of the

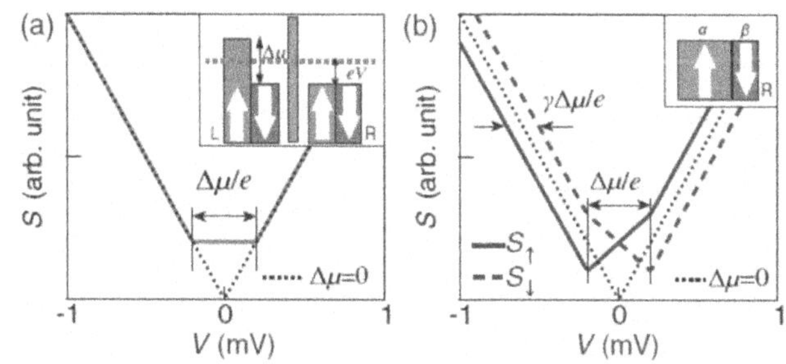

Figure 5.10 Model for the shot noise detection of spin accumulations. Adapted with permission from Arakawa et al. (2015).

structure. Moreover, it was suggested that the current noise may exhibit a sample-dependent, mesoscopic asymmetry under reversal of the electric bias in the presence of SO interaction.

The above-described predictions have been experimentally verified by Arakawa et al. (Arakawa et al., 2015) across a tunneling barrier with one of the leads located in non-equilibrium spin accumulation region, while the opposite ferromagnetic (GaMn)As contact in spin-equilibrium conditions (see sketch 5.9).

For the linear $I - V$ characteristics, at $T = 0$, and in the absence of spin accumulation (i.e., $\Delta\mu = 0$, the current spectral density S due to shot noise is expected to be proportional to the applied voltage V (see dotted line in Fig. 5.10). If one uses tunnel junction as detector with FM electrode opposite to the spin accumulation region (i.e., when $\Delta\mu \neq 0$) being non-magnetic metal, then for $T = 0$ noise power $S(0)$ becomes finite because the noise spectral density vs. bias is now provided by the following relation (Arakawa et al., 2015):

$$S = |eV + \Delta\mu/2| + |eV - \Delta\mu/2|. \tag{5.4}$$

Spin accumulation should then appear as plateau in $S(V)$ at small biases below $\Delta\mu/e$ (see Fig. 5.10). Of cause, as long as the measurements are carried out at some finite temperature, the kinks smear out.

As we have mentioned theoretical model predicts (Meair et al., 2011) that for electric detection of the non-equilibrium spin accumulation via shot noise a ferromagnetic contact, the bias dependence of the expected power density at $T = 0$ changes as long as effective spin polarization of ferromagnetic metal should be taken

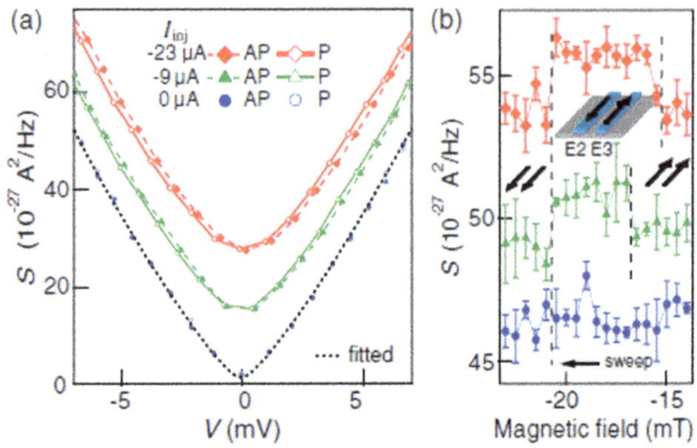

Figure 5.11 Experimental detection of spin accumulation-related shot noise. Adapted with permission from Arakawa et al. (2015).

into consideration. Indeed, in this case,

$$S_{\uparrow/\downarrow} \propto \alpha|eV \pm \Delta\mu/2| + \beta|eV \mp \Delta\mu/2|, \tag{5.5}$$

Here ↑ and ↓ indicate direction of the spin accumulation relative to magnetization direction of the detector contact, while α and β represent normalized conductivities of corresponding channels being $(\alpha + \beta = 1)$.

The resulting $S(V)$ relation becomes asymmetric with respect to the bias as plotted in Fig. 5.10. The spin-up and spin-down noise density $S_{\uparrow/\downarrow}$ become shifted in bias by $\gamma\Delta\mu/e$, with γ being the spin asymmetry $\gamma = \alpha - \beta$.

The experimental detection of the spin accumulated noise (Arakawa et al., 2015) validated the above discussed concept. While the charge-related shot noise as $(S_{\mathrm{AP}} + S_{\mathrm{P}})/2$ (AP and P are the relative alignments of the FM electrode and non-equilibrium spin accumulations), the spin-related shot noise is $(S_{\mathrm{AP}} - S_{\mathrm{P}})/2$ is proportional to $(2e^2/h)TF\Delta\mu$ with T being average between spin up and spin down channel transmissions and F, the Fano factor. Experimental fits reveal that Fano for the spin-related noise is close to 0.7–0.8, which is characteristic of the tunneling through nearly homogeneously distributed defect states in the barrier (compare with similar results obtained for magnetic tunnel junctions (Guerrero et al., 2006b).

Closing up this paragraph, we discuss two other interesting recent suggestions indicating the importance of taking into consideration presence of spin shot noise in the experiments involving spin pumping. One of them is related with spin torque phenomenon due to angular momentum transfer from spin-polarized current to a ferromagnet. Evidently, shot noise of the non-equilibrium charge current should be accompanied by some kind of stochastic force which acts on the ferromagnet adsorbing spin current. Stochastic version of the Landau–Lifshitz–Gilbert equation for a magnetization describing the effect has been solved (Chudnovskiy et al., 2008). It was predicted that the non-equilibrium spin torque-related noise results in a non-monotonic variation of the precession spectrum linewidth on the spin current.

Pure spin current may be pumped by exciting resonance in ferromagnetic films contacting for example normal metal (Tserkovnyak and Brataas, 2002). Kamra and Belzig (Kamra and Belzig, 2016) investigated statistics characteristics of the generated spin current and found that noise spectrum is frequency independent up to the ferromagnetic resonance frequency, and increases linearly above the drive frequency. The low-frequency shot noise is super-Poissonian due to spin transfer in 1 to 2 quanta. This means that ferromagnet may exhibit spontaneous squeezing of magnons with (unconventional) non-integral spin into the adjusted normal metal or thin insulating barrier in MTJs.

Chapter 6

Frequency-Dependent Noise in Single-Barrier Magnetic Tunnel Junctions

The previous chapters introduced the readers to the origin of coherent tunneling and the giant TMR in magnetic tunnel junctions with symmetry filtering as well as provided an introduction to shot noise with direct and sequential tunneling. The present chapter aims on a more detailed discussion of the different physical mechanisms contributing to the noise in MTJs with MgO barriers, including low (below a few kHz) and high (a few GHz) frequency noise, but excluding shot noise, which was discussed in the previous chapter.

6.1 $1/f$ Noise in Single-Barrier Magnetic Tunnel Junctions

The first studies of the dependence of $1/f$ with the applied magnetic field in magnetic tunnel junctions were carried out in systems with AlO_x barriers (Ingvarsson et al., 1999, 2000; Nowak et al., 1998, 1999). It was observed that $1/f$ noise experiences extra (dominating low-frequency noise) magnetic contribution close to the transition between the P and AP states due to fluctuations

Noise in Spintronics: From Understanding to Manipulation
Farkhad G. Aliev and Juan Pedro Cascales

ISBN 978-981-4774-34-5 (Hardcover), 978-1-315-11088-2 (eBook)
www.panstanford.com

arising from the reorientation of the free layer's magnetization (Jiang et al., 2004a). In these conditions, the spectral density of the fluctuations in resistance $S_R(f)$ are expressed as a function of fluctuations in magnetization $S_M(f)$, which in turn is described by the fluctuation-dissipation relation (Hardner et al., 1993b). Resistance and magnetization fluctuations may be related as follows:

$$S_R(f) = \left(\frac{\partial R}{\partial m}\right)^2 S_M(f), \quad (6.1)$$

where m is the magnetic moment and R is the resistance. The expression for $S_M(f)$ corresponds to equation 2.24.

Then, supposing that χ'' is a constant in a ferromagnetic system (Hardner et al., 1993b), and supposing that $S_R(f)$ depends on f as shown in equation 2.22 (Hooge model), we can integrate equation 6.1 in experimentally determined frequency range between f_{min} and f_{max}:

$$\frac{\partial R}{\partial H} = \frac{2m\mu_0}{k_B T \Delta R} \frac{R^2}{A} \ln\left(\frac{f_{max}}{f_{min}}\right) \alpha \quad (6.2)$$

So, close to the P-AP transition, the $1/f$ noise is mainly due to magnetic fluctuations and α is proportional to $\partial R/\partial H$.

The use of crystalline barriers, like MgO, improves the structural quality of the junctions, and in general they present much lower normalized noise than for amorphous barriers (Aliev et al., 2007). MgO barriers also offer an increase (around threefold) in TMR without additional low-frequency noise (Gokce et al., 2006a). Crystalline barriers also influence the tunneling transport due to spin filtering, as was seen in Section 1.2.3. Indeed, the first studies of low-frequency noise in MTJs with MgO(111) barriers showed that the noise may depend quite strongly on the magnetic state. In Refs. (Guerrero et al., 2005, 2006a), the authors found an increase in the normalized low-frequency noise (Hooge parameter) in the AP state for such junctions (see Fig. 4.2). This effect could be due to either spin accumulation or the difference in stress acting on the barrier depending on the magnetic field, referred to as *magnetostriction*. This difference in normalized noise between the magnetic states was further confirmed to exist in MTJs with MgO(100) barriers (Aliev et al., 2007), where record low normalized $1/f$ noise values were obtained.

As mentioned in Section 1.2.3 (and shown in Fig. 1.8), there exists a 3% lattice mismatch between Fe and MgO. Alloying Fe with V changes its lattice parameter, so the interface mismatch between the Fe electrode and the MgO barrier could be effectively reduced. Corresponding lattice mismatch improvement increase the TMR ratio and suppress the non-magnetic (P state) and magnetic (AP state) $1/f$ noise, in junctions with FeV bottom electrodes (Herranz et al., 2010c). Implementation of this strategy will be discussed separately in Chapter 8 on *Tunneling magnetoresistance and noise MTJs with optimized interface mismatch.*

Somewhat different strategy was proposed by Herranz et al. (2012), who suggest that doping MTJs with light elements could be advantageous in reduction of the barrier-related LFN. Indeed, as we shall see below in this chapter, carbon doping of epitaxial MTJs may decrease the $1/f$ noise, and even improve the robustness of MTJs, since it seems that carbon suppresses Fe-O interdiffusion and relaxes the MgO barrier.

Another more straightforward strategy to improve signal to noise ratio in MTJs is annealing. One of the first studies demonstrating essential influence of annealing on the electron and spin transport magnetic tunnel junctions was done in Ref. Feng et al. (2006) for the double-barrier MgO magnetic tunnel junctions with CoFeB electrodes. The junctions show TMR ratios of more than 100% at room temperature, but the TMR presents a highly asymmetric dependence with the voltage. The authors decrease this asymmetry by increasing the annealing temperature, finding that $T = 300°C$ is the optimal value. The temperature is determined by looking for the maximal TMR ratios in the junctions, as well as a minimal asymmetry with the voltage.

The same authors presented a later work, where the influence of annealing on single-barrier CoFeB/MgO/CoFeB magnetic tunnel junctions was investigated (Scola et al., 2007a). The junctions present an impressive TMR up to 235% and the noise is measured for different degrees of crystallization and CoFeB/MgO interface quality depending on the annealing temperature. An extremely low $1/f$ noise, compared to Al_2O_3 junctions, is found although it seems to be independent of the annealing temperature. The authors argue that the origin of the low-frequency noise are the localized charge traps within the MgO barriers.

Series of later works by the authors from Refs. [Stearrett et al. (2010a,b); Wang et al. (2009)] presented several detailed studies on the growth and characterization of CoFeB/MgO/CoFeB MTJs and investigation the dependence of $1/f$ noise with the annealing temperature. The global conclusions extracted from them are (a) the annealing time required for the TMR and noise to reach its maximum and minimum value respectively, depends on the annealing temperature, shorter at higher annealing temperatures, and (b) the noise is attributed to oxygen vacancy defects in the MgO tunnel barrier, and thermal annealing reduces this disorder and the number of defects with levels near the Fermi energy of the electrodes.

6.1.1 Magnetic State-Dependent Low-Frequency Noise in Epitaxial Magnetic Tunnel Junctions with MgO(100) Barriers

Let us now consider in detail the magnetic state dependence of LFN in fully epitaxial MTJs with MgO(100) barriers. We have already mentioned that comparison of fluctuations in the conductance (or resistance) in the P and AP states in MTJs with MgO barriers could provide additional new information on specifics of electron tunneling and electron correlations which are not available on electron transport measurements (Aliev et al., 2007; Gokce et al., 2006b; Scola et al., 2007b). We start with a detailed analysis of the dependence of LFN with the magnetic state of the electrodes followed by the investigation of electron tunneling and noise in fully epitaxial MTJs with MgO(100) barriers (Aliev et al., 2007). The early study of the $1/f$ noise in MTJs with MgO(111) barriers already revealed a possible magnetic state dependence of the normalized (Hooge factor) $1/f$ noise (Guerrero et al., 2005). Similar effects were reported later also for sputtered grown CoFeB/MgO(100)/CoFeB MTJs (Gokce et al., 2006b; Scola et al., 2007b) indicating that the low-frequency noise in MTJs with a MgO barriers cannot be described simply within FDT as long as it is not be determined just by the

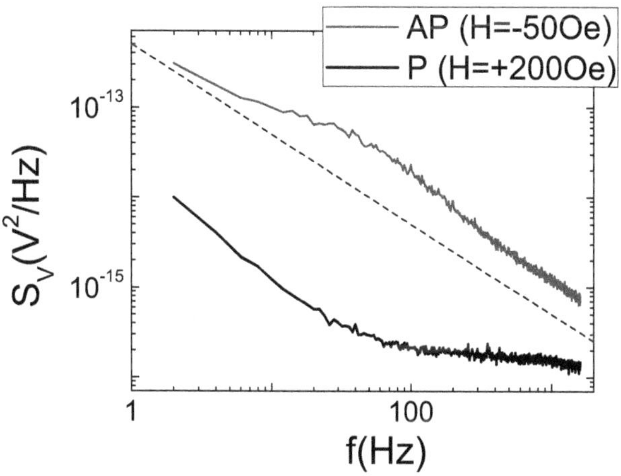

Figure 6.1 Typical noise spectrum measured in carbon-doped MTJs in the P and AP states with a positive bias of 200 mV. The dashed line is a guide for the eye for $1/f$ dependence.

derivative of tunneling resistance on the magnetic field (Mazumdar et al., 2007).

A typical (room temperature) power spectral density $S_V(f)$, below 1 kHz taken in the P and AP states (measured by cross-correlation method and normally at least 100 s after the magnetic field has been changed to ensure quasi-equilibrium conditions) are represented in Fig. 6.1. Besides, these spectra have been measured in the well-defined magnetic states (i.e., far from the magnetization reorientation characteristic fields, namely for $H = 200$ Oe and $H = -50$ Oe corresponding to the P and AP states, respectively). In the P state and at high frequencies ($f > 100$ Hz) one observes a frequency independent noise power spectrum, i.e., white noise (consisting of thermal and the shot noise contributions). The low-frequency part of the noise spectrum, however, is clearly dominated by the $1/f$ noise. Noise power spectrum dramatically changes in the AP state. In the frequency range studied (depending on the applied bias polarity, see below), the voltage noise may consist of the $1/f$ noise background superimposed by additional Lorentzian contribution or pure $1/f$ noise. In order to follow general trends in the noise, we shall analyze the LFN contribution by means of the widely used phenomenological Hooge parameter (α) defined as $\alpha = fAS_V(f)/V^2$, where A is the junction area, f is the frequency and V is the dc voltage applied to the junction (Hooge et al., 1981b).

Let us discuss the experimentally observed difference between LFN in the well-defined P and AP states. At high magnetic field (far from the reorientation transition) and at low biases, the conductance fluctuations are most probably of the purely electronic origin, because the magnetic fluctuations in the electrodes could be neglected. In such conditions, the commonly accepted phenomenological model of the $1/f$ noise suggests a superposition of multiple relaxation processes. As we have discussed previously, in Fe/MgO(100)/Fe MTJs the conductance is governed by electrons with Δ_1 (in the P state) and $\Delta_{5,2'}$ (in the AP state) symmetries at E_F. The conductance in the P state is higher than the AP state, because the propagation of Δ_5 and $\Delta_{2'}$ states are strongly attenuated inside the MgO barrier. The simplest explanation or the excess of the normalized $1/f$ noise in the AP state could be as follows. As the flow of electrons with Δ_5 symmetry substantially reduced in the AP state, the fluctuation of the defects affecting these Δ_5 and $\Delta_{2'}$ conductance channels would have much strong impact on the low-frequency noise (due to the absence of the shunting current otherwise present in the P state). Consequently this strongly enhances LFN in the AP state in comparison with one in the P state, at low bias. In other words, the idea is that scattering on the defect states would produce relatively stronger impact on conductance dominated by less symmetric and more localized Δ_5 and $\Delta_{2'}$ electron waves (AP state) than on Δ_1 states dominating transport in the P state.

One could therefore expect that noise in the P state to be mainly of electronic origin and very weakly affected by the applied voltage (Aliev et al., 2007). This indeed could be seen in Fig. 6.2, where the dependence of the normalized LFN for the carbon-doped and carbon-free MTJs is shown (determined for the frequency range of 1–50 Hz) as a function of bias, at room temperature. The detailed description of how carbon doping of the interface of epitaxial Fe/MgO(100)/Fe MTJs is realized could be found in see also (Tiusan et al., 2007) and its consequences will be also discussed further below.

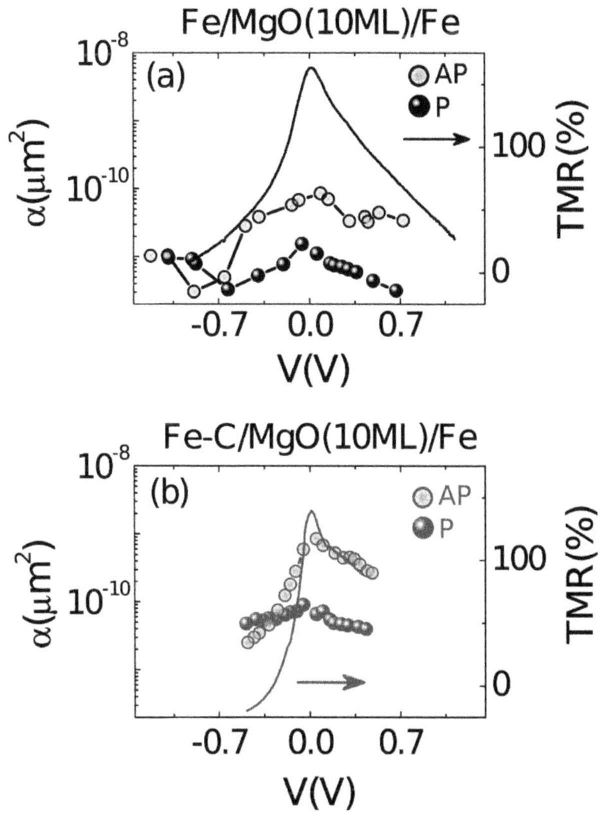

Figure 6.2 Bias dependence of the Hooge parameter (a) for carbon-free and (b) for carbon-doped MTJs in the P and AP states, evaluated for the frequency range of 2 $\sim$ 40 Hz (left axis), and TMR vs. bias (right axis).

The charge and spin transport scenario completely change in the AP state. From the one side, the low-frequency noise becomes highly conditioned by the presence of the magnetic inhomogeneities of the electrodes. From the other side, relative displacement of majority and minority spin bands of the electrodes with the applied bias should influence spin-dependent electron tunneling probability. Thus, the low-frequency noise in the AP state could be determined by a combination of both electronic and magnetic factors.

6.1.2 Interplay between Electronic and Magnetic Noise

While for the spintronic devices without coherent transport (such as metallic SVs) the LFN excess was mainly observed in the transition region between P and AP alignment (see Chapter 4), as we have seen in the previous paragraph, the situation could be very different for fully epitaxial MTJs with electron symmetry-dependent electron transport. Specifically, the dependence of the Hooge factor as a function of bias in the AP state seems to be qualitatively similar to the TMR vs. bias behavior (Aliev et al., 2007) (see Fig. 6.2). Moreover, there is an evident asymmetry in both bias-dependent TMR and LFN. Analyzed in terms of the Hooge factor, one finds that the $1/f$ noise is weakly bias dependent for the positive bias, where TMR is weakly suppressed by the bias. Whereas at the negative biases both the TMR and $1/f$ noise show a much stronger decrease. One of the possible reasons for the TMR vs. bias asymmetry for the Fe-C/MgO(10 ML)/ Fe MTJs is that at negative bias the electrons tunnel from the top (C free) electrode to the bottom electrode (C doped), where a strong interfacial resonant states (IRS) peak in density of states exists contributing to the conductance excess in the AP state (Aliev et al., 2007). The reversal of the TMR approximately coincides with the reduction of the Hooge in the AP state in respect the one in the P state. These observations suggest that the Hooge factor could have a strong dependence on the direction of the current (see Fig. 6.2) opening the possibility to optimize a signal to noise ratio in the AP state by choosing appropriate direction of the applied current or bias (see below). In the carbon-free MTJs, the behavior of the TMR and $1/f$ noise with respect to the bias polarity is more symmetric because the conductance peak from IRS in the AP state occurs at $-V$ only (Zermatten et al., 2008) and consequently, the inversion of the TMR and the $1/f$ noise takes place at much higher biases than for the C-doped MTJs (Aliev et al., 2007).

6.1.3 Detailed Analysis of the Low-Frequency Noise in the AP State

In order to understand in more detail the influence of the applied bias current direction on the LFN (Aliev et al., 2007) we discuss

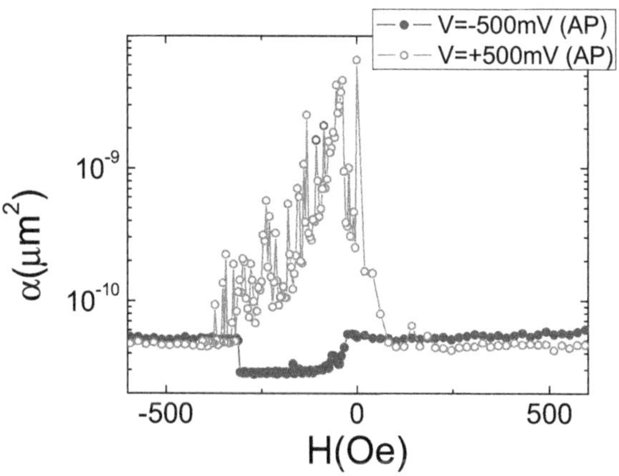

Figure 6.3 Normalized Hooge parameter with respect to magnetic field for ±500 mV in AP state (in carbon-doped MTJs).

below the details of the noise power spectra and relative dispersion in the noise for two fixed but opposite biases (corresponding to ±500 mV applied in the AP state) and measured in the magnetic field from +1 kOe to −1 kOe (Fig. 6.3). In both P states, the Hooge parameter is weakly dependent on the external magnetic fields and it is almost constant as a function of voltage. However, in the AP state normalized in terms of the Hooge factor LFM strongly dependent on the direction of the tunneling current. We recall that the Hooge parameter with respect to magnetic field studied for AlO_x barriers was observed to be proportional to $|dR/dH|^2$ (Ingvarsson et al., 2000; Jiang et al., 2004a) and the LFN excess was attributed to the magnetic instabilities mainly. Guerrero et al. (Guerrero et al., 2005) were the first to report on an excess of the normalized $1/f$ noise in the well-established AP state for the MTJs with MgO(111) barriers. However, in the Fe/MgO(100)/Fe MTJs one also observes an influence of the current direction on the LFN spectra dispersion in the well-defined AP state (not investigated in (Guerrero et al., 2005)). Such strong current asymmetry has been suggested to indicate that one direction of the current stabilizes while the opposite one destabilizes the magnetic fluctuations or magnetic inhomogeneities in the magnetically free electrode (Aliev et al., 2007).

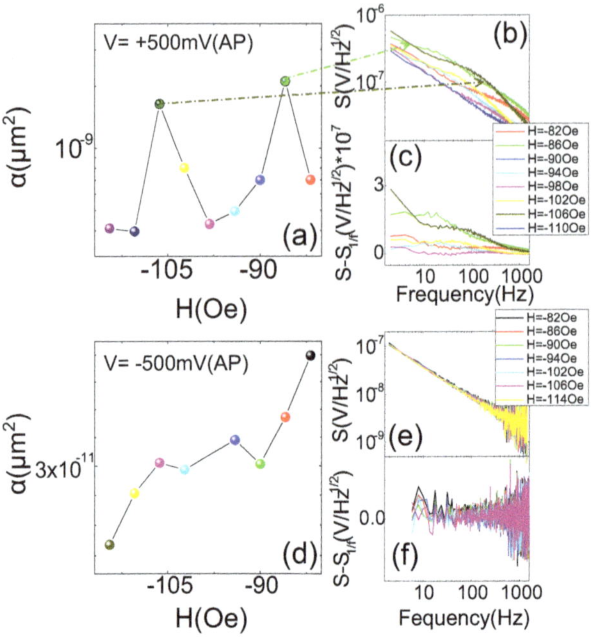

Figure 6.4 (a) The dependence of normalized Hooge parameter with respect to magnetic field for +500 mV in the AP state. (b) Power spectral density in the AP state for different fields (between −80 and −115 Oe. (c) Power spectral density after extracting the $1/f$ background for the same field range of the part (b). Parts (d), (e) and (f) show similar representation to previous figures, but for −500 mV bias voltage in the AP state.

The origin of the asymmetric behavior of the LFN could be better understood from the analysis presented in Fig. 6.4. Figure 6.4a,d zooms the Hooge parameter with respect to magnetic field in the AP state shown before in Fig. 6.3. For the positive bias, when the current flows from the bottom to the top electrode, the value of the TMR is 120%. Decrease of magnetic field followed by its inversion leads to the emergence of the high dispersion in the normalized $1/f$ noise. A detailed study of the power spectral density (Fig. 6.4b) for those fields apparently shows the presence of the random telegraph noise (RTN) contribution superimposed on $1/f$ background for +500 mV in the AP state. At the fields when RTN appears, effective α jumps up. The influence of the RTN could be

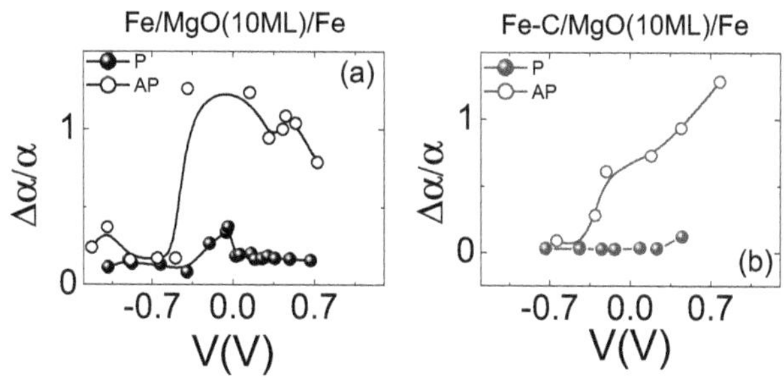

Figure 6.5 Bias dependence of the "noise of the noise" ($\Delta\alpha/\alpha$) vs. bias in the P and AP states in two types MTJ samples (left is carbon undoped and right- carbon-doped sample) at room temperature.

deduced as an additional Lorentzian type noise on top of the $1/f$ noise. In order to verify the Lorentzian contribution, Fig. 6.4c shows the spectrum of the different fields after extracting the $1/f$ noise background. The RTN is probably originated from some two-level fluctuator of magnetic origin. These fluctuations could be related with non-uniform magnetization of the one of the electrodes, for example the presence of 360° DWs. the observed fluctuations in Hooge factor do not persist for all magnetic fields applied in the AP state due to DW pinning/depinning process. Independently of presence of RTN, background $1/f$ noise in the AP state strongly exceeds one in the P state.

On the other hand, as we have already discussed, when the current flows in the opposite direction, the conductance in the AP state increases very fast leading to the TMR inversion. For example, at negative bias of -500 mV, where TMR is about $\sim$20% (Fig. 6.2), the normalized LFN in the AP state becomes substantially lower than one in the P state. This unique for the MTJs situation leads to nearly order of the magnitude TMR signal/noise variation once the applied bias is inverted (Aliev et al., 2007). We note that the strong bias-dependent asymmetry of the Hooge factor is accompanied by even stronger asymmetry in the dispersion of the Hooge.

In order to quantify effects related with sporadic appearance of the RTN as field being varied in the AP state, the LF noise dispersion was investigated (Aliev et al., 2007) as "noise of the noise" parameter for different applied magnetic fields at the same bias. The noise

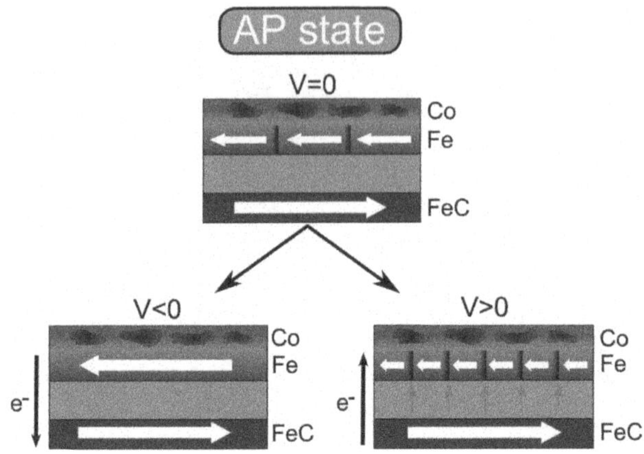

Figure 6.6 Sketch of the possible distribution of the magnetic inhomogeneities in MTJs as a function of the bias polarity.

dispersion parameter (equation 6.3) defines how stable is the LFN when the external magnetic field is changed in steps of a few Oe. The following quantity was defined (Aliev et al., 2007):

$$\frac{\Delta\alpha}{\alpha} = \sqrt{\frac{\langle(\alpha - \langle\alpha\rangle)^2\rangle}{\langle\alpha\rangle}} \tag{6.3}$$

We have mentioned before that the "noise of the noise" in the P state is practically zero (see Fig. 6.5). In contrary, in the well-defined AP state, similar to the LFN, the "noise of the noise" shows a strong dependence on the direction of the current and for biases above the TMR inversion is close to zero (see Fig. 6.3). Interestingly, the dispersion of the LFN remains negligible for the negative biases independently of the TMR values.

Figure 6.6 sketches a possible scenario explaining asymmetric in current magnetic RTN instabilities (Aliev et al., 2007). We know that the top electrode is magnetically hard due to the coupling of Fe layer and the above sputtered Co layer, which is a disordered alloy formed at the Fe/Co interface (Popova et al., 2002). In these conditions, the top FeCo electrode should have more magnetic domains than the bottom one, which is in the contact with substrate. Suppose that a negative current is applied, so that the electrons flow from the bottom to the top electrode. Then in the AP state and far from the transition to the P state, the magnetization of the bottom electrode is well defined, while magnetization of the top electrode could have certain number of magnetic inhomogeneities. In contrary, when the electrons flow from the top to the bottom electrode, the current stabilizes the magnetic inhomogeneities of the top electrode and consequently reduces the level of $1/f$ noise related to magnetic instabilities. Asymmetric character of the RTN generation suggests that a possible contribution of spin torque on the magnetic domain walls in the hard electrode.

Finally, we discuss changes in the bias dependence of the Hooge factor and TMR with carbon doping. The normalized variation of the noise with respect to the noise in the P state at 100 mV, represented in Fig. 6.7 indicates much stronger asymmetry of the noise vs. bias in the carbon-doped MTJ in comparison with undoped MTJs. However, we note that the analysis of the "noise of the noise" is made through normalized noise (α(V)). This important point ensures that the asymmetry of the "noise of the noise" is not directly related with one of TMR. Apart from a practical interest

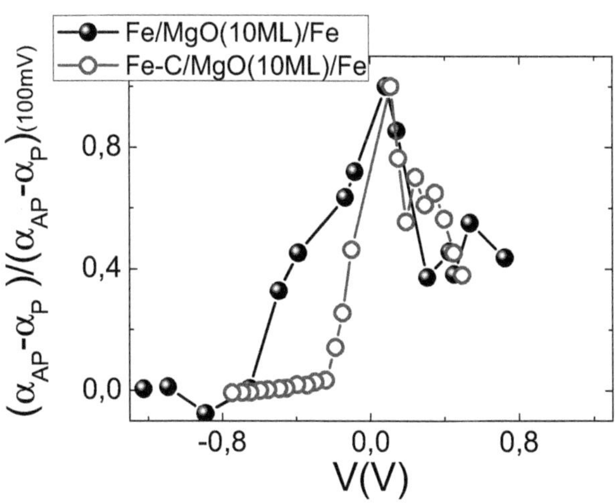

Figure 6.7 Difference between normalized Hooge parameter in the AP state with respect to the P state. In order to compare different samples, the relation of the noise is normalized by the difference at 100 mV.

(i.e., control over RTN by the bias direction), this behavior could provide a new interesting physics, which is sketched in Fig. 6.6. Particularly, a strong asymmetry of $\frac{\Delta\alpha}{\alpha}$ vs. bias could be related to the possible influence of "local topological spin torque effects due to current concentration into DWs in the AP state. Indeed the DW allow local tunneling of current with Δ_1 symmetry in the AP state. Another important aspect is related with the increase/decrease of the stability of DWs and magnetic inhomogeneities vs. bias direction. This means that a careful choice of the polarity and of the currents values could give us a better relation of the signal/noise ratio. For instance, as it is explained before, we could consider ± 500 mV, because the TMR is noticeable ($\sim$20%), but the low-frequency noise is two orders of magnitude smaller, in comparison with the positive bias, and also being reduced even dispersion of the noise (see Figs. 6.2 and 6.7).

Further in this chapter, we shall discuss a qualitatively similar effects in contest of the current induced DW motion in MTJs.

6.1.4 Influence of Carbon Doping on the Low-Frequency Noise

Experimental results (Herranz et al., 2012) evidence that carbon doping could change both dynamic conductance and dramatically decrease $1/f$ noise in MTJs with MgO barrier. Herranz et al. (Herranz et al., 2012) investigated both the conductance and LFN in MTJs with 10 and 12 ML of MgO barriers. Moreover, in the 12 ML MgO barrier the number of defects could be additionally enhanced due to the growth processes. Let us compare the noise power spectra in the carbon-free and carbon-doped MTJs with 12 ML of MgO continue state (Fig. 6.8a). In the frequency range between 1 and 50 Hz, the power spectral density behaves as $S_V(f) \propto 1/f^{\beta}$ (with $0.5 < \beta < 1.5$) (Dutta and Horn, 1981), allowing the description of the noise as $1/f$-type. We also note that the $\alpha(V)$ correlates with TMR(V) (Fig. 6.2), indicating that the origin of $1/f$ noise could be due to electron symmetry-dependent tunneling resistance fluctuations. As could be seen in Fig. 6.8b, the average low-frequency noise power increases approximately as V^2, indicating the dominant $1/f$ noise.

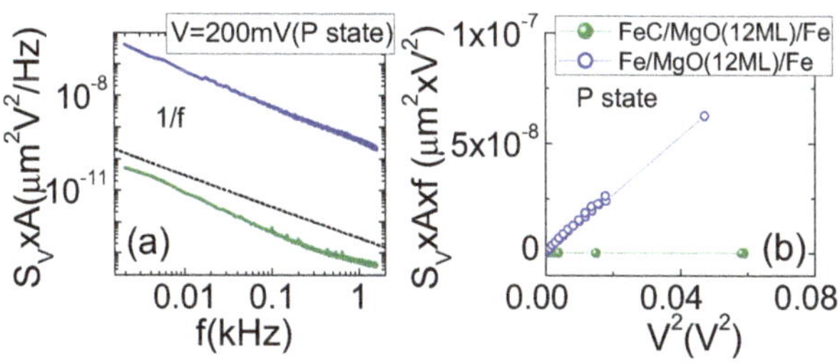

Figure 6.8 (a) Frequency dependence of the typical normalized noise power for carbon-free (dotted) and -doped (full lines) MTJs. Black dotted line represents $1/f$ pure dependence. Part (b) shows proportionality between the noise power and the bias squared for carbon-doped and undoped MTJs.

Figure 6.8 shows a strong suppression of $1/f$ noise with carbon doping in MTJs with 12 ML of MgO barrier.

According to the results by Herranz et al. (2012) the carbon doping substantially diminishes the normalized noise in the P state in this kind of the MTJs. Figure 6.9 summarizes the average value of the TMR and normalized $1/f$ noise (in the range of frequency between 1 and 50 Hz) in the P state at +200 mV as a function of R×A product. One clearly observes the differences in the influence of carbon doping for MTJs with 10 ML and 12 ML-thick MgO barriers. From the one hand, both MTJs with 10 ML MgO barrier show similar values of the LFN and the TMR (Aliev et al., 2007), with an average R × A around 30% higher than for carbon-free MTJs with 12 ML of MgO. In contrast, carbon-doped samples with 12 ML of MgO show much lower $1/f$ noise levels than carbon-free junctions. Besides, the R×A for the carbon-doped MTJs is about two times smaller than for carbon-free MTJs, while the average TMR is somewhat higher for the carbon-doped MTJs. It is important to emphasize the effect of carbon on the MTJs with 12 ML of barrier.

One could observe from Fig. 6.9b that for the MTJs with 12 ML-thick barrier are two well-differentiated groups of samples (representative samples marked with arrows), which seem to have a different concentration of carbon. The samples with lower normalized $1/f$ noise values could probably contain higher concentrations of carbon at the Fe/MgO interface. The dotted line represents

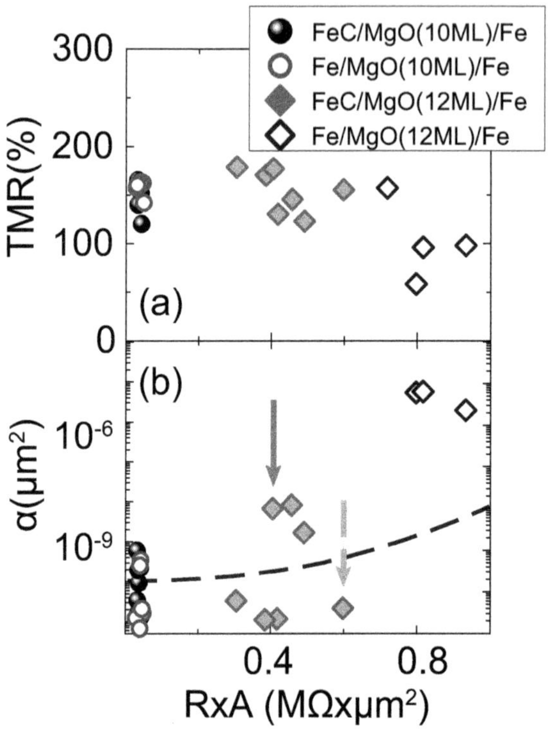

Figure 6.9 Summary of TMR (a) and Hooge factor (b) as a function of R×A product for MTJs with 10- (Aliev et al., 2007) and 12 ML-thick MgO barrier. The dotted line represents the mean empirical value of α reported in Ref. Gokce et al. (2006b). Dotted arrow indicates sample with larger carbon concentration in comparison with other pointed out, which shows reduced carbon concentration as deduced from the bias dependence of conductance (Herranz et al., 2012).

empirical values of the average α, previously reported for MgO and Al_2O_3-based MTJs (Gokce et al., 2006b), clearly indicating that the carbon doping in Fe/MgO (12 ML)/Fe MTJs strongly decreases the $1/f$ noise while increasing or maintaining a high TMR level (Herranz et al., 2012).

Electron symmetry changing at the interfaces could probably account for some part of substantial influence of carbon on the $1/f$ noise for the carbon-doped samples. Band structure calculations

(Lu et al., 2012) suggest the change of symmetry of interface resonant states (IRS) from Δ_5 to Δ_1 with carbon doping. Let us assume that the $1/f$ noise in the P state is originated from SO mediated interaction of the electrons with interface atoms, with uniformly distributed relaxation times, for which SO type scatterers change Δ_1 electrons into the $\Delta_{5\text{IRS}}$ states or vice versa. Then, for carbon-free MTJ, we could roughly estimate the corresponding normalized fluctuations in the conductance as

$$\frac{\langle[\sigma(\Delta_1) - \sigma(\Delta_{5\text{IRS}})]^2\rangle}{\langle\sigma^2(\Delta_1)\rangle} \sim 1 \tag{6.4}$$

(in the conditions $\sigma(\Delta_{5\text{IRS}}) \ll \sigma(\Delta_1)$). At the same time, for the carbon-doped MTJs with IRS dominated by Δ_1 symmetry, fluctuations in the position of interface atoms provide smaller fluctuations of the conductance:

$$\frac{\langle[\sigma(\Delta_1) - \sigma(\Delta_{1\text{IRS}})]^2\rangle}{\langle[\sigma(\Delta_1) + \sigma(\Delta_{1\text{IRS}})]^2\rangle} \ll 1 \tag{6.5}$$

(in the condition $\sigma(\Delta_{1\text{IRS}}) \simeq \sigma(\Delta_1)$).

An alternative or complementary explanation of the strong effect of carbon doping on the reduction of the $1/f$ noise could be that the small concentration of carbon "fills" all the Oxygen vacancies at the interface(Wang et al., 2011b), improving the quality of the barrier, and strongly reducing the low-frequency noise. For those junctions, where the carbon doping of the interface is not sufficient, a different reconstruction (c(3×3)) at Fe-C/MgO interface probably occurs leading only to a incomplete reduction of the $1/f$ noise (Herranz et al., 2012).

6.2 Frequency-Dependent Spin-Flip Noise due to Non-Equilibrium Spin Accumulation

As we discussed in the previous chapter, the non-equilibrium spin accumulation in MTJs could provide some excess shot noise in the AP state (Guerrero et al., 2006b) when spin flips are slow. For the

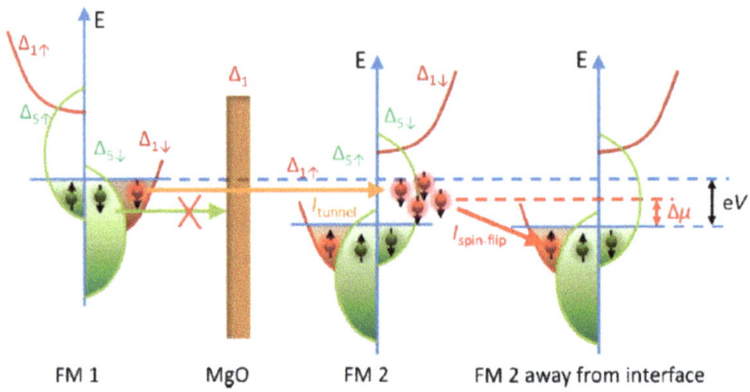

Figure 6.10 Schematics of the spin flips and spin accumulation in the antiparallel state in MTJs with symmetry filtering. As long as there are no available down-spin electron states in the AP state, electron contributing to the current should undergo spin-flips at interface becoming therefore the spin up-spin states. Adapted with permission from Liu et al. (2016).

case of MgO-based MTJs such effects were shown to be proportional to the spin current (Arakawa et al., 2015).

Later, Liu et al. (2016) also considered the spin accumulation scenario due to spin flips in MgO-based MTJs, (i.e., generally different non-equilibrium picture than that of Ref. (Arakawa et al., 2015) and basically related with symmetry filtering by MgO). It has been suggested to treat the LFN in MTJs as an essentially equilibrium phenomenon for which LFN in MTJs would just arise from resistance fluctuations which the applied current I converts into the voltage fluctuations ($S_{\mathrm{V}} = I^2 S_{\mathrm{R}}$). Contrary to the use of standard FDT relations, Liu et al. (Liu et al., 2016) fit the current noise power as

$$S_{\mathrm{I}} f = aI \coth\frac{a}{b} - ab, \tag{6.6}$$

with a, b being constants. Note that the above relation is conceptually similar to the one describing the charge shot noise (see Chapter 2). However, now the current would play a role of the bias voltage in the charge shot noise, while the spin-flip probability would play role of the tunneling probability in non-magnetic tunnel junctions.

A fundamental difference between spin accumulation (frequency independent) shot noise (Arakawa et al., 2015; Guerrero et al., 2006b) and the above-discussed spin-flip frequency-dependent noise (SFN) is as follows. While Ref. (Arakawa et al., 2015) assumes that current noise (charge) and spin-induced excess current noise (spin) could be separated, providing the possibility of a finite current noise at zero bias, spin accumulation in the SFN is directly linked with presence of the charge current, therefore no SFN is expected at zero bias (Liu et al., 2016).

We should mention that it was pointed out back in 2007 that the excess normalized noise in the AP state in fully epitaxial MgO-based MTJs is mainly controlled by the defects inside the barrier (Aliev et al., 2007), which are the source of spin flips in the SFN model (Liu et al., 2016). Besides, electron symmetry flips extended to interfacial resonant states (i.e., from Δ_1 to Δ_5) have been mentioned to explain a strong reduction of the LFN in the P state with carbon doping stimulating appearance of IRS (Herranz et al., 2012). It would be therefore realistic to suggest that role of the second ferromagnet (FM2) in the spin flip noise model (Liu et al., 2016) (see Fig. 6.10) might be played by the IRSs containing Δ_5 symmetry states even in the carbon free MTJs with MgO barrier. In this scenario, the spin flips are mediated by spin orbit coupling (SOC) scattering, which is enhanced between the majority Δ_1 and the minority Δ_5 bands with IRS at the Fe/MgO interface having dominant Δ_5 symmetry (Lu et al., 2012).

6.2.1 Lateral Size Dependence of the Low-Frequency Noise in Sub-Micron MTJs

Reducing the area of magnetic tunnel junctions may be a good approach for obtaining spin-torque-based devices, since the current needed to obtain the necessary current density for STT effects can be notably decreased. The main problem is that decreasing the area of the devices also increases their resistance since $R \propto 1/A$. So, the decrease in the area must be compensated with growth of thinner tunnel barriers, since the resistance scales exponentially with the barrier thickness. When the lateral size of the junction falls below the μm, the matter turn to be technologically difficult, since the barriers must be made thinner than 1 nm to keep the resistance in the hundreds of Ω. The great majority of noise studies in junctions of such size deal with high-frequency dynamics, with their application as microwave oscillators in mind (see the next paragraph) and only limited number of studies are known in the LF spectral window.

In this respect, rather interesting case is investigation of the electron transport and LFN measurements in submicron MTJs where spin torque effects just become emerging. In particular, those junctions with dimensions close to the transition to the single domain regime (Herranz et al., 2011). Those junctions revealed the presence of single magnetic inhomogeneities in the soft and hard layers, appearing as a two-state RTN. It was found that the RTN fluctuations and the related resistance steps of the AP state were due to domain wall motion and magnetic inhomogeneities. These are asymmetrically influenced by the current, which can be used to displace the domain walls in the soft layer by the spin torque effect using relatively low current densities.

Herranz et al. (2011) have shown that the LFN of submicron MTJs could depend on the junctions' lateral dimensions, especially in the regime when the lateral size is of the order of the domain wall width of the magnetically free layer. A detailed study at room temperature of tunneling magnetoresistance and low-frequency noise in elliptic submicron CoFeB/ MgO/ CoFeB MTJs, with 1.1 nm-thick MgO barriers and areas from 0.117 μm^2 to 0.0245 μm^2. The low-frequency noise is strongly affected by magnetic inhomogeneities (MIs) and domain walls (DWs), showing as a function of the area two qualitatively different, robust and reproducible behavior (Herranz et al., 2011). In the smaller area junctions, an unexpected random telegraph noise (RTN1), deeply in the parallel state (P), was found possibly due to stray field induced MI/DWs in the hard layer. The second noise source (RTN2) is observed in the antiparallel state (AP) for the largest junctions. The strong asymmetry of RTN2 related to resistance steps with current, indicate the possible influence of the spin torque, acting on the MI/DWs in the soft layer at very low current densities below 5×10^5 A/cm^2.

Figure 6.11 summarizes the main results of (Herranz et al., 2011). Specifically, the asymmetric in bias RTN2 in relatively large area MTJs on the border to transition into single domain state is explained as due to the influence of spin torque on the stability of the associated magnetic inhomogeneities. Lower part sketches the

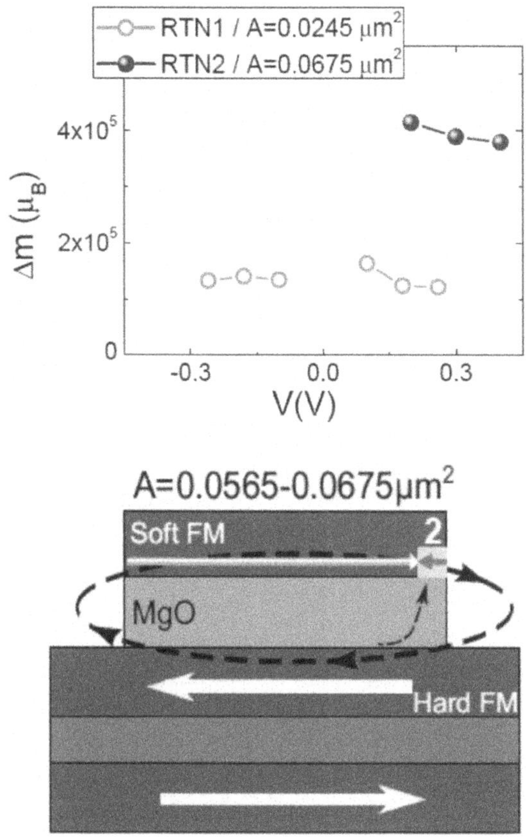

Figure 6.11 (Top) Comparative bias dependencies of estimated fluctuating moments obtained from RTN1 and RTN2. (Bottom) DW or MIs inside hard and soft electrodes indicated as (1) and (2), respectively. Dotted lines inside MgO indicate spin current with Δ_1 symmetry into DW or MI2 in the AP state. The dashed line sketches the stray field.

corresponding proposed model pointing out to Δ_1 tunneling current focusing into the MIs as a main reason for spin torque effects are observed at relatively small total tunneling current densities.

Spin torque acting on domain walls and magnetic inhomogeneities in MTJs presents a novel mechanism for the DW motion under effect of current. The peculiarity of this mechanism is that is related with DW motion under perpendicular tunneling current and with very low current densities (because of the effective concentration of the tunneling current into topologic anomalies). Similar effects have been independently reported from TMR measurements by Chanthbouala et al. (2011). In this paper, a half-ring shape of the MTJ was designed in order to detect the DW motion by vertical current injection. The displacement of DW at low current densities by the tunneling current could be used to create new kind of MRAM operating with lower current densities.

Chapter 7

Interplay between Spin Torque and Noise in Sub-100 nm Size Magnetic Tunnel Junctions

7.1 Introduction to Spin Torque

As we mentioned in Section 2.5, magnetization oscillations known as a ferromagnetic resonance arise from the precessional motion of the magnetization of a ferromagnetic material when an external DC magnetic field with specific (for each external field) frequency is applied in the presence of microwave pump field perpendicular to it. Since magnetic tunnel junctions are composed of ferromagnetic electrodes, it is possible to detect such resonance peaks if the spectra of their voltage fluctuations are measured up to a few tens of GHz.

Oscillations of the magnetization in magnetic tunnel junctions, specially of very small lateral size, could be driven in a different way. The fact that a spin-polarized electrical current flowing non-collinearly to the local magnetization, can apply a large torque to a ferromagnet, through the direct transfer of spin angular momentum (Slonczewski, 1996), offers the possibility of manipulating magnetic elements in devices without the use of magnetic fields. Certain

Noise in Spintronics: From Understanding to Manipulation
Farkhad G. Aliev and Juan Pedro Cascales

ISBN 978-981-4774-34-5 (Hardcover), 978-1-315-11088-2 (eBook)
www.panstanford.com

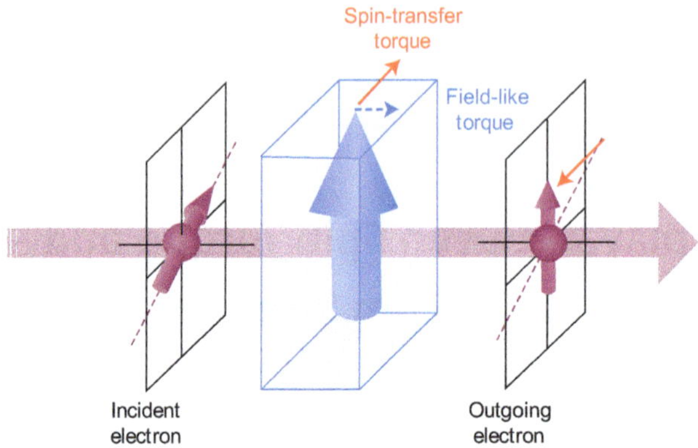

Figure 7.1 Illustration of the current-induced torque of a spin-polarized current on a ferromagnet. Adapted with permission from Ref. Brataas et al. (2012).

oscillatory magnetic modes, unreachable with magnetic fields alone, may be achieved by driving nanomagnets by spin-polarized currents. As we have already mentioned (Section 2.5), the difference in the spin direction of a spin-polarized current and the magnetization of a ferromagnetic layer may change the spin direction of outgoing electrons with respect to incident electrons, and the difference in spin polarization can apply a torque on the ferromagnet. This is schematically shown in Fig. 7.1 and is referred to as the spin-transfer torque effect.

The spin transfer torque effect was independently predicted in 1996 by Slonczewski (Slonczewski, 1996) and Berger (Berger, 1996). They argued that a current flowing perpendicular to the plane in a magnetic multilayer structure could generate a strong enough spin transfer torque to reorient the magnetization of one of the FM layers. These studies sparked interest in the ST effect, and in 1998 the first measurements of current-induced resistance changes in magnetic multilayer devices were reported (Sun, 1999; Tsoi et al., 1998). Spin torque induced magnetization reversal in lithographically patterned devices was observed shortly after (Katine et al., 2000; Myers et al., 1999).

In the regime where the current density applied to the MTJ is low, the resulting *damped oscillatory modes* are due to the external applied magnetic field and thermal fluctuations, referred to as thermal FMR (T-FMR). This effect is typically observed for applied current densities below $J \simeq 10^7$ A/cm^2 (Ralph and Stiles, 2008). The effective damping can be cancelled altogether by the spin torque from a DC current at some critical value of the current density J_C. This results in an auto-oscillation of the magnetization which is often referred to as a steady-state precession.

The study of spin torque effects in magnetic tunnel junctions is not only of fundamental interest but has a huge technological importance as well. The ability to switch the magnetization of a FM layer with only current could pave the way for new, smaller, and faster data storage devices. Using MTJs as data bits could be quite useful since their magnetic state may be switched only by current. Making the MTJs smaller would not only increase the storage density of the devices, it would reduce the amount of current needed to "write" each data bit, since the current density is inversely proportional to the area of the junction. When electrons flow from the free to the pinned electrode, the AP state is favored (and the P state is destabilized) for in-plane MTJs. Conversely, the flow of electrons from the pinned to the free electrode stabilizes the P state. Also, simple magnetic, multilayer structures may act as nanoscale oscillators, converting electrical DC energy into high-frequency magnetic oscillations which could be used as microwave sources and resonators. Although several studies show that a more effective ST effect could be reached with perpendicular magnetization (Houssameddine et al., 2007; Yu et al., 2014), we will focus on magnetic tunnel junctions with **in-plane magnetization**.

A phase diagram of the magnetic state (P or AP) of MTJs as a function of the applied current (or voltage) and external magnetic field may be constructed from experimental magnetoresistance curves, which gives information about the critical switching currents and switching fields. Several authors have reported both experimental (Kiselev et al., 2003) and theoretical (Xiao et al., 2005) studies of such diagrams, which show promise for the application of spin transfer-based devices. One of the first experimental results of such phase diagrams, by Kiselev et al. (2003), confirmed the spin transfer torque mechanism as an effective way to switch FM layers with currents in spin valves.

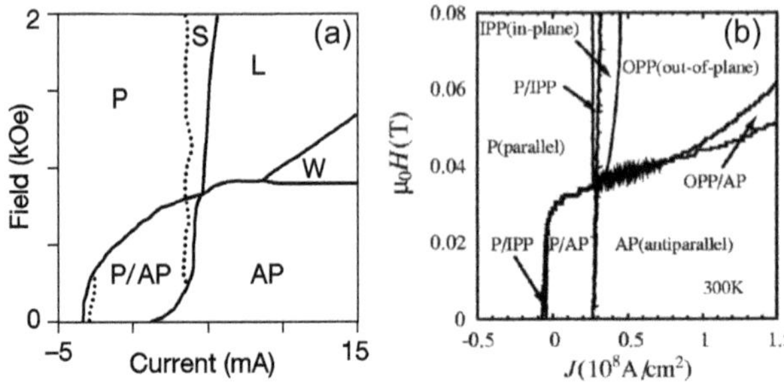

Figure 7.2 Switching phase diagram constructed from (a) experimental measurements of room-temperature current switching in a Co/Cu/Co spin valve adapted with permission from Kiselev et al. (2003) and (b) from the numerical integration of the LLG equation with the Slonczewski term, including thermal fluctuations at $T = 300$ K adapted with permission from Xiao et al. (2005).

Spin valves have a few disadvantages with respect to MTJs, as they present lower resistance and MR ratios. Figure 7.2a shows a experimental current switching phase diagram obtained at room temperature by Kiselev et al. (2003), for Co/Cu/Co spin valves of elliptical cross-sections (130×70 nm^2). In this graph, positive current I denotes the flow of electrons from the free to the fixed layer. As is seen, the state diagram is constructed from the resistance vs. field curves, which indicate at which currents or fields a change in the magnetic configuration occurs. The authors extract the dependence of the modes of the multilayer structure as a function of the current and external field. At low current, the field dependence of the observed oscillations correspond to small-angle precession and can be described by the Kittel formulas (Kittel, 2004). The zones in Fig. 7.2a labeled with S and L indicate *small* and *large* angle precession, respectively. Macrospin simulations reproducing the previous experiment can be found in Ref. Xiao et al. (2005). The model captures the essential features of the experiment when thermal fluctuations are included, as can be seen in Fig. 7.2b.

The first quantitative measurements of the direction and magnitude of the spin-transfer torque effect in magnetic tunnel junctions were carried out by Sankey et al. (2007) and Kubota et al. (2007). In the work by Sankey et al. (2007), the authors measure the bias and angular dependence of the spin-transfer

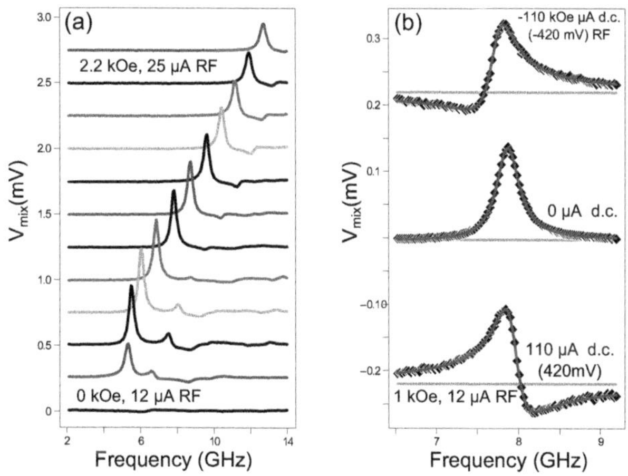

Figure 7.3 Measured by lock-in ST-FMR spectra at room temperature for (a) $I = 0$ for in plane magnetic fields spaced by 0.2 kOe. I_{RF} ranges from 12 μA at low field (high resistance) to 25 μA at high field. (b) Primary ST-FMR peaks at $H = 1000$ Oe and $I_{RF} \sim 12$ μA for different DC biases. Symbols are data; lines are Lorentzian fits. Adapted with permission from Ref. Sankey et al. (2007).

torque in CoFeB/MgO/CoFeB MTJs by using spin-transfer-driven ferromagnetic resonance (ST-FMR) technique (Sankey et al., 2006). The ST-FMR technique consists on biasing the sample with both a direct current I_{DC} and a microwave-frequency current I_{RF} through a bias-tee, and then sweeping the RF frequency. The spin-transfer torque from I_{RF} may excite resonant magnetic dynamics, which causes the resistance to oscillate at the driving frequency, and the resonant voltage response is measured by a lock-in. The fitting of the main ST-FMR peak by an equation derived from a macrospin model gives information on the magnitude and direction of the torque. The authors find that the torque lies in plane at low currents, while an out-of-plane component is found for higher bias. Figure 7.3 shows (a) how the frequency of the resonance peak varies with the magnetic field and (b) how the line-shape resonance peak at a fixed field changes with the bias.

The work by Kubota et al. (2007) used the spin torque diode effect (Tulapurkar et al., 2005) to estimate the characteristics of the spin-transfer torque in MgO-based MTJs. The spin torque diode effect in an MTJ consists on the application of a small radio-frequency alternating current to a nanoscale MTJ, which can generate a measurable DC voltage across the device when the frequency is resonant with the spin oscillations that arise from the spin torque effect. The authors study the spin-transfer torque for biases beyond the critical switching current. They estimate a considerable perpendicular torque at high biases, although the values found for the in-plane torques differ from Ref. Sankey et al. (2007). This could point to the importance of the perpendicular component of the torque for inverting magnetization by current.

A rather complete study of the oscillation modes present in nanoscale MTJs was carried out by Helmer et al. (2010). In this article, the authors study the field dependence of the mode frequency of thermally excited spin waves (T-FMR) in MgO-based, rectangular nanopillar MTJs of lateral sizes between 60×100 nm and 105×190 nm. The experimental results are backed by calculations based on the diagonalization of the dynamical matrix of a system of three coupled, spatially confined magnetic layers. The result is presented in Fig. 7.4, where the calculated dependence of the oscillation modes with the external field is shown for fields applied along the (a) easy and (b) hard axis of magnetization. Modes originating from the free layer or the synthetic antiferromagnet are identified, as each category presents a characteristic shape. This result is important and we will use it to identify the oscillation modes observed in the samples further on.

7.2 Transition from T-FMR to a Steady-State Precession

The next step toward making a practical use STT effects was presented in parallel theoretical (Kim et al., 2008a; Tiberkevich et al., 2007) and experimental studies (Mistral et al., 2006). The transition from the T-FMR regime to an in-plane, steady-state precession (SSP) state is studied in CoFe/Cu/CoFe/NiFe spin valves of nanoscale size (50×100 nm^2) (Mistral et al., 2006). The transition is identified from the sudden decrease in the frequency

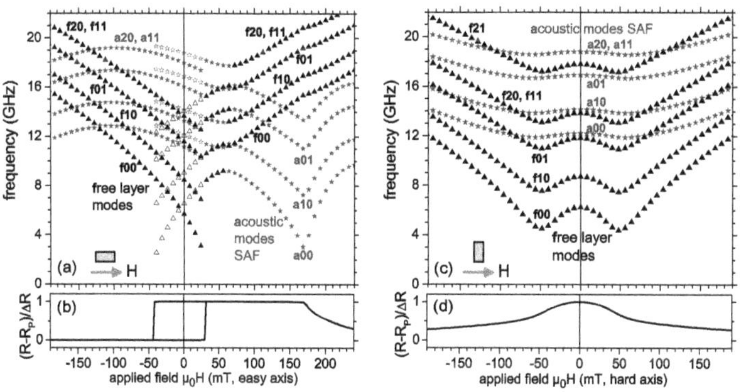

Figure 7.4 Calculated mode frequencies versus magnetic field along (a) easy axis and (c) hard axis for a rectangular nanopillar. (b) and (d) show the corresponding calculated hysteresis loops. Adapted with permission from Ref. Helmer et al. (2010).

and linewidth Δf accompanied by an increase output power P of the emitted microwaves on the applied current (Fig. 7.5a). Petit et al. (2007) found the influence of spin torque on HF fluctuations in nanoscale CoFe/Al_2O_3/CoFe/NiFe magnetic tunnel junctions. The authors study the dependence of the frequency, linewidth and oscillation amplitude for current densities below the critical value for spin torque. They find that the magnetic fluctuations are either reduced or enhanced by the DC current bias depending on its polarity.

Further work on this effect was presented by Houssameddine et al. (2008). The authors study the transition from T-FMR to a steady-state precession in nanoscale MgO tunnel junctions. They find that the devices presenting the clearest transition to the steady-state (steep decrease in frequency, narrower linewidths, high increase in output power) exhibit low resistance and TMR ratios of around 30%. The transition for high (HTMR) and low (LTMR) resistance junctions is shown in Fig. 7.5d,e. The authors report that repeated high-current measurements on HTMR devices may gradually turn them into LTMR devices. These LTMR devices seem to be stable against high-current measurements, and the authors speculate that the difference between sample types could be due to localized reductions in the tunneling barrier (Houssameddine et al., 2008). The authors consider the LTMR samples to have

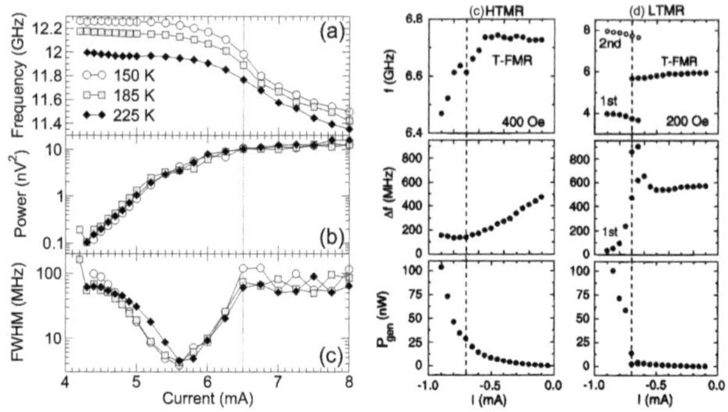

Figure 7.5 (a) Frequency, (b) power, and (c) linewidth of the fundamental mode as a function of applied current in sputtered CoFe/Cu/CoFe/NiFe spin valves. Adapted with permission from Ref. Mistral et al. (2006). Frequency, linewidth, and integrated power vs. current I for (d) the high resistance and (e) low resistance nanoscale MgO MTJs. Adapted with permission from Ref. Houssameddine et al. (2008).

non-uniform tunnel barriers (when the MgO thickness is reduced below 1 nm), so localized high-current densities could potentially create non-homogeneous excitation modes.

In a later publication, Houssamedine et al. investigated the details of the transition from T-FMR to a SSP by time-domain measurements of the magnetization oscillations in LTMR samples (Houssameddine, 2009; Houssameddine et al., 2009). This study presents an important result: the steady-state regime identified in Ref. Houssameddine et al. (2008) from the analysis of frequency-domain results is actually divided into two regimes. These regimes occur in HTMR and LTMR devices alike, as discussed in Ref. Houssameddine (2009), but the transition to the SSP presents a higher and sharper change as the samples tend to the LTMR type. Figure 7.6 presents (a) f and (b) Δf as a function of the current for a LTMR device with a TMR of 50% (Houssameddine, 2009). The first corresponds to currents $|I| > |I_c|$, where I_c is the current which onsets the decrease in frequency, linewidth and increase in microwave power. This regime presents *intermittent* steady-state oscillations which decay every few *ns* ("turn" off and on) (Fig. 7.6c,e) and which give rise to slightly different oscillation frequencies each time. These changes in frequency are responsible for the

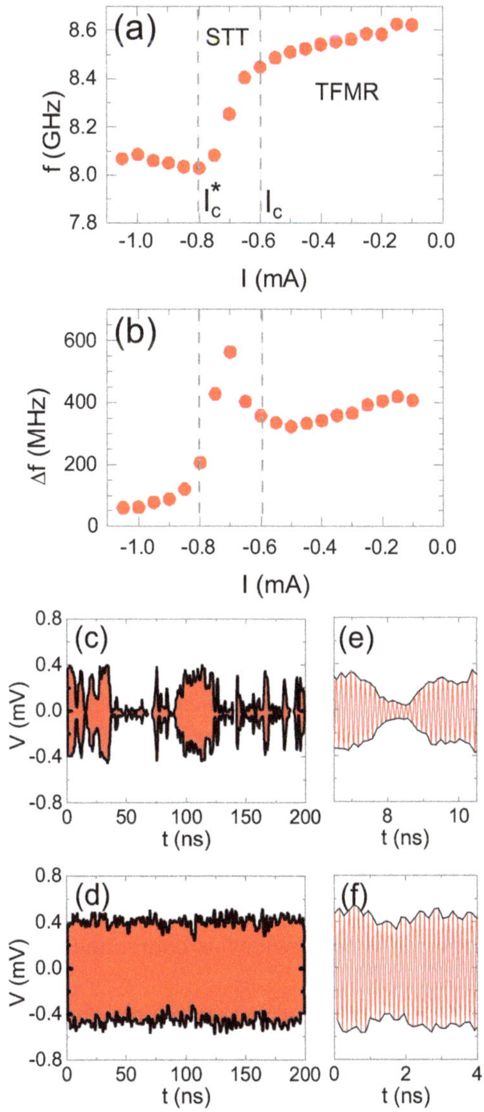

Figure 7.6 Current dependence of the emission frequency in a LTMR (50%) devices presenting a decrease in (a) frequency and (b) linewidth with the current at I_c, which become constant at I_c^*. Adapted with permission from Ref. Houssameddine (2009). 200 ns segment of the full time trace (gain corrected) for (c) $|I_c| < |I| < |I_c^*|$ and (d) $|I| > |I_c^*|$. 4 ns zoom revealing the high frequency oscillations for (e) $|I_c| < |I| < |I_c^*|$ and (f) $|I| > |I_c^*|$. Adapted with permission from Ref. Houssameddine et al. (2009).

magnitude of the linewidth of the resonance peak. The second regime corresponds to $|I| > |I_c^*| > |I_c|$, where I_c^* is a second critical current above which the SSP does not decay (Fig. 7.6d,f). In this regime, the frequency fluctuations (phase noise) yield a linewidth of the order of 25 MHz. The linewidth of a time segment during which the frequency remains constant is of 1.4 MHz.

Returning to pseudo pinholes (or "hotspots") and barrier inhomogeneities, these defects are naturally present in magnetic tunnel junctions with ultrathin barriers (Dimitrov et al., 2009), but they can also be created artificially by a soft breakdown (Herranz et al., 2010a). The presence of pseudo pinholes was suggested to qualitatively modify the magnetoresistance (Vedyayev et al., 2006a) and the magnetization reversal mechanisms in MTJs, even in the absence of ST effects (Finocchio, 2008; Finocchio et al., 2008). Although shot noise could be used to characterize ultrathin barriers in sub-100 nm MTJs, it has not been extensively studied, as they present higher $1/f$ noise amplitudes. The study by Zhong et al. (2011) of shot noise in MgO-based MTJs with areas 10^{-2} μcm^2 reveals Poissonian statistics with barriers 0.95 nm thick.

There has also been some controversy with respect to ST in MTJs with pinholes. While numerical calculations for a single (Zhu, 2005) or for multiple hotspots (Meng and Wang, 2005), point to a decrease of the threshold ST current in comparison with non-broken MTJs, Finocchio et al. (2006) predict an increase of the minimum current to excite microwave dynamics in junctions with pinholes. Thus, the realization that non-homogeneous barriers create local, high current densities could constitute an important step forward in the application of these devices as components of data storage technology.

Information about the current-switching phenomena and the changes in the oscillation modes with the current can also be indirectly obtained by the combined study of TMR and LF noise measurements, and in fact, $1/f$ noise around the transition of a damped oscillation to a steady-state precession has been observed (Devolder et al., 2009) and calculated (Zhu and Zhu, 2004) in previous studies. Possibly, the most complete study of $1/f$ noise and spin torque effects is the work done by Eklund et al. (2014), in which the $1/f$ noise up to 150 MHz and the white noise up to

25 GHz is measured in nano-contact spin valves. The dependence of the $1/f$ and white noise is found to be related to the linewidth of the oscillation. The oscillation linewidth in such devices has been found to be due to the phase noise originated from the non-linearity of the oscillation frequency (i.e., the frequency depends on the oscillation amplitude) (Kim et al., 2008c).

MTJs with ultralow RA are interesting for practical devices (STT-MRAM, STO, read head in hard disk drives). For junctions with low MgO thicknesses (and low RA products, correspondingly), there exists a correlation between the TMR ratio and the MgO thickness (Yuasa et al., 2004). A decrease in TMR is experienced when the thickness of the MgO barrier is reduced, due to local inhomogeneities (or pseudo-pinholes) in the MgO barrier. Obtaining MTJs with low RA products and high TMR ratios, for which homogeneous barriers are needed, is a real engineering challenge. The STT effect requires high current densities, and since MgO barriers only withstand a certain amount of voltage, having a low RA allows high currents to flow through the MTJ without causing the breakdown of the barrier (Alvarez-Hérault, 2010).

The $1/f$ noise and RTN can be used as a tool to quantify or determine the STT-related effects which occur in such MTJs with ultrathin MgO barriers as we will see in the following subsections.

7.3 Influence of Spin Transfer Torque Dynamics on $1/f$ Noise

We have already mentioned early indications on the influence of ST effects on LF noise observed in sub-micron (elliptic few hundreds of nanometer in size) MTJs (Herranz et al., 2011). The study carried out in Ref. Cascales et al. (2015a) presents a comparative investigation of high frequency (HF) dynamics and low-frequency (LF) noise in elliptical magnetic tunnel junctions with lateral dimensions under 100 nm and 0.9 nm thick MgO barriers presenting current-switching phenomena. Previous work comparing FMR linewidth and LFN was done on larger lateral size junctions with Al_2O_3 barriers where ST effects are absent (Sierra, 2009).

The analysis of the high frequency oscillation modes with respect to the current reveals the onset of a steady-state precession regime for negative bias currents above $J = 10^7$ A/cm^2, when the magnetic field is applied along the easy axis of magnetization. Previous studies indicate that the low-frequency tail in the HF noise power could be affected by the transition from a damped oscillation to a steady-state precession (Devolder et al., 2009; Thadani et al., 2008). By the study of low-frequency noise for the same samples, a direct link is shown between changes in the oscillation modes with the applied current and the normalized low-frequency ($1/f$) noise as a function of the bias current.

The devices are embedded in impedance matched RF coplanar waveguides for electrical contacting using special RF probes. The HF and LF measurement system was described in (Cascales, 2015; Cascales et al., 2013a). The representative high- and low-frequency noise measurements shown in this work were carried out in the same 40×80 nm^2 sample with an external field along the easy axis. Positive voltage means that electrons flow from the pinned to the free electrode, promoting the P state. Negative voltage favors the AP state. Figure 7.7a shows changes in the coercive field with the applied current, obtained from resistive transitions measured by sweeping the field positive to negative values. As can be seen in

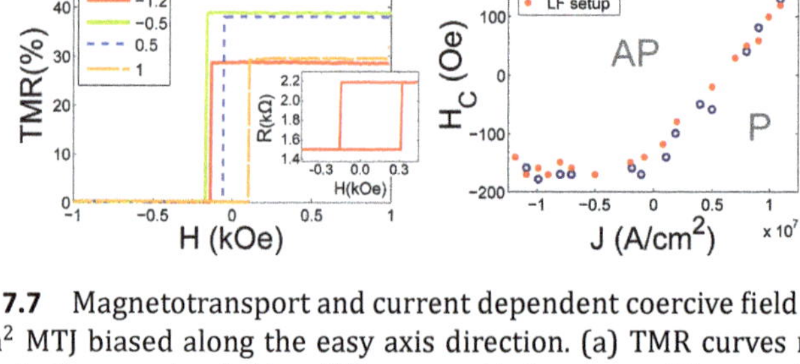

Figure 7.7 Magnetotransport and current dependent coercive field in 40 × 80 nm^2 MTJ biased along the easy axis direction. (a) TMR curves reveal a change in the coercive field at different applied currents. The coercive field of the MTJ changes with increasing positive current. The inset shows a full TMR cycle. (b) Change in the coercive field with the applied current from HF and LF noise measurements. Adapted with permission from Ref. Cascales et al. (2015a).

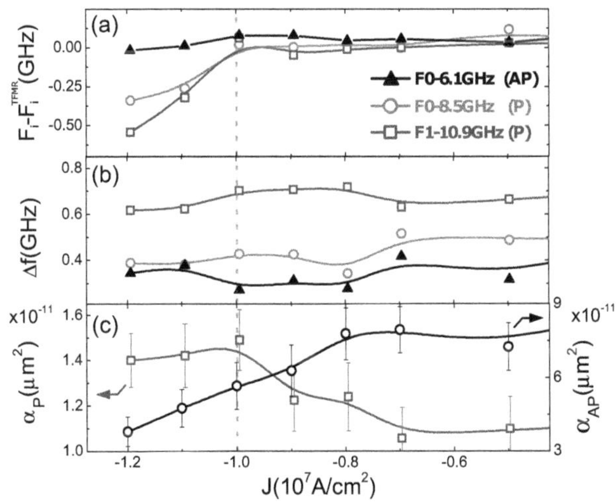

Figure 7.8 High and low-frequency noise measurements in the P (−350 Oe) and AP (250 Oe) states of a 40 × 80 nm^2 MTJ. Dependence of the F0 and F1 modes' (a) frequency shift with respect to the frequency in the TFMR regime, (b) linewidth and (c) average Hooge parameter as a function of the current density in the P and AP states. Adapted with permission from Ref. Cascales et al. (2015a).

Fig. 7.7b, the coercive field H_C is stable for negative currents (the AP state is favored) while it shifts to lower values (favoring the P state) for positive currents.

Six clear modes are detected, which come from oscillations of the free layer (labeled F0, F1, ...) and the SAF structure (A0, A1, and A2). The free layer modes are V-shaped, while the modes not showing a minimum at low fields correspond to SAF modes (Helmer et al., 2010). The analysis of the F0 and F1 modes was carried out in the manner discussed in Ref. Houssameddine et al. (2008). The analysis of these results reveals that for negative currents, a decrease in frequency is observed for the first and second modes in the P state, starting at $J \sim -10^7$ A/cm^2. The change of the oscillation frequency with respect to the TFMR regime and the linewidth of the F0 and F1 modes for the P ($H = -350$ Oe) and AP (250 Oe) states is shown in Fig. 7.8a,b, resembling the transition from T-FMR to SSP reported on similar devices in Ref. Houssameddine et al. (2008).

The minimum linewidth Δf obtained for the F0 mode is around 400 MHz, concluding that the highest negative current takes the sample to an intermittent SSP regime (Houssameddine et al., 2009). The AP state, however, presents no signs of such a transition, and the frequency of the modes in Fig. 7.8a remains constant with the current.

Low-frequency noise measurements were also carried out in the same sample, using the same current and magnetic field values. The low-frequency spectra show $1/f$ noise or Hooge parameter and $0.7 < \beta < 1.5$. The spectra reveal the appearance of random telegraph noise at higher frequencies (Cascales et al., 2013a). The Hooge parameter remains somewhat constant for each magnetic state (P or AP), and it is higher in the AP state than for the P state because the sample presents resistance fluctuations which are proportional to the magnetization noise and $\partial R(H)/\partial H$ (Hardner et al., 1993b). The analysis of the $1/f$ noise data as a function of the current density reveals the signature of effects observed in the HF results. As is shown in Fig. 7.8c, noise in the P state, α_{P}, monotonically increases with the applied current for current densities substantially below $\pm 10^7$ A/cm^2. Only for negative currents, the normalized noise reaches a maximum at around $J = -10^7$ A/cm^2.

The $1/f$ noise in spin torque oscillators has been tied to hopping of the oscillation modes, where each random hopping event leads to a jump in the phase of the oscillator (Eklund et al., 2014; Sharma et al., 2014). As the current is increased toward the critical value for an intermittent SSP (Houssameddine et al., 2009) ($J_{\text{C}} = -10^7$ A/cm^2), these phase jumps increase in number, which is reflected in $1/f$ results. When the intermittent SSP is reached, the hopping events gradually become less frequent until the purely SSP is reached for a second critical current (Houssameddine et al., 2009) J_{C}^* (not reached in the experiment described in this section). The increase in $1/f$ noise is observed only in the P state, as negative currents destabilize this state. Therefore, by comparing the LF and HF results, the authors conclude that signs of spin torque-related phenomena are detected in low-frequency $1/f$ measurements.

7.4 STT Magnetization Switching and Random Telegraph Noise

The work presented in (Cascales et al., 2013a) observed STT switching in low RA junctions, and the analysis of low-frequency random telegraph noise (RTN) effectively evaluates the non-homogeneity of the ultrathin barrier.

As was mentioned above, MTJs of sizes under 100 nm with low TMR (around 30–40%) and low RA products have been suggested to have localized reductions in the tunneling barrier (Houssameddine et al., 2008). The work shows that the fraction of area where the barrier is reduced may be estimated by analyzing the RTN present in the samples. It was found that due to inhomogeneous spin currents, these low TMR MTJs could be switched between the AP and P states using rather low current densities when an external magnetic field is directed along the hard axis. Micromagnetic simulations, with an effective pseudo-pinhole area close to what is estimated from the RTN, support the main experimental observations.

As we have seen, in the absence of external field the free electrode's magnetization M_f and the pinned electrode's magnetization M_p are aligned parallel (P state). When the field is applied along the hard axis, the highest resistance is reached for fields around 1.5 kOe (Fig. 7.9a) and M_f is fully aligned with H. Since M_p remains fixed, the angle between M_f and M_p is around 90°. We will refer to this state by AP_{HA}, the hard axis anti-parallel state. Then we have that $R(P) < R(AP_{HA}) < R(AP)$.

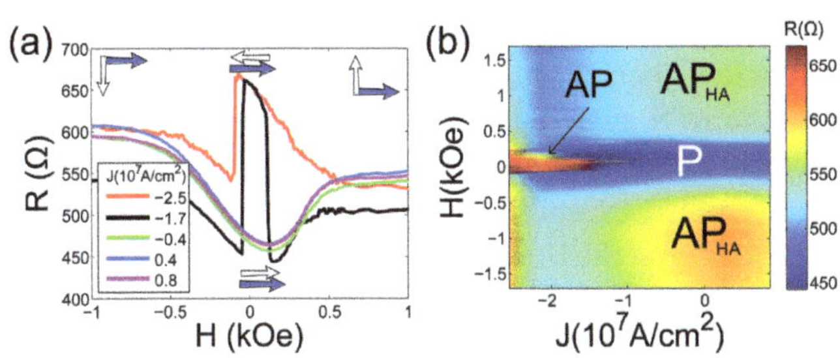

Figure 7.9 50×100 nm^2 MTJ sample measured in the HF setup. The (a) TMR curves and (b) phase diagram show current switching at $J = -1 \times 10^7$ A/cm^2.

As could be seen in Fig. 7.9a, the P state is reached when the field approaches zero for positive currents. However, for the negative currents, when the field is lower than the switching field, i.e., $|H| < H_{switch}$, ST effects overcome the external field and switch the sample to a high resistance state. The resistance of the sample in this state is higher than in the AP_{HA} state, so it is argued that the sample switches to the AP state. Figure 7.9b shows a phase diagram of the magnetic state of the sample constructed from TMR curves at different currents. A region where the sample is switched to the AP state appears at some negative critical current density, and the authors report this behavior was observed in a considerable amount of devices (of similar or different size).

Another notable difference between easy and hard axis measurements is the following: (i) the current switching occurs for the positive currents between the AP/P states in the easy axis case, while (ii) the switching takes place between the P/AP states for negative currents in the hard axis. Therefore, it seems that current switching and steady-state precession occur for opposite signs of the current, and the sign depends on the direction of the applied magnetic field.

The authors imagine the switching process as a double well potential, where for $|H| > H_{switch}$, the minimum of energy corresponds to the typical HA TMR configuration and for $|H| < H_{switch}$, the minimum of energy corresponds to the AP state due to ST. In the vicinity of H_{switch} a bi-stable situation is obtained and thermally activated RTN is detected. The conditions for such a bi-stability should primarily exist in the areas of the soft magnetic electrode which are located close to the pseudo-pinholes, i.e., where the current density is the highest.

Then, the RTN analysis is used to estimate the size of the "hot spot" region of the barrier. The spectra in these samples usually present $1/f$ noise, except for the fields where RTN is present. The field dependence of the Hooge factor α revealed a clear maximum in noise centered around the field value where the resistance switch takes place (Fig. 7.10a). Also, the exponent β (see the $1/f$ noise equation 2.22) approaches 0 for these same field values (Fig. 7.10b), i.e., the curve becomes flat and Lorentzian-like. These features clearly identify the range of H which presents RTN.

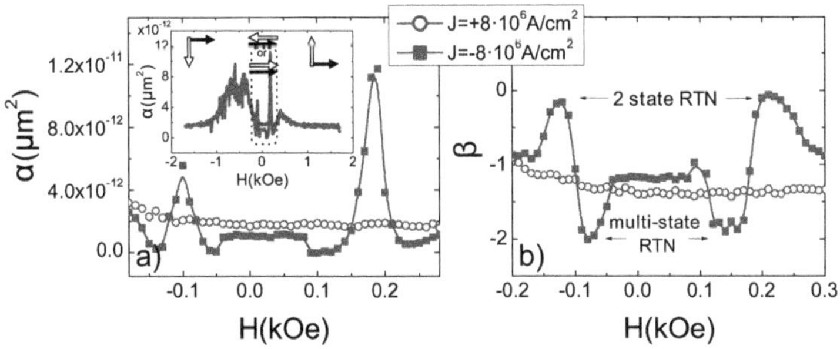

Figure 7.10 Dependence of the (a) Hooge parameter α and the (b) $1/f$ exponent β on the magnetic field. The curves indicate the range of H which presents 2 state ($\beta \simeq 0$) or multi-state ($\beta \simeq -2$) RTN. The inset in (a) plots α for the whole range of external field.

The characterization of the RTN was then carried out by analyzing the spectra and time-series at these fields, following the method explained in section 3.2. Figure 7.11 shows typical RTN features in the LTMR samples, where the field is directed along the HA. Figure 7.11a shows typical time series of the voltage fluctuations for magnetic fields in and outside the field range where a strong RTN is detected. The graph corresponding to $H = +190$ Oe shows well-defined step-like jumps between two voltage levels. The amplitude of these RTN fluctuations is a factor of 10^2 larger than RTN for higher fields. Figure 7.11b shows that for the field values in the range where the magnetization reversal takes place ($H = +190$ Oe) one observes a Lorentzian-like spectrum, typical of RTN. On the other hand, outside the range ($H = +1700$ Oe and $H = +600$ Oe), the spectrum is nearly $1/f$. The graphs shown in Figs. 7.11a,b correspond to $J = -1.2 \times 10^7$ A/cm^2.

For each current, the fluctuating magnetic moment Δm involved in the RTN is estimated in the following way. The fluctuating voltage ΔV is obtained from histogram of the time-series and the spectra are fitted by equation 2.17. Then we calculate $\ln(\tau_{up}/\tau_{down})$ with respect to H for each current, fit by the Arrhenius law $\ln(\tau_{up}/\tau_{down}) = C + 2\Delta m H/k_B T$ (where C is a constant) as shown in Fig. 7.11c. Figure 7.11d shows the estimation of Δm for different currents, which is found to be around $(2 - 4) \times 10^5$ μB, for both resistance switches (AP-P and P-AP) present in each curve. Similar

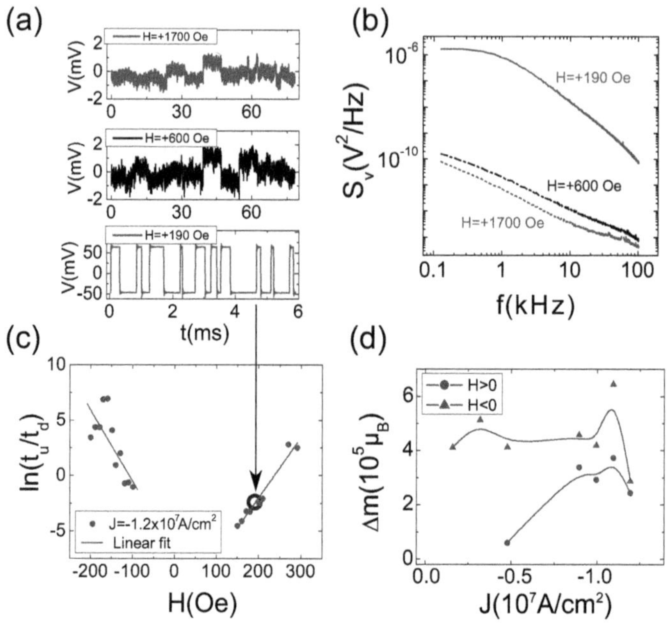

Figure 7.11 (a) Voltage fluctuations for three different applied magnetic fields for $J = -1.2 \times 10^7$ A/cm^2. The time-trace of the magnetization reversal ($H = +190$ Oe) presents RTN fluctuations which are two orders of magnitude higher in amplitude than other values of the field ($H = +1700$ Oe, $H = +600$ Oe). (b) Lorentzian-like spectrum typical of RTN at $H = +190$ Oe and a typical $1/f$ spectrum for $H = +1700$ Oe and $H = +600$ Oe. (c) Linear fit to an Arrhenius-type law of the $\tau_{\text{up}}/\tau_{\text{down}}$ ratio. (d) Estimation of the fluctuating moment Δm with respect to the applied current. Adapted with permission from Cascales et al. (2013a).

effects were observed in several other junctions. Considering the moment per atom in CoFeB to be of 1 μB, its lattice parameter $a = 0.284$ nm, an fcc structure (hence there are 4 μB in a volume of a^3). The authors suppose that the fluctuating moment is only present in the free layer, so the estimated volume which corresponds to $\Delta m = 4 \times 10^5$ μ_B is 23% of the volume of the free electrode, which fluctuates and generates the RTN (Cascales et al., 2013a). This percentage is obtained from the ratio of the volumes $V_{\text{RTN}}/V_{\text{electrode}}$ where $V_{\text{RTN}} = \Delta m a^3/4\,\mu_B$ and $V_{\text{electrode}} = x \cdot y \cdot z$, where $x = 20$ nm, $y = 40$ nm and $z = 3.9$ nm.

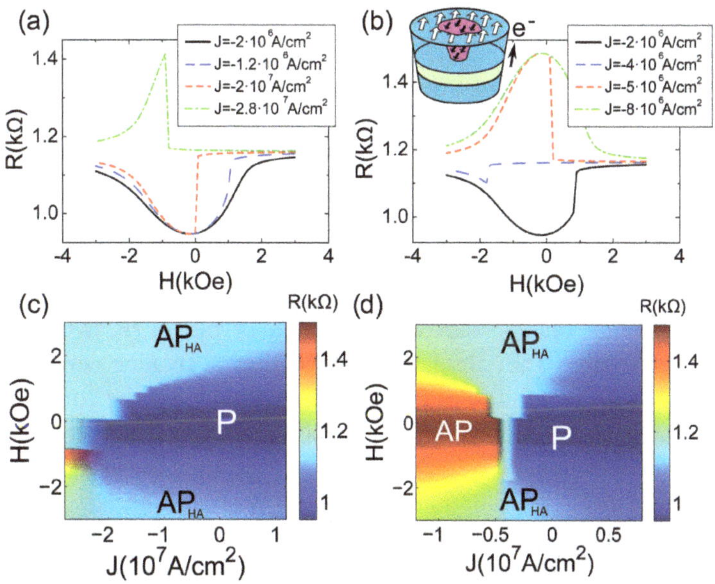

Figure 7.12 (a) Simulated TMR curves for different current values in a junction with a perfect barrier. (b) Simulated TMR curves for different currents with a pin-hole in the barrier. (c) Phase diagram constructed from the TMR curves in (a), where the switching current is found to be $J = -2 \times 10^7$ A/cm^2. (d) Phase diagram constructed from the TMR curves in (b), which presents a much lower switching current density, $J = -4 \times 10^6$ A/cm^2 for $A/a = 5.3$. Adapted with permission from Cascales et al. (2013a).

The authors also carried out numerical simulations, with OOMMF (Donahue et al., 1999), of junctions with and without pinholes described by a simple qualitative model in order to account for the observed phenomena (Cascales et al., 2013a). The reduction of the barrier or the presence of a pseudo-pinhole are modeled as a region of area a in the insulating barrier which concentrates the current going through the structure, as schematically shown in Fig. 7.12b. If J is the current density flowing through the electrodes of area A, then the current density in the pin-hole is $J \cdot A/a$. For negative currents, electrons flow from the free to the pinned layer and this favors an AP alignment of the electrodes' magnetizations. The constants for CoFeB used in the simulations were: spin polarization $P = 0.5$ and saturation magnetization

$M_s = 1150$ kA/m. The TMR curves (Fig. 7.12a,b) have been calculated by obtaining the average angle between the free and pinned layer's magnetizations, and using the expression $R(\theta) = R(0) + \Delta R(1 - \cos\theta)/(2 + \chi(1 + \cos\theta))$(Montaigne et al., 2010). Since H was only applied along the HA in this sample, we have to estimate ΔR. From the data, TMR$(\pi/2) = 23\%$ and $R(0) = 945\ \Omega$, so with $\chi = 0.5$ we obtain that $\Delta R = 0.6R(0)$, i.e., a reasonable TMR $= 60\%$.

The results of the simulation (Fig. 7.12) closely resemble the experimental results with the exception of the AP/P switch for negative fields. The simulations show that a smaller current is necessary to switch the free layer's magnetization if the current is concentrated through some region of the barrier, where there is a higher effective current density, compared to a perfect junction (Cascales et al., 2013a). Figure 7.12a shows that a perfect barrier needs a much higher current to obtain a resistance switch ($J = -2 \times 10^7$ A/cm^2) than junctions with local barrier reductions, as seen in Fig. 7.12b. Figures 7.12c,d show a phase diagram constructed from all the TMR curves for an MTJ with (d) and without (c) a pinhole. Indeed, for the ratio $A/a = 5.3$ (for which close to 20% of the insulating surface concentrates current), a switch to the AP state is obtained for low H at $J = -4 \times 10^6$ A/cm^2. If the fluctuating moments are located in the free layer, and strictly above the area a of the pinhole, this corresponds to 20% of the volume of the free layer, which is remarkably close to the 23% mentioned above. Some disagreement between simulations and experiment could be related to the fact that the simulations are done at zero temperature while the experiments were carried out at 300 K, since it has been demonstrated by simulations that temperature fluctuations facilitate reversal processes (Aurelio et al., 2010).

Chapter 8

Magnetoresistance and Noise in Magnetic Tunnel Junctions with Optimized Interface Mismatch

This chapter discusses lattice mismatch as one of the major noise sources in the fully epitaxial spintronic devices and presents methods of its optimization. The second part explains how the related noise measurements could be converted into band edge spectroscopy allowing obtaining information on the buried interfaces.

8.1 Lattice Mismatch and Interfacial Stress Problems

The seminal MTJs with composition Fe(100) | MgO | Fe were the first to reveal a record high TMR (above 100% at room temperature, due to the high spin polarization of electrons at Fermi level (E_F) with Δ symmetry in Fe bands and the filtering capability of the MgO(100) barrier. However, the experimentally observed TMR values of about $200 - 300\%$ were notably below the theoretically predicted, being

Noise in Spintronics: From Understanding to Manipulation
Farkhad G. Aliev and Juan Pedro Cascales

ISBN 978-981-4774-34-5 (Hardcover), 978-1-315-11088-2 (eBook)
www.panstanford.com

approximately 1000% (see previous chapters). The understanding of the origin of the TMR limitation has been a fundamental issue to control the spin-polarized tunneling. Among the possible reasons is oxidation that occurs at the Fe/MgO interface during the growth process (Velev et al., 2007; Waldron et al., 2006). Another important source of TMR limitation could be a small structural lattice mismatch between the MgO barrier and the Fe layer. In the Fe (100)/MgO(100)/Fe junctions, the Fe lattice is rotated by 45° with respect to the MgO (Tiusan et al., 2007). With a bcc crystalline structure, the lattice parameter of the MgO is $a_{\mathrm{MgO}} = 0.421$ nm. Consequently, in the [110] direction, the interatomic distance in bulk MgO is $\frac{a_{\mathrm{MgO}}}{\sqrt{2}} = 0.298$ nm, close to Fe in the (100) direction ($a_{\mathrm{Fe}} = 0.287$ nm). As could be estimated from lattice parameters shown in Fig. 8.1, the he lattice mismatch between both elements is around to 3.9%, and this fact should naturally induce a stress at the Fe/MgO interface.

This strain is partially relaxed during the MgO growth via interfacial dislocations (Tiusan et al., 2007). The lattice mismatch could both partially suppress the TMR and contribute to the defect-related $1/f$ noise (Aliev et al., 2007; Guerrero et al., 2005). In order to search for the alternative ferromagnetic electrodes to improve the TMR and suppress the noise, it is critical to choose those ferromagnetic electrodes which satisfy several specific criteria.

- A bcc structure for electrode is necessary to maintain the sensitivity of the tunneling to states with Δ symmetry to magnetic states (P vs. AP) (Butler et al., 2001).
- The lattice parameter of the electrode should be close to that of MgO.
- The spin polarization of the ferromagnetic material has to be large, in particular, the spin polarization of tunneling Δ electrons.

In the direction of improving the TMR, recently ultrathin Co electrodes emerged, as an alternative to the Fe, with a high spin polarization of the Δ bands at the E_{F}. For example, bcc Co (Yuasa et al., 2005) fulfill the condition of high effective spin polarization. The advantage of using Co, in contrast to Fe, is that (at small biases)

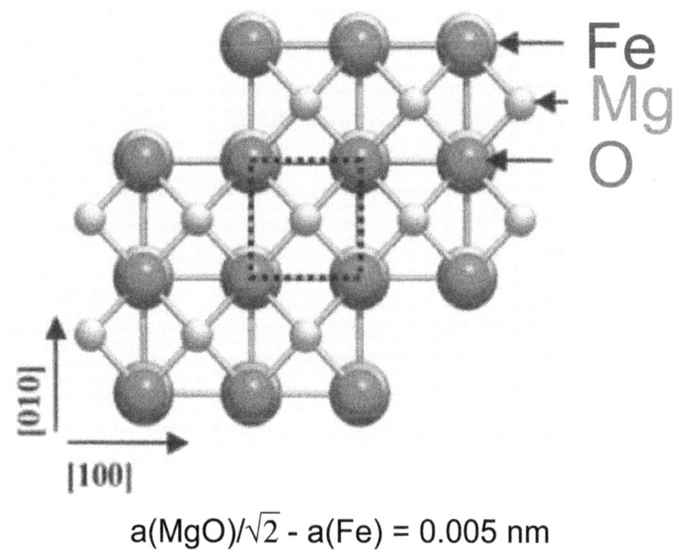

Figure 8.1 Schematic representation of the MgO and Fe crystallized in a cubic structure.

only electrons with Δ_1 symmetry contribute to the transport at E_F for the majority-spin band, because the E_F level is well above of $\Delta_{5,2'}$ bands. Consequently, in the antiparallel (AP) state, it is expected that all the states are reflected at $\vec{k}_{\parallel} = 0$. Based on this hypothesis, Yuasa et al. (Yuasa et al., 2005), grew Co (100)/ MgO/ Fe and Co (100)/ MgO/ Co MTJs, reaching values of TMR at room temperature of 270% and 410%, respectively. However, substitution of Fe by Co even worsens the lattice mismatch. Indeed, the lattice parameter of the Co is known to be lower than one of Fe ($a_{Co} = 0.251$ nm).

With the aim to search for the ferromagnetic materials with similar lattice parameter as that of the MgO and the reduced stress of the barrier, a new material combination has been probed. For example, amorphous CoFeB crystallizes in a bcc structure in the (001) direction and depending on the temperature of the annealing one can obtain similar lattice parameter to these of the MgO barrier (Ikeda et al., 2008). The CoFeB (001)/ MgO/ CoFeB MTJs has reached TMR ratios about 600% at RT due to the highly oriented crystalline structure (Ikeda et al., 2008).

A different strategy has been tested by Bonell et al. (2009a) trying to improve the tunneling transport with the control of the order/disorder at the interface electrode/insulating interface using Fe electrode with oxidized surfaces. But at the end, the obtained experimental results show lower TMR values than those expected for control MTJs. An alternative approach would be tuning the lattice parameter of the barrier itself though creation of the small lattice mismatch at the Fe–MgAlOx interface (Tao et al., 2015).

In this chapter, we shall discuss electron transport and noise in MTJs with effectively improved interface mismatch following original proposal (Bonell et al., 2009b) to create ferromagnetic electrodes in form of Fe alloys with different concentrations of vanadium. Importantly, the substitution of Fe with vanadium both decreases the magnetic damping (Scheck et al., 2007) and increases the lattice parameter of the electrode and therefore reduces the misfit (Bonell et al., 2009b) with the MgO barrier. This type of interface engineering could also be expected to enhance the electronic stability of MTJs through suppressed defect-related noise as long as the quality of the metal-insulator interface in MTJs determines the $1/f$ noise.

The detailed description of the room temperature electron transport and noise in such MTJs is provided in (Herranz et al., 2010b) and PhD thesis by D. Herranz (Herranz, 2012), who carried out a study of a total of 89 MTJ samples at room temperature, 61 of which had a MgO thickness of 9.5 ± 0.5 ML and 28 had a 13 ML MgO thickness. The electron transport and low-frequency noise were measured in fully epitaxial $Fe_{(1-x)}V_x$/MgO/Fe and Fe/MgO/$Fe_{(1-x)}V_x$ MTJs. Summarizing the main observations to be discussed below, the Vanadium doping provided an increase of TMR of about 10% for the thinner barrier and of 35% for the thicker barrier, with a relatively small vanadium doping of the bottom electrode (x about 0.1). What is most relevant for the present book, a remarkable reduction in nearly two orders of magnitude of the normalized $1/f$ noise (Hooge factor) in the non-magnetic (parallel) state was observed. The sample growth and composition details could be found in Refs. Bonell et al. (2009b) and Herranz (2012).

8.2 Optimization of Tunneling Magnetoresistance through Improved Lattice Mismatch

It has been shown that the low concentration of vanadium improves the misfit at the Fe-V/MgO interface and that due to structural processes of the growth, the density of dislocations is lower when the Fe electrode containing vanadium in the bottom electrode (Bonell et al., 2009b) and (Herranz, 2012). However, the higher critical concentrations of vanadium decreases the total spin polarization and affects the Δ_1 and Δ_5 majority bands. In order to verify these effects, the tunneling transport and low-frequency noise in two types of MTJs (using Fe-V alloy in the bottom or top electrode) and with different thickness of the MgO barrier (9.5 ± 0.5 and 13 ML) have been investigated and the results were compared with the reference samples (Fe/MgO/Fe) (Herranz, 2012).

Figure 8.2 shows the typical TMR measured in zero bias limit by using a low amplitude (with excitation below 10 mV) square current excitation. Part (a) shows the TMR of the MTJs with the thinner (9.5 ± 0.5 ML) barrier. For these MTJs, the reference sample has a value of TMR close to 185% at room temperature. When one uses $Fe_{90}V_{10}$ alloys in the top electrode, the TMR decreases to about 130%. This is because the density of dislocations at the Fe/MgO interface is large (respect to Fe-V/MgO) when the MgO barrier is grown on the Fe electrode. Besides, the top electrode is formed by a Fe-V alloy, where the total spin polarization is lower than the bottom electrode and, consequently, the TMR decreases respect to the TMR value of the reference samples. In contrast, for the samples with $Fe_{92}V_{08}$ in the bottom electrode, the density of dislocations at the interface should significantly decrease, which is pointed out by an improvement of the tunneling conductance and the increase of TMR values up to 210%. For this concentration of vanadium, the total spin polarization of the bottom electrode is lower than of the top Fe electrode but not yet significantly decreased. Due to the competition between the mismatch and the spin polarization, for the restricted range of small concentrations of vanadium, the first term is more relevant and the TMR increases with vanadium.

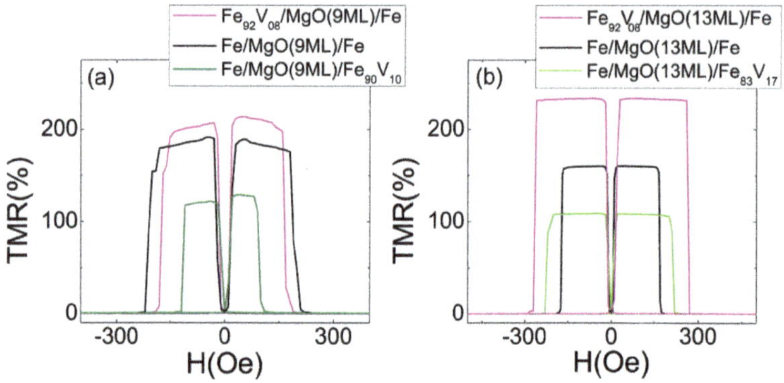

Figure 8.2 (a), (b) TMR of various $Fe_{1-x}V_x$/MgO (9.5±0.5 and 13 ML)/ $Fe_{1-x'}V_{x'}$ MTJs versus applied magnetic field at RT and 0 applied bias, respectively.

Similar effects were observed in the tunneling transport when the thickness of the barrier is higher (confirmed by the preliminary report (Bonell et al., 2010)). When one uses the $Fe_{92}V_{08}$ alloy as the bottom electrode, the value of the TMR increases in a 60% in comparison with the reference sample Fig. 8.2b). The difference in the TMR for the thicker barrier could be due a plastic relaxation that occurs after to growth 5–6 ML of MgO barrier (Bonell et al., 2009b; Herranz, 2012). In this range, an increase of the thickness of the barrier increases the density of defects. Therefore, when one uses pure Fe as the bottom electrode, the defects density is higher and they propagate through all the structure. As a consequence, the effective tunneling conductance and TMR could be reduced (compare Figs. 8.2a,b). In the contrary, if the bottom electrode is a Fe-V alloy, it will have reduced the mismatch improving the effective tunneling transport through the barrier and increasing the TMR with respect to thinner barrier due to Δ_5 filtering of the MgO barrier.

Further below, following (Aliev et al., 2014), we discuss in more details the TMR in MTJs with different concentrations of vanadium. As one could see in Fig. 8.3, for MTJs with 9.5 ± 0.5 ML thickness of the MgO barrier with Fe-V alloys as the top electrode, the TMR values are monotonically decreasing as a function of vanadium concentration. On the other hand, for Fe-V alloys as the bottom

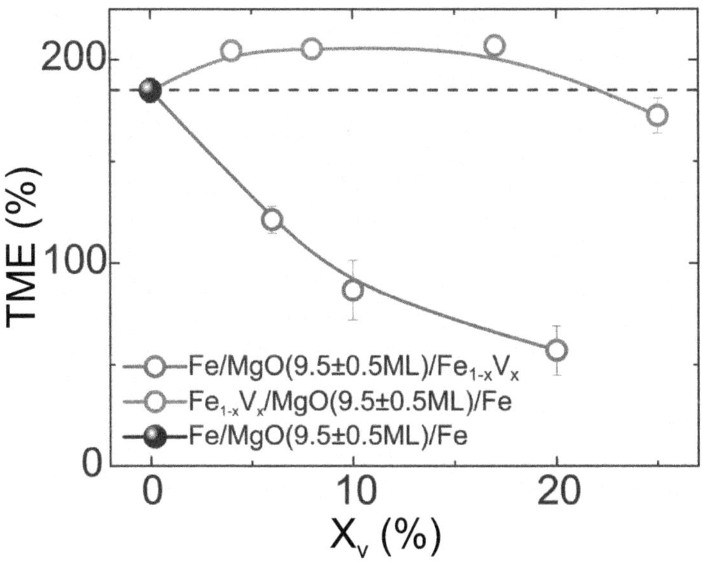

Figure 8.3 TMR variation at RT versus vanadium concentration x for the different series of MTJs with thickness of the MgO 9.5±0.5 ML.

electrode (for $x < 20\%$) the TMR is higher than the reference sample (via a decrease of the number of the dislocations) and for the $x > 20\%$ TMR values decrease with respect to control samples (due to the decrease of the total spin polarization of the bottom electrode). A similar behavior has been observed for thicker barrier (Bonell et al., 2010). Given that the density of defects is related to the low-frequency noise, we shall discuss below in more details the noise as a function of vanadium concentration.

8.3 Low-Frequency Noise

In order to simplify the analysis of the variation of the LFN, we use again a Hooge model. More details on the TMR and noise vs. vanadium doping could be found in Ref. Herranz et al. (2010b).

In Fig. 8.4a comparative typical noise power spectra times area ($\alpha \times$A, for which Hooge factor is being analyzed) are represented in the P state for the junctions with undoped, bottom or top

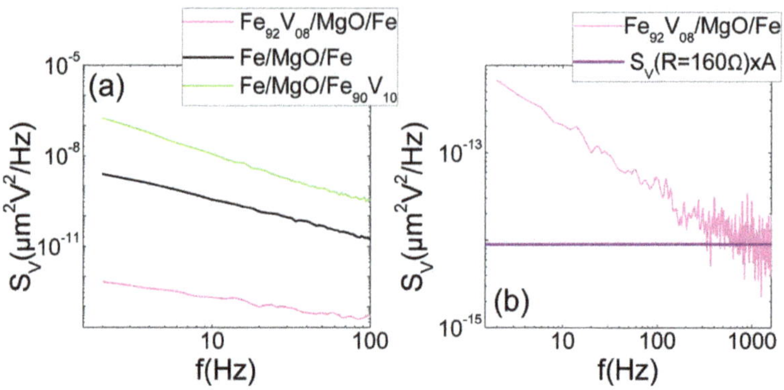

Figure 8.4 (a) Voltage noise power spectral density times area measured on the junctions with undoped, bottom doped or top doped electrodes with bias of 200 mV in the P state at RT. (b) Power spectral density (up to 2000 Hz) for the junction with bottom doped electrode. The purple horizontal line marks the noise power times area expected level of $Fe_{92}V_{08}$/MgO(9.5 $\pm$ 0.5 ML)/Fe with resistance of 160 Ω and direct electron tunneling processes (Shot noise + thermal noise).

Fe-V electrodes. Clearly, vanadium doping of the bottom Fe electrode trends to reduce the low-frequency contribution to the noise, while vanadium doping of the upper electrode trends to enhance the normalized $1/f$ noise. Figure 8.4b is an expansion of the spectra up to 2000 Hz. It is shown that noise for MTJ with bottom doped electrode ($Fe_{92}V_8$/MgO (9 ML)/Fe) may be accounted by the junction resistance in the conditions of the direct electron tunneling (i.e., Poissonian shot noise + thermal noise).

Let us now discuss the influence of vanadium concentration on the $1/f$ noise. For low concentrations of vanadium, as we noted in the figure above, the power spectral noise at low frequencies decreases when using Fe-V alloy as the bottom electrode due to the smaller number of dislocations in the interface between electrode and barrier. We now extend this study to all concentrations of vanadium for which we measured the dynamic conductance vs. bias. Figure 8.5 shows the variations of the zero bias TMR and normalized low-frequency noise (Hooge parameter analyzed between 1 and 50 Hz)), averaged for each set of samples of ($Fe_{(1-x)}V_x$/MgO(9.5 $\pm$ 0.5 and 13 ML)/$Fe_{(1-x')}V_{x'}$), as a function of vanadium content (x). The error bars represent the dispersion of the corresponding values measured within each junction sets. The convention used is negative x values correspond to a bottom $Fe_{(1-x)}V_x$ electrode, whereas positive x values correspond to a top $Fe_{(1-x)}V_x$ electrode. First, we study the influence of vanadium for thinner (9.5 ± 0.5 ML) barrier (Fig. 8.5a. In the latter case (top $Fe_{(1-x)}V_x$ electrode), the TMR systematically decreases with x being lower than for control (standard) Fe/MgO/Fe MTJs (185% at room temperature). As was shown before, the TMR decreases due to the reduced total spin polarization of Δ states in Fe-V alloys (Bonell et al., 2009b). In the presence of a bottom Fe-V electrode and for $x < 0.17$, we observe an increase of TMR from 185% to 207%, whereas the TMR again decreases for larger vanadium contents. Similar trends are observed with a thicker MgO barrier (13 ML) as shown in Fig. 8.5b, the TMR reaching up to 240% at RT in the latter case (Bonell et al., 2009b), due to a better spin filtering when increasing the MgO barrier. It was shown that the optimum TMR results from the competition between the reduction of the electrodes polarization (detrimental to the TMR) and the structural improvement of the barrier (beneficial to the TMR) (Bonell et al., 2009b). Indeed, when used as the supporting bottom electrode, Fe-V alloys reduce the dislocations density and therefore the strain of the barrier and its roughness.

The main finding of (Herranz et al., 2010b) is the observation of a strong (nearly two orders of magnitude) decrease of the Hooge factor measured in the P state (i.e., defect-related noise) with vanadium doping of the bottom Fe electrode for the thinner (9.5 ± 0.5 ML) barrier (Fig. 8.5a). We note also that the minimum of the normalized non-magnetic noise (α_P) roughly coincides with the maximum TMR of about 207%. This corresponds to an enhancement of the signal to non-magnetic noise ratio by more than two orders of magnitude in $Fe_{(1-x)}V_x$/MgO/Fe ($0.08 \leq x \leq 0.17$) MTJs, in comparison with the reference Fe/MgO/Fe one. Although additional vanadium doping further decreases the lattice mismatch at the Fe-V/MgO interface, contrary to expectations, the normalized noise values start to increase when x exceeds 17%. One possible reason could be the increased chemical disorder at the Fe-V/MgO interface, and the related spatial fluctuations of the potential. In contrast, vanadium doping of the upper electrode in Fe/MgO/Fe-V MTJs results in an increase of the non-magnetic noise. Contrary to

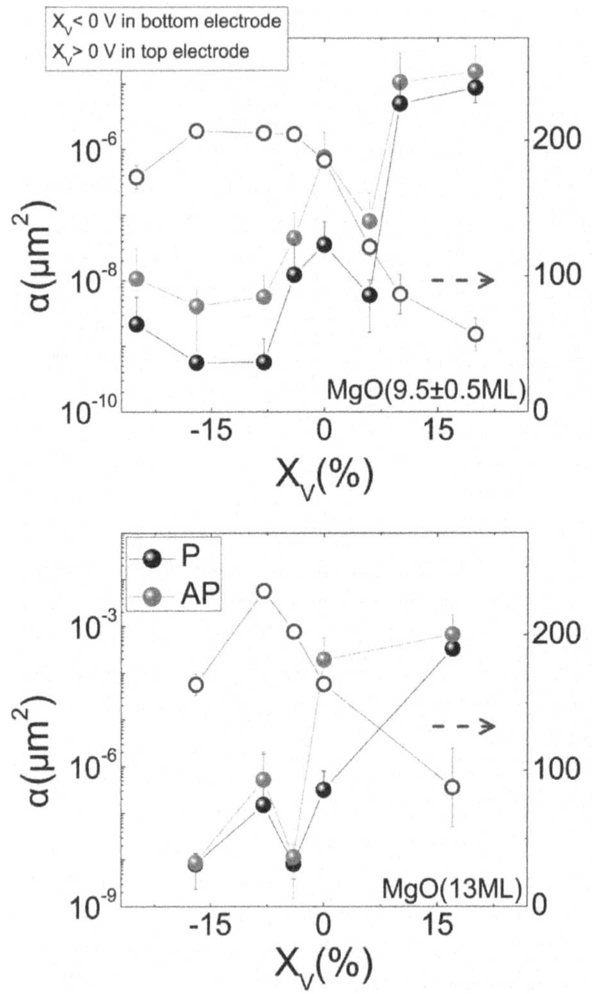

Figure 8.5 Dependence of zero bias TMR and normalized noise (Hooge factor), averaged over each set, as a function of vanadium content in the bottom ($x < 0$) and upper ($x > 0$) electrodes and obtained with two different MgO barrier thickness as indicated in each of figures. All measurements have been done at room temperature. The upper figure has been adapted with permission from Ref. Herranz et al. (2010b).

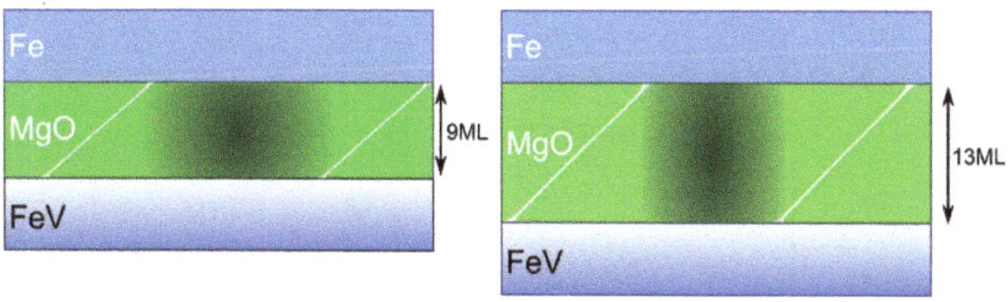

Figure 8.6 Schematic representation of the coherent tunneling for samples using Fe-V alloys in the bottom electrode with different thickness of the barrier. The shadow area represents the defect-free junction area.

the case of a bottom Fe-V electrode, a top Fe-V one acts much less on the strain of the barrier because dislocations mainly nucleate during the growth of MgO. Here, the noise variations could reflect the increasing of chemical disorder.

Qualitatively, similar behavior of $1/f$ noise and TMR as a function of vanadium content is observed for thicker MgO barrier (13 ML) (Herranz, 2012). But now, contrary to what is observed in tunneling transport, the reduction of the normalized low-frequency noise as a function of concentration of vanadium is lower than for the thinner barrier. One of the possible scenarios is shown in Fig. 8.6. The tunneling transport is affected by the mismatch and the filtering of the MgO barrier. For the thicker barrier the filtering of the Δ_5 is higher and consequently also the TMR (Butler et al., 2001). On the other hand, the low-frequency noise measurements are more sensitive to the tunneling process than the dynamic conductance. As we increase the number of layers of the MgO barrier, due to the oblique orientation of the dislocation the effective tunnel area is reduced (see sketch of Fig. 8.6) and the low-frequency noise will be affected.

The normalized difference of the $1/f$ noise between AP and P states defined as $(\alpha_{\text{AP}} - \alpha_{\text{P}})/\alpha_{\text{P}}$ (see Fig. 8.7) shows an interesting trend as a function of vanadium alloying, which may be important for applications of these MTJs. One observes that an enhanced TMR with vanadium alloying is accompanied by a reduced relative noise in the AP state. This effect was attributed (Herranz et al., 2010b) to the minority Fe-V band hybridization (Johnson et al., 1987). As long as the minority Δ bands control the conductance in the AP state

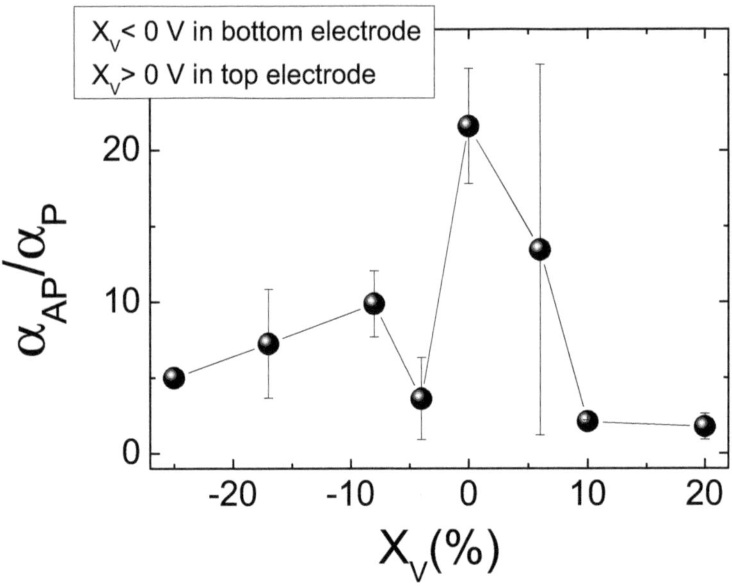

Figure 8.7 Normalized relative variation in the noise between AP and P alignments, as determined from Fig. 8.5 as a function of vanadium content.

(Butler et al., 2001), the Fe-V minority band hybridization could reduce the corresponding $1/f$ noise level produced by the defects through enhanced charge screening.

8.4 Band Edge Noise Spectroscopy

Buried metal (semiconductor)/insulator interfaces are found at the heart of electronics (Kroemer, 2001). The current in tunneling devices is determined by the bias, barrier and density of states of the electrodes (Tiusan et al., 2007; Yuasa and Djayaprawira, 2007). Electron states not allowed in bulk could become permitted at the surface leading to topological (Hasan and Kane, 2010; Schlenk et al., 2013) or interface resonant states (Belashchenko et al., 2005). For metallic structures the scarce knowledge on the interface bands is mainly obtained by indirect methods such as ballistic electron emission spectroscopy (Bell and Kaiser, 1988) or high-resolution

X-ray spectroscopy (Bonell et al., 2012). The possibility of having a reliable and down-scalable emphin situ method for investigating interface electron bands remains centrally important (Berger et al., 2011).

Tunneling magnetoresistance (Julliere, 1975; Miyazaki and Tezuka, 1995; Moodera et al., 1995) is extremely sensitive to the band structures of ferromagnet/insulator (FM/I) interfaces (Bowen et al., 2001; Butler et al., 2001; De Teresa et al., 1999; Faure-Vincent et al., 2003; Mathon and Umerski, 2001; Parkin et al., 2004; Stewart, 2010; Yuasa and Djayaprawira, 2007; Yuasa et al., 2004). Despite recent attempts to understand the nature of the electron bands which contribute to electron transport in spintronic devices (Harada et al., 2012; Rungger et al., 2009; Zermatten et al., 2008), the issue remains unsettled. The main tool to characterize interfaces or barriers has been inelastic electron tunneling spectroscopy (IETS) (Harada et al., 2012; Jaklevic and Lambe, 1966; Teixeira et al., 2011) analyzing the derivative of the *conductance as a function of bias*. The resulting IETS signals depend on the tunneling density of states (DOS) and inelastic scattering (Tiusan et al., 2007; Wortmann et al., 2005; Yuasa and Djayaprawira, 2007) which could obscure the detection of the band edges in the presence of interface disorder. The bias dependence of *the conductance and its low-frequency fluctuations* could be an alternative way to study the interface or electron confinement (Nikolić and MacKinnon, 1994; Xu et al., 2010a) DOS.

As we will discuss in detail below, a commonly accepted phenomenological approach relates the excess low-frequency noise (LFN), which often presents $1/f$ dependence, with electrons scattering from defects characterized by a broad distribution of relaxation times with energy (Dutta and Horn, 1981). If dominant defect states are located close to the interfaces, they could create interface band edge tails. Therefore, when the tunneling is tuned to some specific band edge in the opposite electrode, the current could acquire an extra LFN due to multiple relaxations originating from defect states contributing to the formation of the band edge tails.

In this chapter, we investigate the bias dependence of conductance and LFN in single-barrier tunneling devices in order to determine emphin situ the energies of the band edges of the

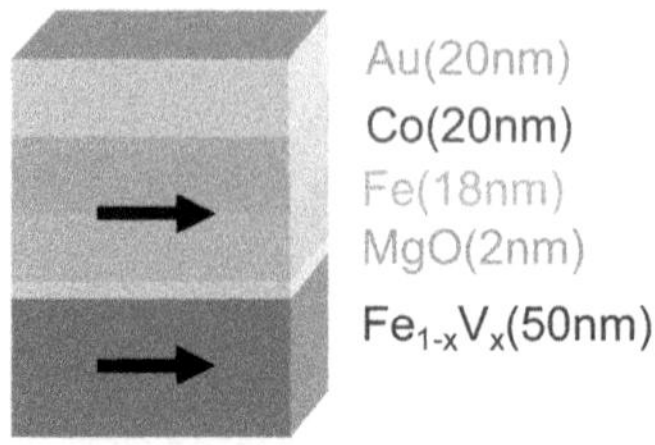

Figure 8.8 Sketch of the sample layer structure with reduced lattice mismatch.

buried interfaces. We unambiguously demonstrate the validity of the *band edge noise spectroscopy* (BENS) concept by studying seminal Fe/MgO/Fe MTJs with partial doping of the bottom electrode (Fe) with vanadium (V). Such substitution has been shown to reduce defect states inside the MgO barrier due to improved interface matching between $Fe_{1-x}V_x$ and MgO in $Fe_{1-x}V_x$MgO/Fe MTJs. (Bonell et al., 2009b, 2010; Herranz et al., 2010c). Numerical simulations confirm that tunneling of band-tail electrons, influenced by spin orbit interactions, are responsible for the observed LFN anomalies. Summary of the BENS method could be found in Ref. Aliev et al. (2014).

8.4.1 Growth and Characterization of the Samples

The samples for which we shall discuss implementation of band edge noise spectroscopy as same for which room temperature noise optimization has been discussed above. Only those samples for which $1/f$ noise is reduced by vanadium doping (i.e., obtained by vanadium alloying the bottom electrode) and where effects are relatively strongest will be discussed.

As can be seen in Fig. 8.8, the layer sequence of the studied MTJs is

$$\text{MgO//Fe}_{1-x}\text{V}_x\text{(50 nm)/MgO(2 nm)/Fe(18 nm)/Co(20 nm)/Au(20 nm)}.$$

The single-crystal multilayers were deposited on MgO(100) substrates by molecular beam epitaxy (MBE) at a typical base pressure of 10×10^{-10} mbar. Before the deposition, the substrates are outgassed at 875 K. A 10 nm MgO layer is grown at 725 K in order to prevent the diffusion of residual carbon to the bottom electrode (Sicot et al., 2006). The bottom electrode, magnetically the softest, is grown using Fe-V alloys with different concentrations of vanadium, obtained by Fe and V co-evaporation. The film is then annealed at 825 K for 20 minutes to reduce its roughness. Afterwards, a 2 nm-thick MgO barrier is deposited and controlled by reflection high-energy electron diffraction (RHEED) intensity oscillations. Then, an 18 nm-thick Fe layer is grown over the MgO barrier and annealed for 10 min at 475 K. A 20 nm-thick Co is grown over the Fe layer, forming a magnetically harder electrode by the coupling between layers. The layered structure is protected from oxidation by a 20 nm Au film. The MTJs were patterned by UV photolithography and Ar etching into pillars with square cross sections of lateral size from 10 to 50 μm. Additional details about the fabrication of the samples can be found in Refs. Bonell et al. (2009b, 2008).

8.4.2 Experimental Results

The typical noise power spectra (S_V) in the antiparallel (AP) or parallel (P) states reveal the presence of $1/f$ noise in the frequency range between 1 and 100 Hz (depending on the sample type), where $S_V(f) \propto 1/f^{\beta}$ (with $0.8 < \beta < 1.5$), as can be seen in Fig. 8.9. The fact that the $1/f$ noise is only observed up to such low frequencies is

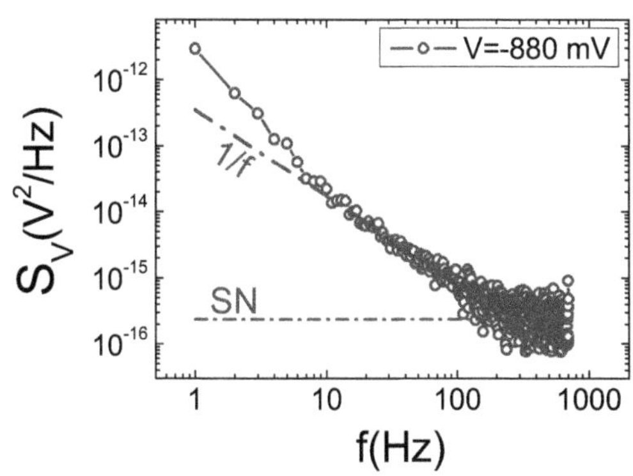

Figure 8.9 Typical low-frequency noise spectrum presenting $1/f$ and shot noise contributions for a sample with a $Fe_{0.75}V_{0.25}$ bottom electrode.

indicative of a high structural quality. The bias dependence of the LFN has been determined through the Hooge factor (α) from the phenomenological expression (equation 2.22) $S_V(f) = \alpha \cdot V^2 A \cdot f$, where V, A and f indicate bias voltage, area of the sample and frequency, respectively. Qualitatively similar results have been obtained by analyzing integrated LFN. Shot noise was obtained from the frequency independent part of the spectra for temperatures below 10 K, as explained in Section 3.2.

We begin by analyzing the electron transport and SN behavior at $T = 4$ K, which is summarized in Fig. 8.10. As can be seen in Fig. 8.10a, the zero bias TMR as a function of V content shows a maximum value for a vanadium content of 8%, confirming a reduction of the interface mismatch reported previously at room temperature (Bonell et al., 2009b, 2010; Herranz et al., 2010c). In the same graph, it can be seen that the tunneling statistics of these junctions are Poissonian (within the experimental tolerance) presenting an average Fano factor of $F = 1 \pm 0.05$. This indicates that the samples have a very high structural quality and present direct (or very nearly direct) tunneling processes.

Figure 8.10b shows the bias dependence of the Hooge factor with the bias in both P and AP states for a Fe/MgO/Fe MTJ used as reference or control sample. An excess LFN is seen for biases below 200 mV, where FeO (Du et al., 2010) and Fe/MgO (Tiusan et al., 2007) interface defect states have been predicted to influence the conductance. For the MTJ with a non-optimized Fe/MgO interface

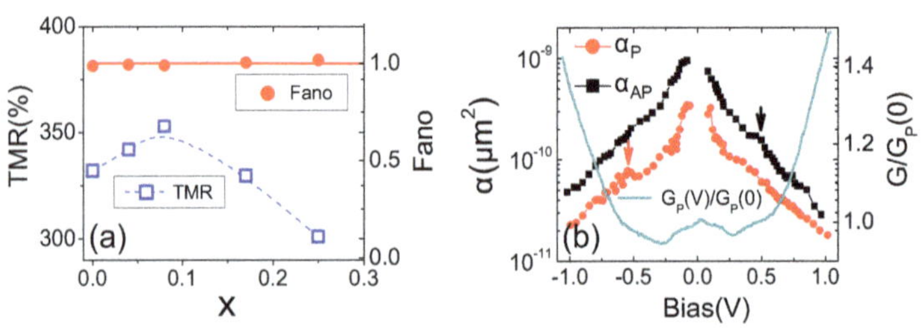

Figure 8.10 (a) Dependence of the zero bias TMR and the Fano factor at $T = 4$ K as a function of V content. (b) Bias dependence, at $T = 4$ K, of the dynamic conductance in the P state, and the Hooge factor α for both P and AP states for Fe/MgO/Fe junctions. Arrows indicate weak peaks. Adapted with permission from Ref. Aliev et al. (2014).

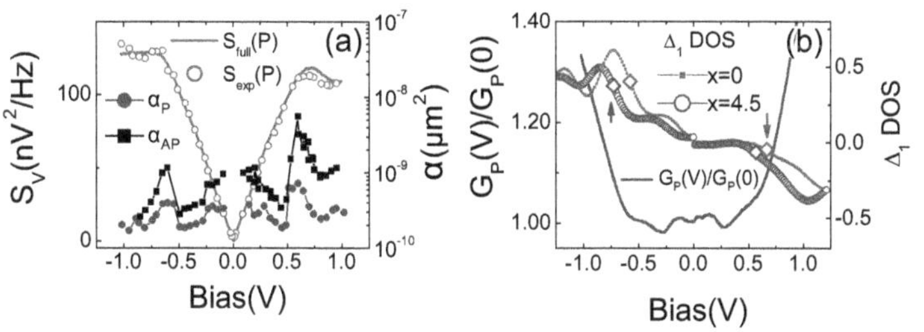

Figure 8.11 (a) Bias dependence at $T = 4$ K of the Hooge factor and SN for Fe $_{0.96}$ V$_{0.04}$ /MgO/Fe MTJ. (b) Dependence of the conductance with the applied voltage at $T = 4$ K combined with the calculated Δ_1 DOS as a function of energy with respect to E_F. Inflection points (open dots) indicate Δ_1 DOS band edges for 4% vanadium for $V < 0$ and pure Fe ($x = 0$) for $V > 0$. Adapted with permission from Ref. Aliev et al. (2014).

one observes a strong suppression of LFN with bias with weak anomalies in α around 0.5 V (indicated by arrows in the graph).

The doping of Fe with V improves the interface mismatch and decreases the Fe/MgO interface defect states density (Bonell et al., 2009b, 2010; Herranz et al., 2010c), which allows the implementation of the BENS method. Figure 8.11a shows the dependence of the shot noise and the Hooge factor α with the applied bias, for $Fe_{0.96}V_{0.04}$/MgO/Fe MTJs. The shot noise analysis yields a Fano factor close to one, indicating that direct tunneling takes place in the bias range under study (Fig. 8.11a). In contrast to what is observed for the reference sample (Fig. 8.10), the LFN shows a clear enhancement (factor of 2) of conductance fluctuations around ± 0.6 V. Yet a stronger enhancement of the LFN close to 0.6 V is observed in the AP configuration. Along with the enhancement of $1/f$ noise, as can be seen in Fig. 8.11b, the dynamic conductance in the P state shows an upturn around 0.6 V. The upturn is much clearer in the P state, so the AP state not shown for simplicity. Numerical calculations of the tunneling electron DOS indicate that the upturn in conductance and the noise enhancement could be related with the opening of a new transmission channel when the Fermi level of one magnetic electrode crosses one of the band edges of the other magnetic electrode, indicated by arrows in Fig. 8.11b. Details of the calculations are discussed further on.

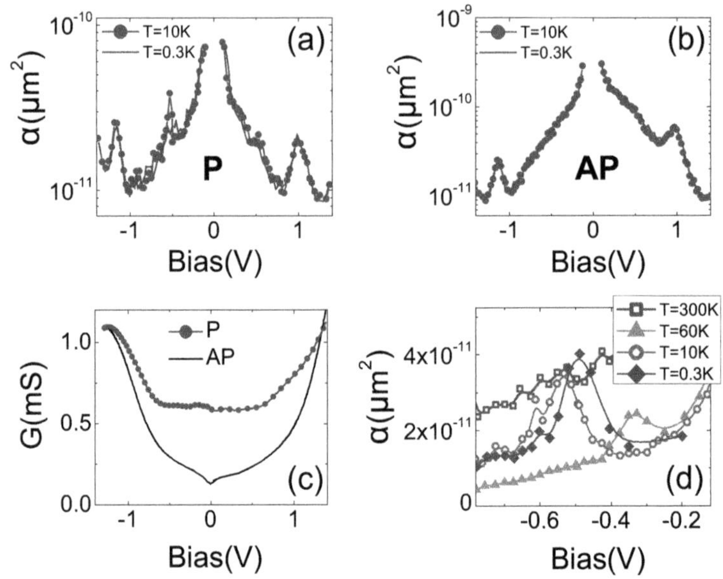

Figure 8.12 Bias dependence at $T = 4$ K of the Hooge coefficient for the (a) P state and (b) AP state in $Fe_{0.92}$ $V_{0.08}$ /MgO/Fe MTJs. (c) Bias dependence at $T = 4$ K of the dynamic conductance for the P and AP state. (d) Low-frequency noise peaks gradually disappear with increasing temperature. Adapted with permission from Ref. Aliev et al. (2014).

Even clearer signs of the band edges in LFN are seen when the V content reaches 8%, for which the lowest background LFN and the maximum TMR (Fig. 8.10) are achieved. Figure 8.12 shows the dependence of α with the applied bias in $Fe_{0.92}V_{0.08}$/MgO/Fe MTJs were the optimum relation between two competing effects is reached: (i) the relaxation of the FM/I interface and (ii) a non-essential suppression of the magnetization and the induced Fe-V structural disorder (Bonell et al., 2009b, 2010; Herranz et al., 2010c).

It has been found (Aliev et al., 2014) that the $Fe_{0.92}V_{0.08}$/MgO/Fe MTJs show clear anomalies in the Hooge factor for biases around 1 V and around 0.6 V for the P state only, as shown in Fig. 8.12a,b. Figure 8.12d demonstrates how the anomaly in the P state around 0.6 V gradually disappears with temperature, probably due to thermal excitations.

Qualitatively similar effects were seen for $Fe_{0.83}V_{0.17}$/MgO/Fe and $Fe_{0.75}V_{0.25}$/MgO/Fe MTJs with the latter being the most robust to electrical breakdown (standing up to 2.5 V) (Aliev et al., 2014). In the high V content range, the LFN is strongly influenced by random telegraph noise at positive biases around 1 V, reflecting a stronger asymmetry in interface defect states previously visualized with scanning electron microscopy for $Fe_{0.8}V_{0.2}$/MgO/Fe MTJs (Bonell et al., 2010).

8.4.2.1 BENS method and phenomenological 1/*f* model

Let us now explain the peaks observed in $1/f$ noise with the concept of *band edge noise spectroscopy* (BENS) which is qualitatively sketched in Fig. 8.13a. If the tunneling through the barrier is coherent, the main source of LFN is conductance fluctuations due to atomic defects affecting Δ_1 and Δ_5 interface states. The resulting localized states close to the band edges (Fahy et al., 2006) could

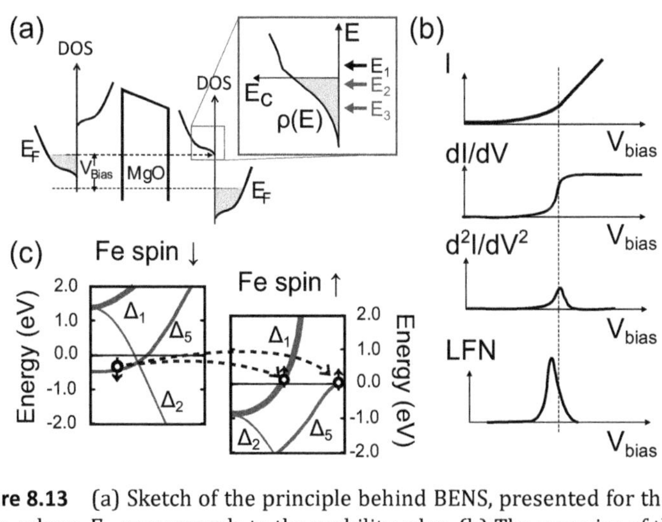

Figure 8.13 (a) Sketch of the principle behind BENS, presented for the AP state, where E_C corresponds to the mobility edge. (b) The energies of these defect states can be inferred from the I–V curve of the sample, and its first (dynamic conductance) and second (IETS) derivatives, but they are detected in a much clearer way though low-frequency measurements.(c) Sketch of a band edge (Δ_1, Δ_5) contribution to the tunneling at ~ -1.2 V. Adapted with permission from Ref. Aliev et al. (2014).

contribute, as reported for bulk semiconductors (Borovitskaya and Shur, 2001; Jayaraman and Sodini, 1989), to the enhanced LFN. The key new feature of the BENS is the versatility in displacing the Fermi level (E_F in Fig. 8.13a) of the ejector electrode with respect to the different band edges (or mobility edge, E_C in Fig. 8.13a) by simply varying the applied bias. The right panel shows how the conductance and its derivatives are expected to change when a new electron channel with a band edge opens at E_F. In order to clearly detect inelastic relaxation through IETS, some well-defined defect states should relax energy through coupling to a well-defined set of phonon energies.

A simplified physical picture (see also Ref. Aliev et al. (2014)) explaining the variation of LFN when tunneling to three different energies $E_{1,2,3}$ around E_C (Figs. 8.13a and 8.14a) is as follows. When electrons tunnel to energies $E_1 > E_C$, their relaxation time is fast due to the delocalized character of the band states near E_1 with a correspondingly small contribution to LFN. For tunneling to electron states $E_3 < E_C$ the LFN is also expected to be small due to low probability of these tunneling events. However, when electrons tunnel to the energies $E_2 \lesssim E_C$, the tunneling current could be affected by multiple trapping-type relaxations originating from shallow defect states contributing to the formation of the band edge tails. We estimate that the LFN peak width is roughly determined by the energy difference between the mobility edge and the bottom of the band tail.

A commonly accepted phenomenological model (already discussed in Section 2.4) attributes the occurrence of $1/f$ noise to the scattering of electrons from defects, where each of these defects is characterized by a certain relaxation time τ. The total $1/f$ noise is described as the sum of the contribution of multiple defects presenting a broad distribution of relaxation times with energy (Dutta and Horn, 1981). If the dominating defect states are located close to the interfaces, they may modify the edges of interface bands, creating tails. Defect states with different relaxation times are sketched in Fig. 8.14a. These defects create a band edge tail, as shown in Fig. 8.13b. The sum of the Lorentzian-type spectra from the relaxation of each defect state, characterized by τ_i, all contribute to the total $1/f$ noise.

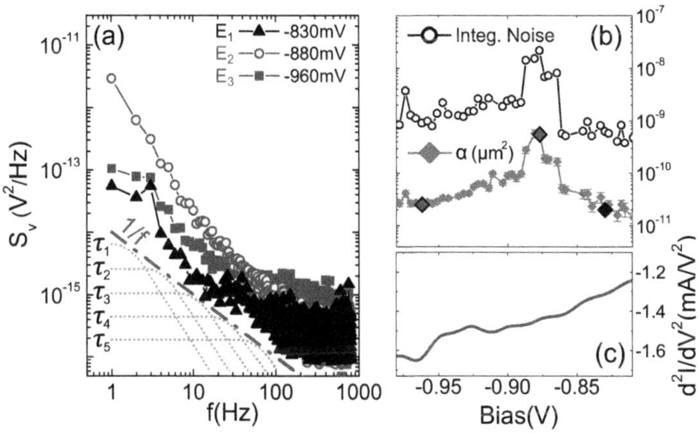

Figure 8.14 (a) Spectra at $T = 0.3$ K corresponding to the three energy states from a junction with 25% vanadium in the P state, with a maximum $1/f$ noise at the expected band edge energy. (b) The peak is clear in the dependence bias of the Hooge factor as well as of the integrated noise (normalized by the voltage squared, and therefore, adimensional). (c) The noise peak is not reflected in the IETS curve. Adapted with permission from Ref. Aliev et al. (2014).

The following arguments accompanied by rough experimental noise power spectra explain the nature of the LFN anomalies found in the tunnel junctions with different vanadium concentrations (Aliev et al., 2014). If the Fermi levels of the ferromagnetic leads are adjusted (by applying a bias) so the Fermi level of one electrode aligns with a certain band edge from the other electrode (as shown in Fig. 8.13c of the manuscript), an extra LFN may appear due to multiple relaxations which originate from the defect states which create the band edge tails. These band edge tails contain localized states below a mobility edge E_c (Fig. 8.14b). When electrons tunnel to energies above E_c, ($E_1 > E_c$) their relaxation time is fast because of the delocalized character of these band states and their contribution to the LFN is small (in the frequency window where conductance fluctuations are experimentally measured). The low-frequency noise is also expected to be small when tunneling to electron states much below the mobility edge ($E_3 < E_c$), which corresponds to a small density of states due to the low probability of these events taking place. However, when electrons tunnel to energies in the proximity or just below the mobility edge ($E_1 \lesssim E_c$), the tunneling current may be affected by the defect states which form the band edge tails, and provide an excess contribution to the low-frequency noise. Spectra for these three situations are presented in Fig. 8.14a. Note that due to their high barrier and interface quality, the single crystalline magnetic tunnel junctions show mostly flat spectra (shot noise) above 100 Hz even at high bias ($\sim$1 V).

The LFN peak width is roughly determined by the energy difference between the mobility edge and bottom of the band tail. The band tail width by itself could be determined by the influence of the different type of defect states (O, Mg, Fe, vacancies or interface steps) on the position of the energy bands. Previous numerical calculations (Du et al., 2010) evaluated energies of O and Mg defects states inside an MgO barrier to be below 150 mV.

In order to demonstrate that the position of the observed LFN anomalies depends only weakly on the method of analysis, Fig. 8.14b compares the dependence of α and the integrated LFN noise (which is adimensional because it is normalized by the square of the bias) with the bias in the P state, at $T = 0.3$ K, of a tunnel junction with 25% vanadium. Both methods provide qualitatively similar results.

As could be seen in Fig. 8.14c, the noise peak is not reflected in the IETS curve. We believe that the random interface potential and the absence of well-defined defect states smear out the IETS signals. Tunneling to the band tail weakly influences IETS reflecting only the derivative of the DOS close to E_C. On the other hand, much stronger changes in LFN vs. bias are seen due to a strong change of excited defect relaxation times (Borovitskaya and Shur, 2001) when tunneling close to E_C, activating an excess of the low-frequency conductance fluctuations. Therefore, interface defect states dominate the LFN, and not the derivative of the conductance (Fig. 8.14c). We argue that the LFN mainly originates from disorder/defects close to the FM/I interface because: (i) the tunnel through the barrier is direct (Fig. 8.10); (ii) the metallic nature of the electrodes, with resistance a few thousand times below the barrier resistance, ensures that electric signals and their fluctuations mainly come from regions in the barrier and interfaces; (iii) by analyzing LFN at higher biases we avoid the direct resonant excitation of localized FeO or O interface defect levels predicted below 200 mV (Du et al., 2010).

In the MTJs under study, electron tunneling mainly occurs between polarized bands with different Bloch state $\Delta_{1,5}$ symmetries spin filtered by the MgO barrier (Bowen et al., 2001; Butler et al., 2001; Faure-Vincent et al., 2003; Mathon and Umerski, 2001; Parkin et al., 2004; Yuasa et al., 2004). This allows a rough estimation of the interband mixing at the interface by analyzing variation of BENS response with relative alignment of the electrodes. Let us discuss qualitatively the reasons why BENS could provide LFN peaks in both P and AP states (Figs. 8.11 and 8.12). For simplicity, we shall use the majority and minority Fe electron bands tunneling in Fe/MgO/Fe junctions (Fig. 8.13c). When the MTJ is in the AP state, then in accordance with BENS arguments $\Delta_{5\uparrow} \Rightarrow \Delta_{5\downarrow}$ and $\Delta_{1\uparrow} \Rightarrow \Delta_{1\downarrow}$ band edge tunneling could provide a peak in the low-frequency noise of the AP state at different biases from 0.4 to 1.3 V if conductance fluctuations originate from elastic scattering events. Experimentally, however, we observe LFN peaks in the P state as well (Fig. 8.11a), which we link with the presence of spin orbit coupling induced $\Delta_{1(\uparrow\downarrow)} \Longleftrightarrow \Delta_{5(\downarrow\uparrow)}$ interband mixing at the Fe/MgO interface (Lu et al., 2012). Indeed, large lateral momentum transfer and interband scattering could be dominant only close to the interfaces (Tsymbal and Zutic, 2011). Within such scenario, the relation between amplitudes of the peaks LFN(P)/LFN(AP) provides an evaluation of the degree of interband mixing between majority $\Delta_{1\uparrow}$ band and the minority $\Delta_{5\downarrow}$ of roughly 0.2–0.3.

8.4.3 Numerical Calculations

In order to examine quantitatively the applicability of the model, numerical calculations have also been carried out (Aliev et al., 2014). The simulation consists of emphab initio calculations of $\sqrt{2} \times \sqrt{2}$ unit cell of $Fe_{1-x}V_x$/MgO ($x = 0$, 0.045, 0.091, 0.182) with five monolayers (ML) of MgO and 11 ML of $Fe_{1-x}V_x$. The first-principles calculations are based on density functional theory (DFT) as implemented in the Vienna emphab initio simulation package (VASP) (Kresse and Furthmüller, 1996a,b; Kresse and Hafner, 1993) within the framework of the projector augmented

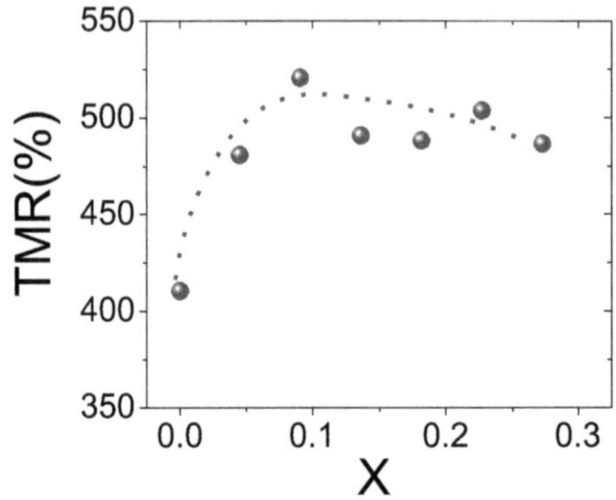

Figure 8.15 Dependence of the calculated TMR values as a function of the vanadium content. The symbols correspond to the estimated TMR values and the dotted line is a guide for the eye. Adapted with permission from Ref. Aliev et al. (2014).

wave (PAW) potentials (Blöchl, 1994; Kresse and Joubert, 1999) to describe electron–ion interaction and generalized gradient approximation (GGA) (Wang and Perdew, 1991) for exchange-correlation interactions. A $13\times13\times3$ K-point mesh was used in the calculations. A plane wave energy cut-off equal to 500 eV for all calculations was used and is found to be sufficient for the system under study.

The TMR is calculated from the numerical calculations using the Jullière model (Julliere, 1975), as shown in Fig. 8.15. A maximum value is reached for 9% of V, which is quite similar to the experimentally observed 8% value.

Figure 8.16 compares the experimentally observed LFN anomalies in the P state (open dots) with the band edge positions (closed dots) estimated from inflection points in the DOS simulations for the majority and minority Δ_1 and Δ_5 states of $Fe_{1-x}V_x/MgO$ ($x =$ 0, 0.045, 0.091, 0.182) structures (as indicated by arrows in the Fig. 8.11b). We have also indicated by horizontal dotted lines the estimated positions of the band edges of the Fe/MgO structure.

A reasonable agreement between simulations and the experiment is observed, especially for the vanadium content between ($0.04 < x < 0.17$) with reduced lattice mismatch, the lowest background LFN and the highest TMR.

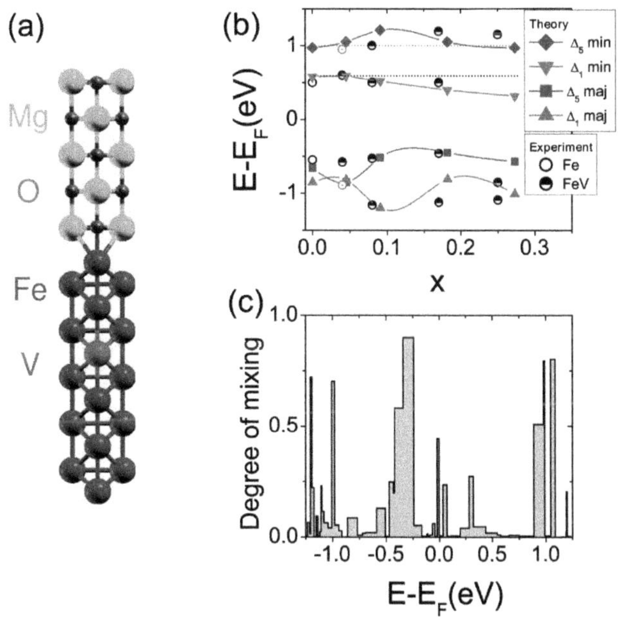

Figure 8.16 (a) Schematic of the calculated crystalline structure for a $\sqrt{2} \times \sqrt{2}$ unit cell of $(Fe_{1-x}V_x)_{11}/MgO_5$. (b) Calculated changes in the energies of the band edges in $Fe_{1-x}V_x$ compared to the experimental data of low-frequency noise anomalies for the P state. Fully open experimental points indicate a weak peak (increase of noise in less than 10%). (c) Calculated degree of mixing between Δ_1 and Δ_5 interface Bloch state character in $(Fe_{1-x}V_x)_{11}/MgO_5$ for $x = 0.091$. Adapted with permission from Ref. Aliev et al. (2014).

A few factors could contribute to some difference between experimental results and calculations. First of all, calculations do not consider the presence of dislocation induced mismatch as well as the structural disorder difference between bottom and top interfaces (Bonell et al., 2010). On the experimental side, measurements on MTJs with the least vanadium were done below 1 V due to their vulnerability, making them difficult to compare with the calculation results above 1 V.

Finally, in order to understand the influence of spin mixing at the interface, the authors (Aliev et al., 2014) have also analyzed the Bloch state character of the interfacial Fe atom in the presence of SOI as a function of the energy difference to E_F. Figure 8.16c presents this analysis for Δ_1 and Δ_5 interface states in $Fe_{0.909}V_{0.091}/MgO$ structure, mainly participating in the electron tunneling through MgO. When the degree of mixing at certain energy is equal to zero, it means that there is no mixing between different Δ channels and there is only one Δ Bloch state character that dominates the tunneling at this energy tunneling. The channel mixing is more pronounced at biases around $-(0.4-0.5)$ V and not above ± 1 V, i.e., close to the intervals where LFN anomalies of different magnitude were observed in both magnetic states (Fig. 8.16c). Authors (Aliev et al., 2014) believe that $\Delta_{5\uparrow} \Rightarrow \Delta_{5\downarrow}$ and $\Delta_{1\uparrow} \Rightarrow \Delta_{5\downarrow}$ mixing could be due to surface induced band crossings and explains the appearance of peaks in LFN in both P and AP states.

Chapter 9

Noise in Double-Barrier Magnetic Tunnel Junctions

9.1 Introduction to Double-Barrier Magnetic Tunnel Junctions

A double-barrier magnetic tunnel junction (DMTJ) is a spintronic device composed of two or three ferromagnetic layers separated by tunneling barriers (see Fig. 9.1). With two ferromagnetic layers, the non-magnetic layer (normal or superconducting) could be located in the middle (case a) or at the edge (case b) of the DMTJ. In the first case one would expect sequential tunneling to occur with Coulomb blockade (see further below) being negligible. The second case, still to be investigated, would have spin injection controlled by the relative magnetic alignment of the two electrodes in the MTJ, from the ferromagnet to the normal metal or superconductor.

A double-barrier magnetic tunnel junction (DMTJ) is a spintronic device composed of two or three ferromagnetic layers separated by tunneling barriers (see Fig. 9.1). With two ferromagnetic layers, the non-magnetic layer (normal or superconducting) could be located in the middle (case a) or at the edge (case b) of the DMTJ. In the first case one would expect sequential tunneling to occur with Coulomb

Noise in Spintronics: From Understanding to Manipulation
Farkhad G. Aliev and Juan Pedro Cascales

ISBN 978-981-4774-34-5 (Hardcover), 978-1-315-11088-2 (eBook)
www.panstanford.com

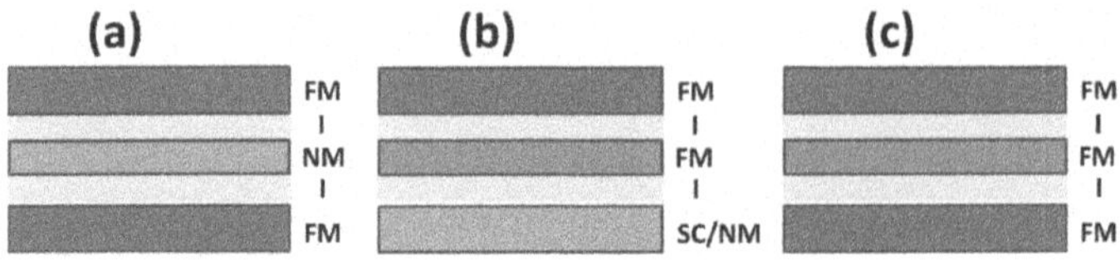

Figure 9.1 Sketch showing three possible double-barrier magnetic tunnel junctions involving middle non-magnetic electrode (a), asymmetrically situated non-magnetic or superconducting electrode (b) or all three ferromagnetic electrodes (c).

blockade (see further below) being negligible. The second case, still to be investigated, would have spin injection controlled by the relative magnetic alignment of the two electrodes in the MTJ, from the ferromagnet to the normal metal or superconductor.

One could expect that the type (c) DMTJs to have enormous TMR ratios, since two tunnel barriers come into play and contribute to the resistance of the antiparallel (↑↓↑) state. What is found instead is that the TMR values obtained at low temperatures barely reach 200% (Li et al., 2013), quite far from the 604% (Ikeda et al., 2008) record in single-barrier junctions mentioned above. This could be due to spin-relaxation in the central electrode, so the current loses some spin polarization and the TMR values observed are not as high as initially expected.

It has been shown that these structures could find application in different electronic and spintronic devices such as, for example, spin transistor (see Fig. 9.2, or memory cells in MRAMs (Ikegawa et al., 2003) or as spin-current diode (Iovan et al., 2008). One may also go further and think about these junctions as data bits who could store twice as much information as a normal MTJ since they can show 4 distinct magnetic states (Colis et al., 2003).

The growth and characterization of sputter-grown DMTJs with MgO barriers and $Co_{40}Fe_{40}B_{20}$ ferromagnetic electrodes is presented in Ref. Feng et al. (2009). The conductance at each magnetic state, and thus the TMR are observed to have an asymmetric bias voltage dependence with annealing only at low temperatures. The authors show that the TMR asymmetry decreases when the junctions are annealed at higher and higher temperatures. The structural improvement is shown in Fig. 9.3. The bias dependence

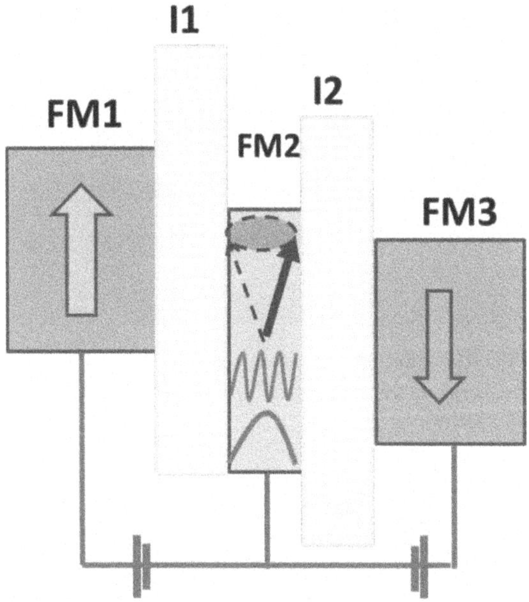

Figure 9.2 Sketch describing versatility of DMTJ-based spin transistor. With magnetically free central layer, due to presence of quantum well states (QWSs), the DMTJ dynamics (spin torque effects) and related noise could be tuned by the gate voltage.

becomes almost symmetric after annealing at 350°C. A maximum TMR ratio of 120% is obtained for the DMTJs, and its somewhat low value is attributed to an amorphous middle CoFeB layer, which does not crystallize after annealing.

Recent study by Guo with co-authors (Hui-Qiang et al., 2015) showed that Hooge is proportional to magnetoresistance-sensitivity-product $(\frac{\Delta R}{R^2})(\frac{dR}{dH})$ only in the field range where the middle free layer rotates. This points out consistency with the fluctuation dissipation theorem for thermal magnetization fluctuations mainly in the field region where magnetization could fluctuate under thermal agitation.

The work in Ref. Wilczynski and Barnaś (2000) considers the dependence of the TMR as a function of the central electrode's thickness in FM-NM-FM DMTJs. At certain thicknesses of the middle electrode, periodic oscillations of the TMR may be observed

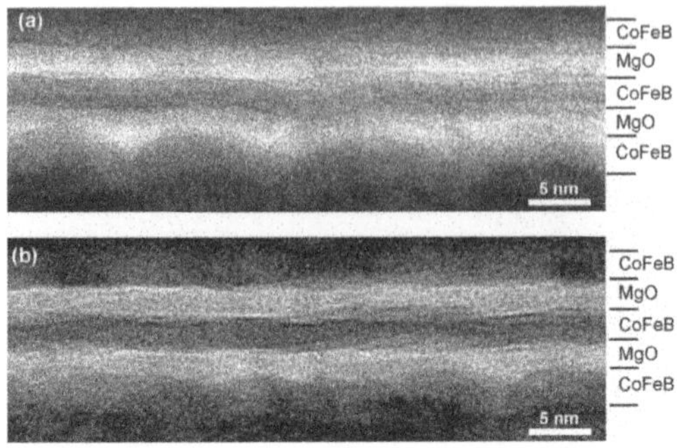

Figure 9.3 Electron micrographs from the barrier region of DMTJs (a) as deposited and (b) after annealing at 350°C. Adapted with permission from Ref. Feng et al. (2009).

which are related to quantum well states formed in this electrode (Moodera et al., 1999b) and to resonant tunneling through the whole structure. The authors present a theoretical study of coherent tunneling in a FM-NM-FM system within the free-electron approximation. The TMR, which depends on the relative orientation of the magnetic moments of the ferromagnetic electrodes, shows an oscillatory behavior with pronounced peaks as a function of the central electrode's thickness. These oscillations are related to the resonant tunneling through the whole system.

Evidence of tunneling through quantum well states in the middle free electrode was found by Herranz et al. (2010a). Oscillations in resistance were observed at room temperature in both magnetic configurations and for both bias polarities. The quantum well states were observed due to a specific feature of junctions: they experienced a *soft* breakdown due to a nitrogen doping of the barriers. Quantum well states in the central electrode may also be well observed in nitrogen-free DMTJs as we shall discuss further below. Moreover, the recent experiments reveal presence of QWSs even at room temperature in DMTJs with relatively thick central electrode when lattice mismatch is minimized through MgO barrier alloying (Tao et al., 2015).

A model describing the spin-dependent current and TMR in an all-ferromagnetic DMTJ (FM-NM-FM-NM-FM) is presented in Ref. Useinov et al. (2011). The theory is based on the two-band model and quasiclassical equations in order to describe the transport phenomena in the ferromagnetic metals. The TMR is studied for the resonance (through quantum well states) and non-resonance tunneling regimes. In the case of non-resonant tunneling, the DMTJs are modeled in two different ways: as a single system presenting coherent (ballistic transport with no scattering) or as two single-barrier MTJs connected in series, presenting a consecutive tunneling system. By the use of both these models, including a barrier asymmetry the authors are able to reproduce the asymmetric TMR curves with the voltage which are observed in experiments (Feng et al., 2009).

9.2 Shot Noise in DMTJs with Two Ferromagnetic Layers

Historically the first theoretical consideration of the shot noise in double-barrier magnetic tunnel junctions was made in application to the type (i) DMTJs, with a normal metal central electrode (FM-NM-FM), by Tserkovnyak et al. (Tserkovnyak and Brataas, 2001). The authors study the case of elastic and inelastic regimes, in the case of diffusive (high electron scattering) and ballistic (negligible electron scattering) junctions. The Fano factor is calculated as a function of the angle between the magnetization of the electrodes (i.e., the magnetic state), at different values of spin polarization. In the case of ballistic junctions with a large polarization, the Fano factor sharply drops from 1 to 0 when the junction approaches the antiparallel state. Diffusive junctions, on the other hand, present much lower Fano factors ($F = 1/3$ for a diffusive, single-barrier junction) which depends non-monotonically with the angle.

Figure 9.4 compares calculated shot noise for symmetric double-barrier junction in diffusive and tunneling regimes. As we have noted above, in the diffusive regime for two possible collinear configuration the Fano has the universal value of 1/3 and Fano decreases for non-collinear configuration due influence of highly coherent electron transport in the normal region. In the tunneling regime dependence of Fano on the relative angle of the electrodes is

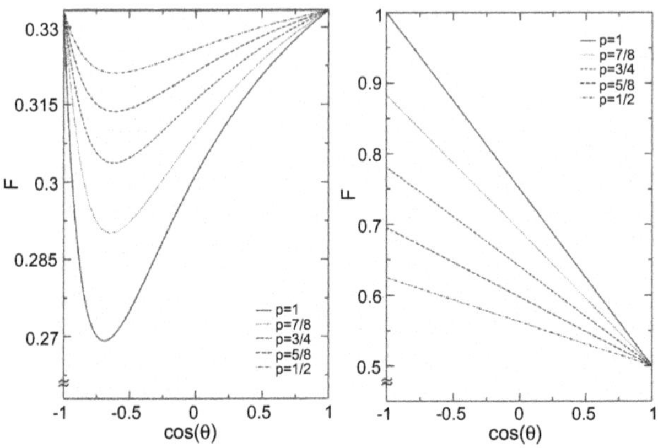

Figure 9.4 Calculated angular dependence of the Fano factor for F-N-F junctions in the diffusive (left) and tunneling (right part) regimes. Adapted with permission from Ref. Tserkovnyak and Brataas (2001).

more simple and could be described by an analytical dependence for any polarization:

$$F(\theta) = 1/2 * (1 + p^2 \sin^2 \theta/2) \quad (9.1)$$

Corresponding angular dependence of Fano is just a monotonic interpolation between the two different Fano factors in the collinear configurations. We note that spin-flip processes are disregarded, which is justified if the central electrode is thinner than the spin-flip scattering length. Spin-flip scattering in the normal metal electrode is expected to make the angular dependence of the Fano factor quantitatively weaker, since the effective polarization and mixing conductance are reduced.

9.3 1/*f* Noise in DMTJs with Three Ferromagnetic Electrodes

Although there have been several reports on the fabrication and characterization of DMTJs only, few of these deal with low-frequency noise in this type of junctions.

Ref. Yu et al. (2011) presents a comparison of low-frequency noise between sputtered CoFeB/MgO DMTJs and single MTJs with

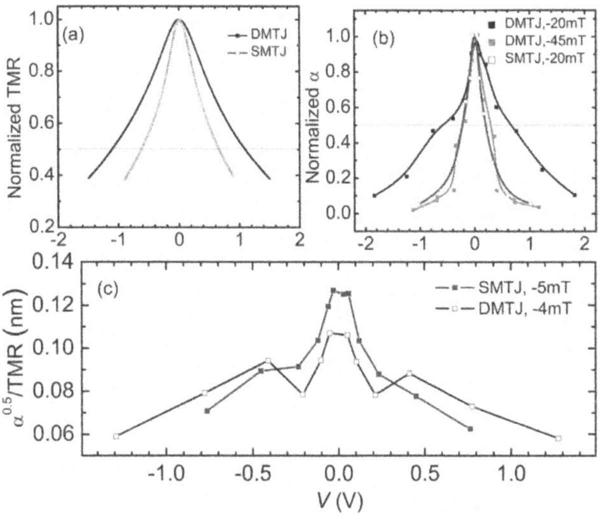

Figure 9.5 Bias dependence of (a) TMR, (b) the noise magnitude parameter in the antiparallel state and (a) the $\alpha^{1/2}$/TMR value during free layer switching for an SMTJ and a DMTJ with symmetric MgO layers. Adapted with permission from Ref. Yu et al. (2011).

high TMR ratios. From a practical point of view, DMTJs could be expected to show a reduced $1/f$ noise if the tunneling is not coherent. The junctions were annealed at temperatures between 250–375°C and have sizes of 20 × 20 and 50 × 150 μm^2. The authors find that while the normalized noise is slightly lower for DMTJs in the P state ($\alpha \sim 10^{-10}$ μm^2), α in the AP state for DMTJs decays slowly with the voltage compared to that single MTJs, as is shown in Fig. 9.5a,b. The authors argue that the bias dependence of the TMR and α points to the DMTJs behaving like two single MTJs in series, since there is also no sign of coherent/resonant tunneling. Figure 9.5c presents $\alpha^{1/2}$/TMR (which is proportional to the field detection capability of MTJs) as a function of the voltage for a DMTJ and a simple MTJ. The results indicate that for a low bias operation, the DMTJs could be used as more sensitive detectors than single-barrier MTJs. In summary, double-barrier tunnel junctions have been successfully fabricated on numerous reports, but although there exist several theoretical descriptions of the tunnel through this type of devices, there is yet no systematic investigation of the transport mechanisms in them, and particularly of shot noise.

9.4 Conductance and Shot Noise in Epitaxial DMTJs with Three Ferromagnetic Electrodes

Double-barrier magnetic tunnel junctions are spintronic elements which could, in the future, replace single-barrier junctions as components in hard-disk drives or as magnetic sensors. First of all, single-barrier MTJs are more susceptible to dielectric breakdown than DMTJs.

DMTJ devices also present a high versatility for controlling their electrical resistance (or TMR). They can present lower shot noise values than single MTJs, and consequently have better signal-to-noise ratios. As we will show in this chapter, engineering DMTJs with certain barrier and electrode thicknesses gives control over the shot noise value of the device, which can be switched with the different magnetic states.

As was mentioned in the introductory chapter, MgO-based junctions became important elements of spintronic devices after spin dependent transport was observed in Fe/MgO/Fe magnetic tunnel junctions (Parkin et al., 2004; Yuasa et al., 2004). The use of MgO for an effective spin injection in MTJs has allowed reducing spin relaxation effects which occur because of conductivity mismatch (Fert and Jaffrs, 2001; Schmidt et al., 2000). Previous studies regarding spin coherence and shot noise in MTJs revealed that junctions with Al_2O_3 barriers present suppressed shot noise values ($0.7 < F < 1$) due to sequential tunneling (Guerrero et al., 2006b). Serial arrays of magnetic tunnel junctions also present a suppressed SN (Gokce et al., 2011).

On the other hand, MTJs with MgO barriers present full shot noise values ($F = 1$), which is independent of the magnetic state. This was observed in Fe/MgO/Fe MTJs (Guerrero et al., 2007). The study of junctions with ultrathin MgO barriers (less than 1 nm thick), yielded a Fano factor of $F \simeq 0.92$ in the parallel state (Arakawa et al., 2011; Liu et al., 2012).

Double-barrier magnetic tunnel junctions have been fabricated using nanoparticles (Nozaki et al., 2006a; Wang et al., 2006) or a continuous magnetic layer as the central electrode (Tiusan et al., 2007). DMTJs present some advantages in comparison with single-barrier MTJs, like higher TMR ratios (Faure-Vincent et al., 2002; Tiusan et al., 2007). Also, DMTJs present a smaller decrease of the TMR with the applied voltage in comparison with single MTJs. For example, epitaxial DMTJs with room temperature TMR ratios of 140% only show a decrease down to 75% for voltages around 1 V (Tiusan et al., 2006). Having high values of TMR at high voltages is essential for applications.

Incidentally, the dependence of the TMR with the applied bias reveals oscillations induced by quantum well states (QWSs) (Herranz et al., 2010a; Wang et al., 2006). Another advantage is that spin accumulation in the central layer is expected to notably enhance spin torque (Berger, 2003; M. Wilczyński, 2009; Theodonis et al., 2007; Vedyayev et al., 2006b). Resonant tunneling in DMTJs may have advantages in comparison with standard MTJs due to resonant spin-torque effects (Theodonis et al., 2007; Vedyayev et al., 2006a; Watanabe et al., 2008). Also, a current driven magnetization reversal may occur at lower current densities in DMTJs (Watanabe et al., 2008).

The control of shot noise has been extensively studied in the area of photonics, and is currently exploited to improve detection methods (Quinlan et al., 2013). The precise control of shot noise in spintronics still remains an important and unexplored issue.

Thus, being able to control the spin diffusion and coherence in spin-polarized tunneling could be important for the further development and technological application of hybrid magnetic nanostructures [10]. In this chapter, we report on the investigation of shot noise in seminal, epitaxial Fe/MgO/Fe/MgO/Fe DMTJs.

9.4.1 Growth Method and Junction Types

As can be seen in Fig. 9.6, a double-barrier tunnel junction consists of three ferromagnetic electrodes, separated by tunnel barriers. The resistance of the DMTJ corresponds to the sum of the resistance of each tunnel barrier, which depends on the thickness of the barrier. If

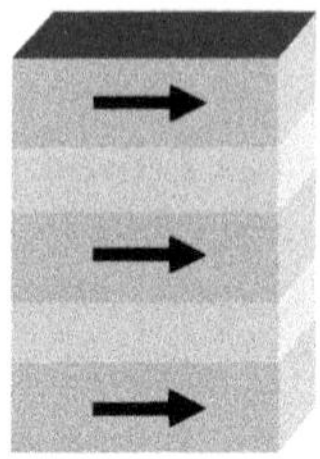

Figure 9.6 Sketch of a DMTJ. The thickness of the barriers determines the resistances R_1, R_2, and the thickness of the FM layers the coercive field of each magnetic electrode.

the coercive fields of the FM electrodes are different, then up to four different magnetic states can be obtained in a DMTJ. The different coercive fields are achieved by either growing FM layers of different thickness or coupling an FM electrode to an antiferromagnetic layer. The growth of the junctions was carried out by the group of Michel Hehn and Coriolan Tiusan and the growth method is described in details in Ref. (Tiusan et al., 2007).

Three different types of DMTJs, with different combinations of electrode and barrier thicknesses, were grown at room temperature by molecular beam epitaxy (MBE) (Cascales et al., 2012). The junctions were grown on MgO(100) substrates under ultrahigh-vacuum conditions (UHV) at a base pressure of around 10^{-11} mbar. In MBE, the source materials used in the growth are heated in evaporation cells. The sublimation of the source materials creates a gaseous beam of the required substance, which then condenses on a wafer or substrate. The substrate may be heated if necessary, as well as kept continually rotating to obtain a homogeneous growth. The slow deposition rate of the MBE technique allows the films to be grown epitaxially. UHV conditions are necessary for the epitaxial growth to take place in a clean environment with a minimal contamination.

The MgO(100) substrates were annealed at 600°C for 20 min before the growth in order to remove carbon impurities. The diffusion of the remaining C impurities to the Fe bottom electrode is prevented by growing a 10–15 nm-thick MgO buffer layer at 450°C on the substrate (Tiusan et al., 2007). A control of the growth down to the monolayer level was achieved by monitoring in-

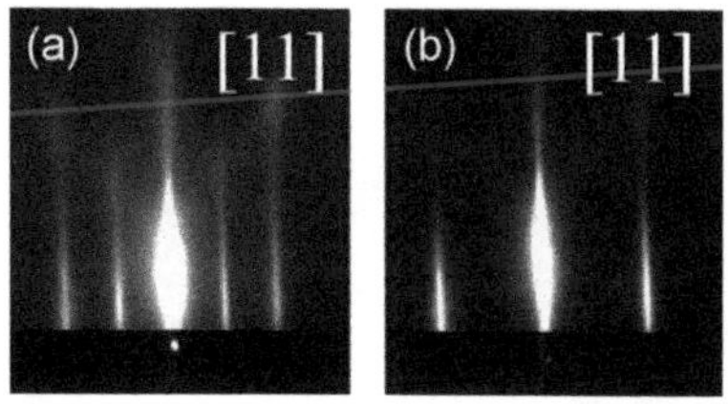

Figure 9.7 RHEED patterns along the [110] crystallographic direction of the Fe bottom layer of samples (a) without or (b) with an MgO seed layer between the substrate and the Fe layer. The buffer layer prevents the diffusion of carbon to the Fe electrode, so evidence of the $c(2\times2)$ reconstruction, related to C, is seen in (a) but not in (b). Adapted with permission from Ref. Tiusan et al. (2007).

situ the intensity oscillations in the reflection high-energy electron diffraction (RHEED) pattern along the [100] direction during the layer-by-layer growth of the MgO barriers. Figure 9.7 shows such a RHEED pattern of the bottom Fe layer where the MgO buffer layer is (a) absent or (b) present. The sample with an MgO buffer layer does not show the $c(2\times2)$ surface reconstruction related to carbon, which is seen in the sample with no buffer layer.

DMTJs with lateral sizes from 10 to 30 μm were patterned by a combination of UV lithography and Ar ion etching, controlled by Auger spectroscopy (Tiusan et al., 2007). The square-shaped pillar is fabricated in the following way. After the deposition of the different layers, the surface of the wafer is cleaned and covered with a photosensitive resin (or resist) (Fig. 9.8b). The resist is then exposed to ultraviolet light through a mask with opaque and transparent regions with the desired square-shaped pattern. The chemical properties of the exposed resist change, and such regions are removed with the appropriate solvent. The remaining resin protects the surface underneath from the Ar ions, so only the resin-free surfaces are etched, leaving a square-shaped pillar MTJ. Once the etching reaches the bottom electrode (Fig. 9.8c), the bottom electrode is patterned down to the substrate using a different mask following the process we just described.

A silica layer is sputtered over the whole wafer, isolating the top and bottom electrodes to avoid short-circuits. Then, "windows" are opened in the silica onto the top and bottom electrodes by

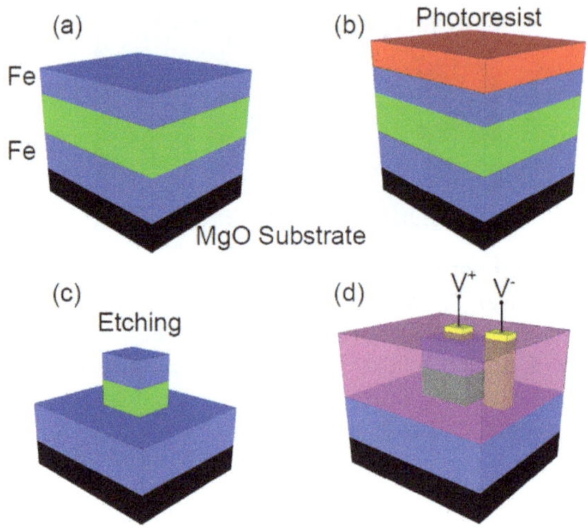

Figure 9.8 Diagram of the photolithography process. (a) Unpatterned sample. (b) Deposition of the resin. (c) Etching of the top electrode and MgO barrier. (d) MTJ pillar with the deposited contact pads.

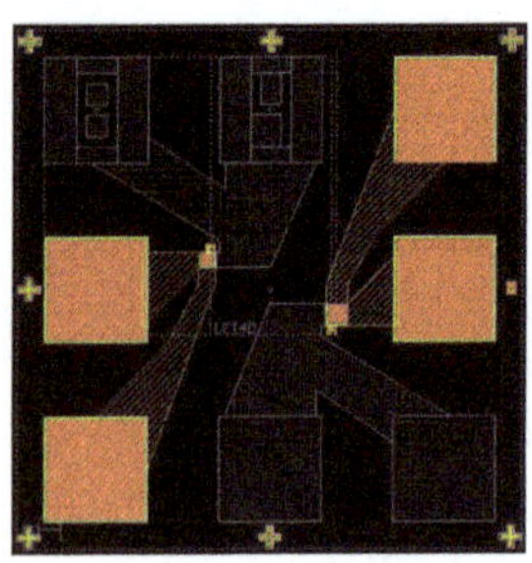

Figure 9.9 Diagram of the contact pads in DMTJs (courtesy of C. Tiusan).

lithography (Fig. 9.8d). Finally, aluminum contacts of the type shown in Fig. 9.9 are deposited.

The first type of junctions (DMTJ1) have the structure:

$$\text{MgO//MgO(15)/Fe}_1\text{(45)/MgO(11ML)/Fe}_2\text{(10)/MgO(3ML)/}$$
$$\text{Fe}_3\text{(20)/Au(20)},$$

where numbers in parenthesis are thicknesses in nm, while ML stands for monolayer. These junctions have a strong asymmetry between barriers, as one is considerably thinner than the other.

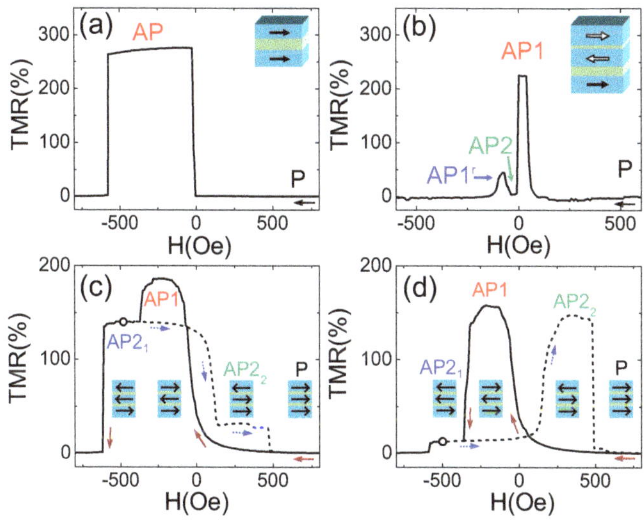

Figure 9.10 TMR curves at $T = 0.3$ K with a field sweep from positive to negative, adapted with permission from Ref. Cascales et al. (2012). (a) Single-barrier MTJ with an area of 100 μm^2. (b) DMTJ1 with a very thin barrier, resulting in coupled top and central layers. (c) DMTJ2 junctions with a thick barrier 2. The resistance difference between AP1 and $AP2_1$ is higher than between $AP2_1$ and P. (d) TMR in DMTJ3 with a thick barrier 1. This results in a bigger resistance jump from $AP2_1$ to P, than from AP1 to $AP2_1$. The circle in (c) and (d) indicates when the field sweep changes direction in order to obtain the $AP2_2$ state.

The TMR as a function of the magnetic field is shown in Fig. 9.10b. In agreement with previous findings (Faure-Vincent et al., 2002; Tiusan et al., 2007), the antiferromagnetic coupling between two ferromagnetic layers across the thinnest MgO barrier in DMTJ1 results in the presence of two different AP1 (↑↓↑) and $AP1^r$ (↓↑↓) states for which the central layer is aligned opposite to the neighboring ones. The difference in resistance between the AP1 and $AP1^r$ states is attributed to the possible influence of domain walls formed in the synthetic antiferromagnet in the latter state. This option is supported by the fact that the normalized $1/f$ noise observed in the $AP1^r$, as is shown in Fig. 9.11. The TMR dip in-between corresponds to the AP2 (↓↓↑) state.

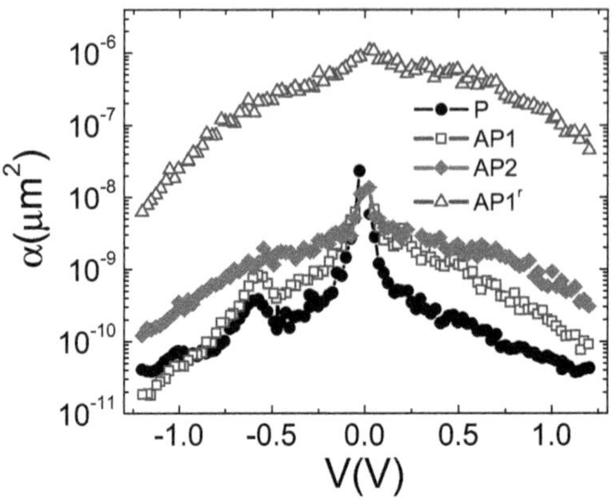

Figure 9.11 Normalized $1/f$ noise at $T = 4$ K in a DMTJ1 sample in four different magnetic configurations.

The second and third type of DMTJs, DMTJ2 and DMTJ3, have the following layered structure:

$$\text{MgO//MgO(10)/Cr(42)/Co(10)/Fe}_1\text{(5)/MgO(10 or 8 ML)/}$$
$$\text{Fe}_2\text{(5)/MgO(9 ML)/Fe}_3\text{(10)/Co(30)/Au(10)}$$

The wafers for DMTJ2s and DMTJ3s were grown simultaneously. The 2 ML difference in thickness of the bottom barrier is controlled by a shutter. The asymmetry between the tunnel barriers in these junctions is much lower than for DMTJ1s because both barriers in these junctions differ only slightly. The barrier asymmetry of DMTJ2 ($R_2 > R_1$) samples is opposite to that of DMTJ3 ($R_1 > R_2$), where the topmost barrier is labeled as number R_1 and the bottom barrier as R_2.

The dependence of the TMR on the magnetic field for DMTJ2 and DMTJ3 samples is shown in Fig. 9.10c,d. Both TMR graphs present the features expected from the barrier thicknesses. DMTJ3s show a larger resistance jump between the AP1 (↑↓↑) and $AP2_1$ (↑↓↓) states than between the $AP2_1$ (↑↓↓) and P (↓↓↓) states. DMTJ2 samples show the opposite behavior.

The $AP2_2$ state (↑↑↓) is present in both DMTJ2 and DMTJ3 samples, although reliable shot noise measurements in this state were only obtained for DMTJ3s. Indeed, one can observe in

Fig. 9.10c,d, that if the magnetic field sweep is stopped when the sample is in the $AP2_1$ state (↑↓↓), and then the field sweep goes back to positive values, then the central layer is inverted and we obtain the $AP2_2$ state (↑↑↓).

As can be seen, the central Fe layer in DMTJ1 samples has a thickness of 10 nm, while for DMTJs 2 and 3 the thickness is 5 nm. The thinnest continuous Fe layer which could be grown by molecular epitaxy (system at the University Henri Poincarè of Nancy) is of 5 nm. By making the central layer as thick as 10 nm, the Fe layer is guaranteed to be epitaxially flat, which is necessary for the growth of the thin 3 ML MgO top barrier in DMTJ1s. On the other hand, for DMTJ2 and DMTJ3 samples, the central layer is grown as thin as possible so the spin flip processes are minimized.

9.4.2 Experimental Results

The main experimental findings show that the shot noise in weakly asymmetric samples (DMTJ2s and DMTJ3s) is suppressed below the classic value ($F = 1$). The Fano factor was obtained as explained in Section 3.2. This approximation is justified above 100 mV, where $eV/k_B T > 100$. Even though the noise and electron transport measurements were done at 0.3 K and 4 K, the real temperature for the less resistive samples was somewhere below 10 K at the highest biases (around 1 V). This was estimated by comparing the I–V curves measured at 0.3 K, 4 K and 10 K. Measurements at 60 K were also carried out, for which the contribution of thermal noise to the frequency independent part of the noise spectrum was estimated and subtracted.

9.4.2.1 Control MTJs

The shot noise measurements are compared with control measurements carried out in single-barrier MTJs with the following layer structure: MgO//MgO(10 nm)/Fe(45)/MgO(2.5)/Fe(10)/Co(20)/Pd(10) /Au(10) with the numbers indicating thickness in nm. These junctions were studied in a previous work by Guerrero et al. (2007). The TMR of a 10×10 μm^2 MTJ is shown in Fig. 9.10a. Shot noise measurements carried out at $T = 4$ K in epitaxial

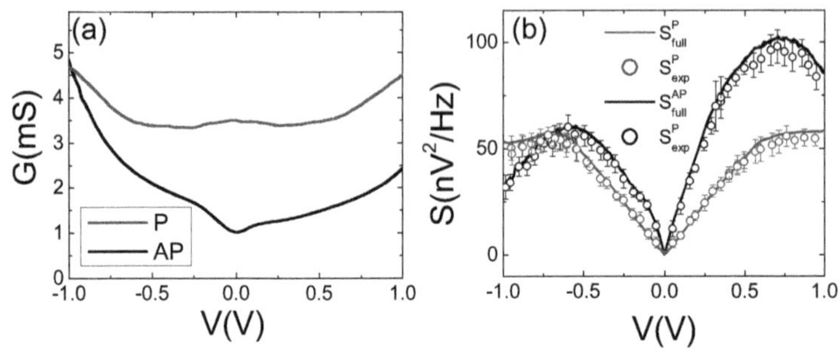

Figure 9.12 (a) Conductance in the P and AP states of a single-barrier control MTJ at $T = 4$ K. (b) Experimental (circles) and full (lines) shot noise in the P and AP states on a single-barrier MTJ with an area of 100 μm^2. These control measurements yield $F \simeq 1$, suggesting a good structural quality of the junctions. Samples from the study are shown in Ref. Guerrero et al. (2007).

MTJs with a single, 2.5 nm MgO barrier, present nearly Poissonian statistics ($F \simeq 1$), in agreement with previous results (Guerrero et al., 2007). Figure 9.12 shows that shot noise measurements in the P and AP states yield the Poissonian value. These single-barrier junctions serve as control for the shot noise measurements in the DMTJs. The fact that $F \simeq 1$ in the control MTJs indicates that the epitaxially grown samples are of good structural quality.

9.4.2.2 Shot noise in DMTJs with high barrier asymmetry

This type of junctions present a signature of QWSs in the central layer in the dependence of the conductance with the applied bias. The oscillations are clearly observed for two of the magnetic states (P and AP2), as shown in Fig. 9.13a.

The shot noise measurements carried out in DMTJ1 samples revealed that shot noise in these structures is only weakly suppressed ($F \sim 0.9$) and is nearly independent of the magnetic state. The thicker barrier dominates the transport statistics because of the large difference between barriers. This is evident in the TMR curve (Fig. 9.10b), which only deviates from the P state resistance when the central layer's magnetization is opposed to the bottom layer's magnetization.

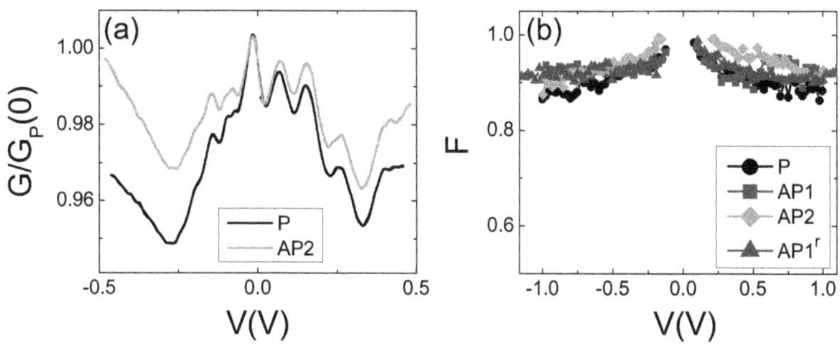

Figure 9.13 (a) Oscillations in conductance for two magnetic states. Each conductance curve is normalized by its value at zero bias. (b) Bias dependence of the normalized shot noise for a DMTJ1 sample. The Fano factor is found to depend weakly on the magnetic state.

9.4.2.3 DMTJs with low barrier asymmetry

In DMTJ2 and DMTJ3 samples, which have a low barrier asymmetry, the slight difference between barriers becomes important and the shot noise measured depends quite strongly on the magnetic state of the DMTJ. Depending on the magnetic state and bias, the Fano factor varies in the range of $F \in (0.5 - 0.9)$. Figure 9.14 shows shot noise measurements at $T = 0.3$ K in a DMTJ3 sample in all four magnetic states.

The solid curves represent the estimated full SN from the $I-V$ curves while the dots indicate the experimental shot noise obtained from the spectra. As can be seen, the shot noise is sub-Poissonian in all four magnetic states. The dependence of the shot noise with the magnetic state can be better appreciated when the normalized shot noise is plotted. Figure 9.15 presents the Fano factor for (a) a DMTJ2 sample (three magnetic states shown) and (b) the DMTJ3 sample from Fig. 9.14. As can be seen, the asymmetry of the barrier determines the value of the shot noise for each magnetic state. For the DMTJ3 sample ($R_1 > R_2$) we have that $F_P > F_{AP2_2} > F_{AP1} > F_{AP2_1}$ while the order is reversed in DMTJ2s ($R_2 > R_1$). Also, an enhancement of the shot noise occurs at voltages higher than 500 mV for certain values of the energy, as can be seen in Fig. 9.15b. The relative magnitude of these quasi-periodic anomalies increases with the applied bias and is weakly dependent on the magnetic state, and

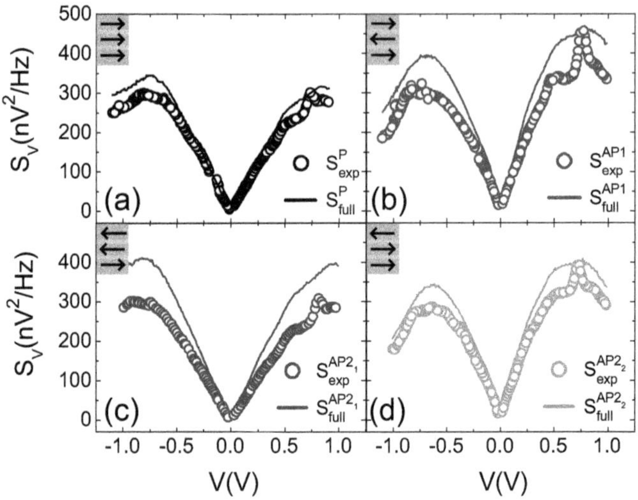

Figure 9.14 (a)-(d) Typical bias dependence of the shot noise for DMTJ2 measured in four different magnetic configurations at $T = 0.3$ K. The experimental data (points) is compared to full shot noise (lines). Adapted with permission from Cascales et al. (2012).

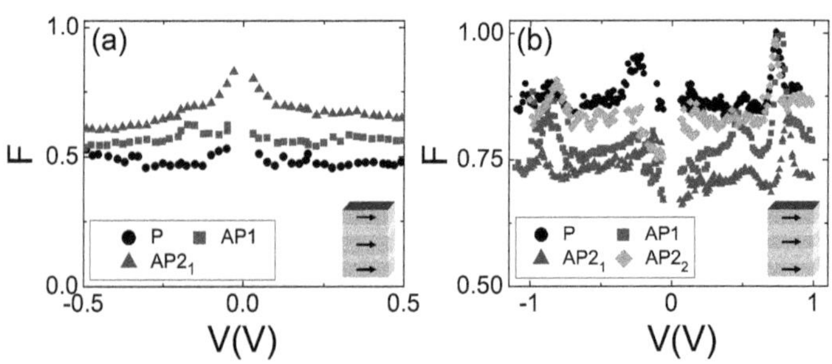

Figure 9.15 Dependence of the Fano factor with the magnetic state as a function of the applied bias for (a) DMTJ2 and (b) DMTJ3 samples. Adapted with permission from Cascales et al. (2012). The normalized shot noise is determined by the magnetic configuration of the junction.

they seem to originate from resonant tunneling through quantum well states.

The shot noise was also measured at a fixed current, while sweeping the magnetic field in order to achieve all four magnetic states of a DMTJ3 sample (Cascales et al., 2013b). As can be seen

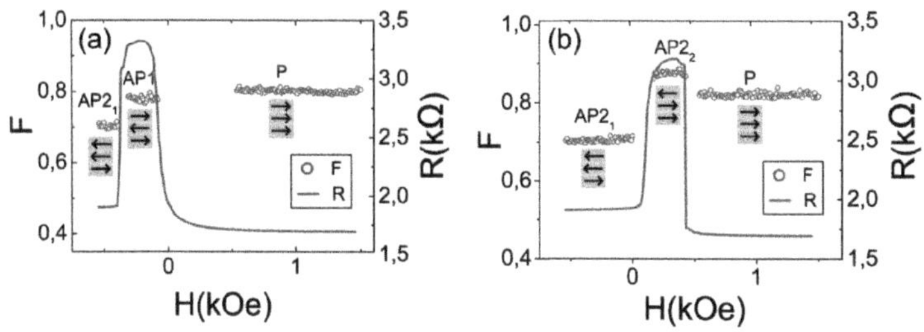

Figure 9.16 Shot noise and resistance at a fixed current during a magnetic field sweep at $T = 0.3$ K (a) from positive to negative values and (b) from the $AP2_1$ at negative fields back to positive fields. © 2013 IEEE. Reprinted, with permission, from Cascales et al. (2013b).

in Fig. 9.16, the Fano values match the results carried out at fixed fields while varying the bias. The main disadvantage of this method is having control over the applied bias, since the resistance of the junction changes with each magnetic configuration under constant current.

9.4.2.4 Tunneling through quantum well states in DMTJs

Quantum well states may form in the central electrode of a DMTJ. Evidence of the presence of such states has been detected as oscillations in the conductance of all three types of DMTJs shown here.

The differential conductance, after subtracting the parabolic background G_0, for DMTJ3s is compared with the Fano factor at different temperatures (Cascales et al., 2012) in Fig. 9.17a,b. The well-defined oscillations of the differential conductance indicate the presence of resonant transmission through QWSs formed in the central Fe_2 layer.

Figure 9.17c compares the conductance oscillations for each type of DMTJ. As can be seen, DMTJ1 samples present more oscillations in the range of energy shown than DMTJ2 or DMTJ3 samples. The appearance of QWSs in the central electrode of a DMTJ can be compared to the finite potential well problem in quantum mechanics. In this simplified situation, the number of bound states increases with the width d of the well as $N \propto d^2$, while the states become more closely spaced in energy. Indeed, the samples with

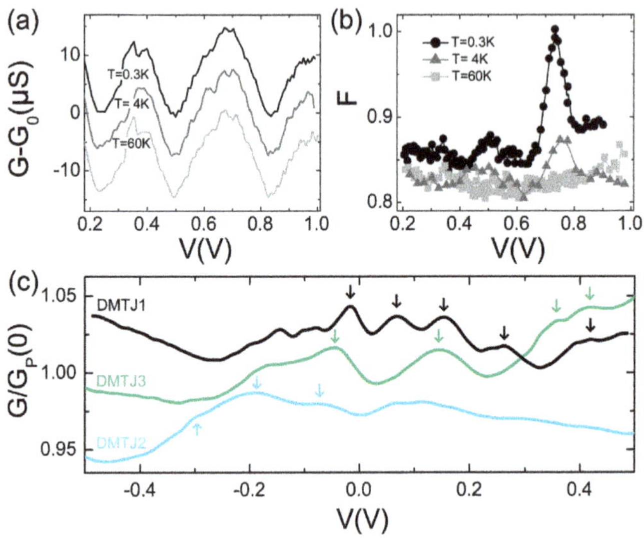

Figure 9.17 (a) Oscillations in the conductance of a DMTJ3, which are not affected by temperature (© 2013 IEEE. Reprinted, with permission, from Cascales et al. (2013b)). (b) Shot noise in the P state at 0.3 K, 4 K, and 60 K (© 2013 IEEE. Reprinted, with permission, from Cascales et al. (2013b)). The effect of quantum well states is gradually suppressed with increasing temperature. (c) Oscillations in the P state conductance due to QWSs (indicated by arrows) (Cascales et al., 2012). The oscillations are more pronounced for $V > 0$ in DMTJs 1 and 3, and for $V < 1$ for DMTJ2s. The curves have been offset for better observation.

a 10 nm central Fe electrode (DMTJ1s) present a larger number of conductance oscillations than those with a 5 nm thick central layer (DMTJ2s or DMTJ3s) for the same bias range (Fig. 9.17c). Moreover, the QWSs calculated in Ref. (Wang et al., 2006) for Fe/MgO/Fe/MgO/Fe DMTJs qualitatively agree with the oscillations obtained for DMTJ3s with 5 nm Fe central layers. QWSs are more pronounced for positive biases (when electrons tunnel from the bottom to the upper electrodes) for DMTJs1,3, and, as could be expected from the barrier asymmetry, are more pronounced for negative bias for DMTJ2s as can be seen in Fig. 9.17c. Indeed, the voltage distribution across an asymmetric DMTJ shows that QWSs affect the conductance mainly when electrons tunnel from the contact's Fermi level to the central layer through the thicker barrier

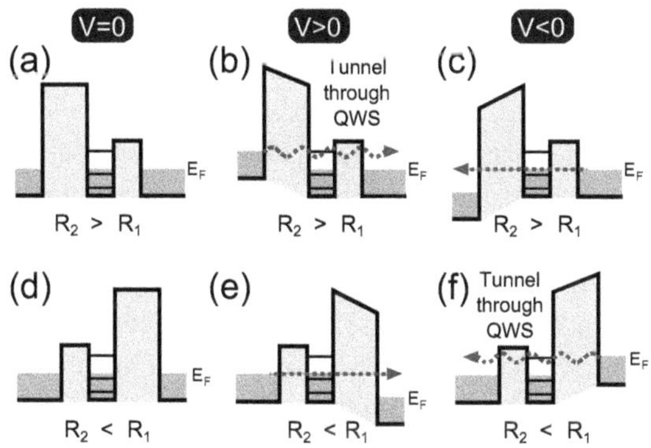

Figure 9.18 Diagram of resonant tunneling through QWSs in a DMTJ structure. DMTJs 1 and 3 correspond to (a), (b), and (c) and DMTJ2s to (d), (e), and (f).

(Chshiev et al., 2002; Greullet et al., 2007). A diagram representing this effect is shown in Fig. 9.18. Also, the oscillations in conductance due to QWSs are more significant, in general, in DMTJ1 and DMTJ3 samples. This is explained by the fact that the barrier through which the tunnel to QWSs occurs has a better structural quality for DMTJ1 and DMTJ3 samples (bottom barrier, R_1) than for DMTJ2s (top barrier, R_2).

Depending on the degree of coherence involved in the transmission through QWSs, SN is expected to show a shallow dip in the Fano factor due to coherent resonant transmission followed by a resonant enhancement in the negative differential conductance regime due to Coulomb interactions (Blanter and Büttiker, 2000a; Iannaccone et al., 1998; Nam Do et al., 2007). These anomalies in Fano are more pronounced for DMTJ3s and for positive bias, where QWSs in the conductance are more clearly observed (Fig. 9.15b). We also remark that the observed periodic anomalies in the conductance and Fano factor cannot be attributed to Coulomb blockade effects, where the Fano factor has periodic minima decreasing in amplitude with the applied bias (Hershfield et al., 1993).

9.4.3 Theoretical Model

The following theoretical model was developed by groups of Józef Barnaś and Vitalii Dugaev, with the aim of understanding the variation of the Fano factor with the barrier asymmetry and magnetic state of the system. The model is described briefly in Refs. (Cascales et al., 2012, 2013b) along with the results shown above, and explained in more details in Ref. (Szczepański et al., 2013).

The shot noise through a double-barrier junction with ferromagnetic electrodes is calculated using a model of sequential tunneling. In the model, DMTJs are described by the parameters α, β_1 and β_2, as well as the spin relaxation parameter g. These correspond to the asymmetry (α), which takes into account the difference between the barriers, as well as the two spin filtering parameters (β_1 and β_2, one for each barrier). The influence of resonant tunneling is not taken into consideration. The parameters have been evaluated by fitting the experimental Fano factor for all magnetic configurations as well as the TMR ratio, showing a good agreement the experimental observations in the case of strong spin relaxation.

9.4.3.1 Calculation of shot noise in the absence of spin relaxation

If spin-flip transitions can be discarded (or are absent), the two spin channels $\sigma = [\uparrow, \downarrow]$ can be considered as fully separated (see Fig. 9.19). The shot noise of each spin channel, S_σ, can be then calculated from the expression for spinless particles (Beenakker and Büttiker, 1992b):

$$S_\sigma = \frac{R_{1\sigma}^2 S_{1\sigma} + R_{2\sigma}^2 S_{2\sigma}}{R_\sigma^2}, \tag{9.2}$$

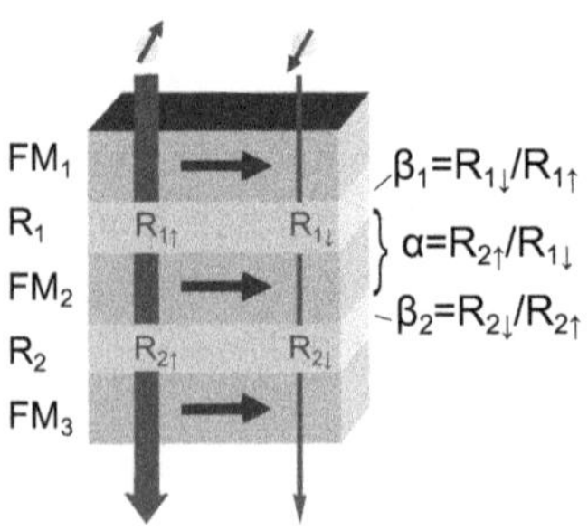

Figure 9.19 Diagram of a DMTJ illustrating the two-current model and the parameters α, β_1 and β_2.

where $R_{i\sigma}$ is the resistance of the i-th junction in the spin-σ channel, R_σ is the total resistance of the spin-σ channel, $R_\sigma = R_{1\sigma} + R_{2\sigma}$, and $S_{i\sigma} \simeq \frac{2eV}{R_\sigma}$ (valid if the transmission probability is small for each barrier). V is the external voltage applied to the system, and the index $i = 1, 2$ refers to the two tunnel barriers of the DMTJs: $i = 1$ for the left and $i = 2$ for the right barrier. Including both spin channels when calculating the shot noise S of the double-junction structure one obtains $S = S_\uparrow + S_\downarrow$.

The average value of charge current driven by the voltage V is

$$I = V\frac{R_\uparrow + R_\downarrow}{R_\uparrow R_\downarrow}. \tag{9.3}$$

From equations 9.2 and 9.3, the Fano factor in the absence of spin relaxation is given by

$$F \equiv \frac{S}{2eI} = \frac{\left(R_{1\uparrow}^2 + R_{2\uparrow}^2\right) R_\downarrow^3 + \left(R_{1\downarrow}^2 + R_{2\downarrow}^2\right) R_\uparrow^3}{R_\uparrow^2 R_\downarrow^2 \left(R_\uparrow + R_\downarrow\right)}. \tag{9.4}$$

In the case of a symmetric structure, $R_{1\sigma} = R_{2\sigma} = R_\sigma/2$, from equation 9.4, one finds $F = 1/2$. In order to describe the numerical results, we introduce the following parameters (see Fig. 9.19):

$$\alpha = R_{2\uparrow}/R_{1\uparrow}, \qquad \beta_i = R_{i\downarrow}/R_{i\uparrow}. \tag{9.5}$$

Let us now consider the case of a DMTJ with three ferromagnetic electrodes. The Fano factor in each magnetic configuration can be obtained as a function of α, β_1 and β_2 by using equations 9.4 and 9.5. Using as a reference the resistance of each barrier in the P state, $R_{\sigma\uparrow}^0$, the coefficients take the following values for each state:

P	$R_{1\sigma}^P = R_{1\sigma}^0$	$R_{2\sigma}^P = R_{2\sigma}^0$
AP1	$R_{1\sigma}^{AP} = \sqrt{R_{1\uparrow}^0 R_{1\downarrow}^0}$	$R_{2\sigma}^{AP} = \sqrt{R_{2\uparrow}^0 R_{1\downarrow}^0}$
AP2$_1$	$R_{1\sigma}^{AP2_1} = R_{1\sigma}^0$	$R_{2\sigma}^{AP2_1} = \sqrt{R_{2\uparrow}^0 R_{1\downarrow}^0}$
AP2$_1$	$R_{1\sigma}^{AP2_2} = \sqrt{R_{1\uparrow}^0 R_{1\downarrow}^0}$	$R_{2\sigma}^{AP2_2} = R_{2\sigma}^0$

As an example, let us consider the AP1 state, for which the resistance of each junction is highest. The corresponding resistance $R_{i\sigma}^{\mathrm{AP1}}$ (for $i = 1, 2$) can be expressed by the resistances in the P configuration as $\sqrt{R_{i\uparrow}^0 R_{i\downarrow}^0}$.

9.4.3.2 Calculation of shot noise in the presence of spin relaxation

The previous model proves to be insufficient to describe the experimental data, since for example, the Fano in the P state is symmetric with the parameter α while in the experiment that is not the case. Thus, spin relaxation in the central electrode is included in the theory in order to obtain a better comparison.

Therefore, taking the spin relaxation into account, one writes the relevant equation for spin density fluctuations δS_z (Szczepański et al., 2013):

$$\Delta J_z^{(2)} - \Delta J_z^{(1)} = -\frac{\delta S_z}{\tau_s}, \tag{9.6}$$

where $J_z^{(i)}$ is the z-component of spin current in the i-th barrier and τ_s is the spin relaxation time in the central electrode. In the limit of strong spin relaxation, instead of equation 9.4 one finds (Szczepański et al., 2013):

$$F = \frac{R_{2\uparrow}R_{2\downarrow}(R_{1\uparrow}+R_{1\downarrow})^2 + R_{1\uparrow}R_{1\downarrow}(R_{2\uparrow}+R_{2\downarrow})^2}{[R_{1\uparrow}R_{1\downarrow}(R_{2\uparrow}+R_{2\downarrow}) + R_{2\uparrow}R_{2\downarrow}(R_{1\uparrow}+R_{1\downarrow})]^2}. \tag{9.7}$$

The Fano factors in all four magnetic configurations can be calculated in the same fashion as the case without spin relaxation. In a general case, the authors take into account the spin fluctuations via equation 9.6. However, the corresponding formulas are cumbersome and will not be presented here and may be found in Ref. Szczepański et al. (2013). The key element of the theory is the dependence of the SN on the spin density fluctuations. These fluctuations are described by equation 9.6 and depend on the spin relaxation. The spin relaxation is conveniently described by the parameter $g = d/v_F\tau_s$, with d being the thickness of the central layer, and v_F the Fermi velocity.

Since the resistance of each state can be calculated as a function of the parameters, the TMR for one of the AP states (AP1, $AP2_1$ or $AP2_2$) may be defined by

$$\text{TMR} = \left(\frac{R_{AP}}{R_P} - 1\right). \tag{9.8}$$

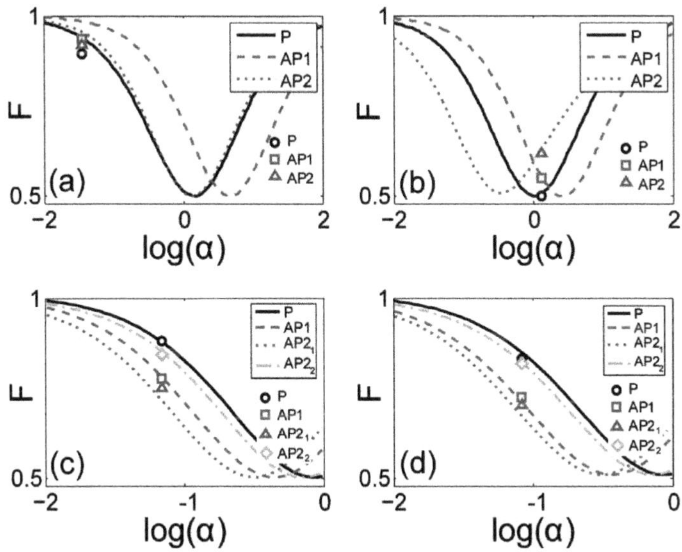

Figure 9.20 Comparison of theory (solid lines) and experiment (points) of the Fano factor measured for three different DMTJs. (a) DMTJ1 at $T = 0.3$ K ($\alpha = 0.035, g = 4.9, \beta_1 = 42, \beta_2 = 2$) (Cascales et al., 2012). (b) DMTJ2 at $T = 0.3$ K ($\alpha = 1.3, g = 0.3, \beta_1 = 75, \beta_2 = 11$) (Cascales et al., 2012). DMTJ3 (c) at $T = 0.3$ K ($g = 100, \alpha = 0.07, \beta_1 = 5.7, \beta_2 = 27.8$, and (d) $T = 60$ K, with $g = 100, \alpha = 0.08, \beta_1 = 3.5, \beta_2 = 28.3$. (c) and (d) © 2013 IEEE. Reprinted, with permission, from Cascales et al. (2013b).

The maximum TMR occurs in the AP1 configuration, which is given by

$$\mathrm{TMR_{AP1}} = \frac{(\sqrt{\beta_1} + \alpha\sqrt{\beta_2})(1 + \alpha + \beta_1 + \alpha\beta_2)}{2(1+\alpha)(\beta_1 + \alpha\beta_2)} - 1\,. \qquad (9.9)$$

By using the experimental Fano values for the magnetic states and the TMR in the AP1 state, we have been able to carry out a comparison between experiment and theory.

9.4.3.3 Comparison between experiment and theory

As can be seen in Fig. 9.20, the experimental data can be accounted for rather well by tuning the parameters α, β_1, β_2 and the spin relaxation parameter g (Cascales et al., 2012, 2013b; Szczepański et al., 2013). To compare the theoretical results with the experimental data we have used average Fano values in the range between 0.2 and 0.5 V in order to avoid the possible influence of defect states in the barrier below 200 mV (Du et al., 2010) or of QWSs observed above 0.5 V. Figures 9.20a–c show the comparison, at $T = 0.3$ K, of the calculated Fano factors as a function of the asymmetry parameter α with the Fano in the measured magnetic states in a DMTJ1,2 and 3 sample, respectively. Figure 9.20d presents the same comparison at $T = 60$ K of the DMTJ3 sample of graph (c). There is a good qualitative and quantitative agreement with the experimental results. The combined TMR and SN provide an evaluation of the three independent parameters α, β_1, and β_2 as well as the spin relaxation g.

Figure 9.21 shows the estimated parameter g and the barrier asymmetry α for DMTJs at $T = 0.3$ K. It is interesting to note that the best fits to the theory for two measured DMTJ3s (see Fig. 9.20c) appear with relatively low g (i.e., large τ_s, estimated to be around 10^{-12} s for $v_F = 10^4$ m/s). On the other hand, SN in both measured DMTJ2s is best described with $g \sim 100$ (Cascales et al., 2012) (i.e., short τ_s) as seen from Fig. 9.21a. We relate shorter τ_s in DMTJ2s with an increased density of oblique defects as the epitaxial MgO is grown above the critical thickness for the plastic relaxation of MgO on Fe (Tiusan et al., 2007). These defects could

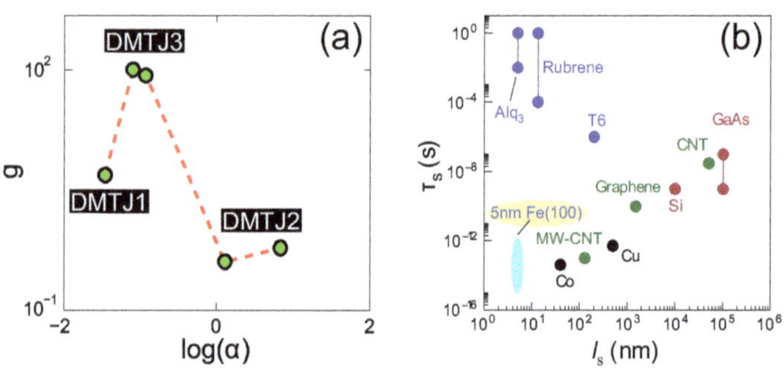

Figure 9.21 (a) Minimum values of the spin relaxation parameter g vs. α which yield the experimental shot noise and TMR values for all three types of DMTJs. (b) Spin diffusion length l_S as a function of the spin diffusion time τ_S for various spintronics materials, which includes the rough estimation for DMTJ2 and DMTJ3 samples. Adapted with permission from Ref. Szulczewski et al. (2009).

be "imprinted" on the central electrode, increasing its defectiveness and, in agreement with the Overhauser–Elliott–Yafet model (Žutić et al., 2004), strongly reducing τ_S. The spin relaxation time data estimated from the theoretical fit is plotted along with the results for other materials (Szulczewski et al., 2009) in Fig. 9.21b.

We note that the model neglects other possible sources of noise, like $1/f$ and thermal noise. Also, the SN is calculated neglecting spin coherent resonant tunneling. Moreover, the model does not include any deviation of the angle between magnetizations from 0 or π, which may influence the Fano factor (Tserkovnyak and Brataas, 2001) or the influence of QWSs, disorder and interfacial states, which may reduce the Fano factor (Liu et al., 2012). All these contributions can be responsible for the deviation of the theoretical curves from the experimental points in Fig. 9.20. The strongest deviation in the case of DMTJ1 (Fig. 9.20a) could be attributed to the presence of exchange coupling through the thin (3 ML) MgO barrier (Faure-Vincent et al., 2002; Tiusan et al., 2007) and the possible formation of domain walls in the central Fe electrode.

9.5 Summary and Perspectives

In summary, double-barrier MTJs are rather prospective and currently under-explored spintronic devices. The main challenge seems to be related with the reduction of structural defects accumulation once thick stack is grown. On the other side, the presence of few barriers provides much more freedom to control electronic transport and noise by exploring QWSs and magnetization degrees.

For example, having control over the shot noise in DMTJ devices could be used for the controlled injection of spins between ferromagnets, as well as between FM and other materials. Furthermore, when other materials are used as a central electrode, shot noise could prove to be a novel method to study the spin relaxation time in different materials. From a technical point of view, DMTJs present fundamentally different from single-barrier MTJs with the most fundamental noise source in electronics could be controlled and reduced. This scheme could be useful for vertical (e.g., spin current injection in semiconductors through double MgO barriers) or lateral (e.g., quantum dots) electronic structures. DMTJs also seem the preferred option for reducing the intrinsic noise of magnetic field sensors (Yu et al., 2011), in detriment of serial arrays of MTJs (Guerrero et al., 2009) which may experience a higher chance of failure.

An important but yet under-explored issue is high-frequency noise study in DMTJs where spin torque effects could be activated at rather low current densities. Theoretical calculations (Chen et al., 2008) predict spin transfer torque to be oscillating with the bias voltage due to presence of QWS in the middle FM (Herranz et al., 2010b; Nozaki et al., 2006b). Critical currents to activate STT are predicted to be rather low ($10^5 - 10^6$ A/cm^2).

Alternative way to develop DMTJs is to embed magnetic nanoparticles into a single-barrier. Such approach may further induce the valuable reduction of the critical current density for current-induced magnetization switching (Useinov et al., 2015).

Chapter 10

Low-Frequency Noise in Organic Electronics and Spintronics

Downscaling of inorganic electronics has currently become increasingly difficult due to both physical limitations such as modification of electronic structure due to increasing relation between surface and volume as well as increasing fabrication costs of top-down structures. Since last decades, especially after Moore law has shown saturation trend, interest has been gradually building up in alternative concepts providing strong reduction of the electronic devices at reasonable prices. Originated in the seventies (Mann and Kuhn, 1971), molecular electronic concept of single molecules or small groups of molecules conducting electrical current between two electrodes is among the most promising candidates to create future devices capable substitute traditional Si-based electronics (Nitzan and Ratner, 2003). It has been realized; however, that electron transport in molecular or organic tunnel junctions (OTJs) is much more complex than initially thought (Galperin et al., 2007).

Effective action on ferromagnetic layers on the electron tunneling over inorganic and organic (molecular) barriers could be rather different (Droghetti et al., 2015). This is mainly because of resistive,

Noise in Spintronics: From Understanding to Manipulation
Farkhad G. Aliev and Juan Pedro Cascales

ISBN 978-981-4774-34-5 (Hardcover), 978-1-315-11088-2 (eBook)
www.panstanford.com

chemical and atomic mass mismatch between molecules and metallic electrodes at interfaces.

Indeed, a variety of factors determine electron transport through OTJs, which as a consequence makes it much more difficult to understand their noise. Among them the most important factors, in addition to the band states of the electrode and molecule electronic structure are molecular vibrations (Galperin et al., 2007), particularly those corresponding to the bonds modified by the proximity to the interfaces (Brumme et al., 2012a). Besides, electron spin may become relevant for both transport in organic magnetic tunnel junctions (O-MTJs) (Dediu et al., 2002). Magnetic state of the interfacial molecules could also be qualitatively changed by the electron coupling to magnetic or even to non-magnetic electrodes at *spinterfaces* (Sanvito, 2010).

In realistic organic electronics renormalized electron structure of the interfacial molecules could acquire to some (Gaussian) distribution in the position of the interfacial molecular energy levels with evident consequences for the electron transport (Shi et al., 2015) and probably low-frequency noise. Besides atomic mass mismatch and unavoidable roughness of the interfaces (both in the case of macroscopic and in some cases atomic junctions) should increase interface related contribution to $1/f$ noise.

The above-discussed complexity of the booming organic electronics explains why noise studies in molecular-based electronic and spintronic devices is currently emerging area of the research. This chapter starts with the application of noise measurements to characterize organic electronics. The second part summarizes so far scare knowledge on the noise in organic spintronics.

10.1 Single Molecular Imaging and Identification through the Low-Frequency Noise

Noise imaging and identification of the molecules is a new direction in the surface science which emerged after 2010 (Tsutsui et al., 2010). Before discussing these methods in detail, let us first review the main physical mechanisms which could contribute to the noise in conductance when metallic tip approaches the surface covered

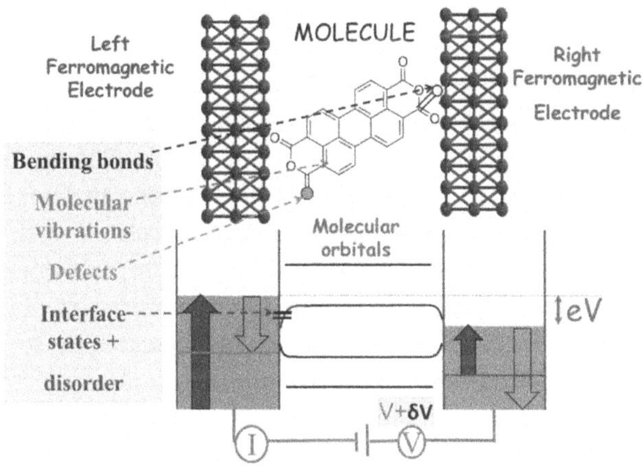

Figure 10.1 Sketch describing relevant noise sources in the molecular spintronic devices with two ferromagnets contacting molecules. Left (yellow background) window lists different factors contributing to electron transport through molecules.

by the molecules or when molecule is sandwiched between two ferromagnets (see Fig. 10.1). One should clearly distinguish between two qualitatively different regimes which may be tuned by changing tip position in the first case or varying growth conditions in the second (OMTJs). First, experiments with STM imply transition from tunneling to contact regime, when the noise in conductance may become rather large due to stochastic interruption of current path to molecule (Brumme et al., 2012a; Néel et al., 2011) or single atom (Sperl et al., 2010).

Second, different LFN sources could be dominant in almost established conductance regime. Such situation is resumed in Fig. 10.1. In this case the excess noise could appear due to molecular bonds fluctuations or vibrations, electron-electron (electron-phonon) correlations controlled by the surface states, or defects related noise among others. Interface states could filter out electron spin (Sanvito, 2010). The origin of conductance fluctuations in the molecular junctions could also be a consequence of spontaneous motion of molecule as whole. Corresponding process could be described as similar to one occurring in electron nanomechanical shuttle (Novotný et al., 2004). Spontaneous molecule vibrations in the form of conformational changes could be stimulated by inelastic tunneling process (Kaun and Seideman, 2005) with emission or adsorption of phonon with some specific energy.

Another possible source of noise in electron transport over molecular barriers could be a consequence of charge trapping/detrapping on defect states (Kim et al., 2010a). Among the most informative contributions to the noise in OTJs, which will be discussed in more details below, is related with above mentioned stochastic bond fluctuation (vibration) from molecular attachments to metallic electrodes (Ramachandran et al., 2003) Finally, when electrons (holes) are injected resonantly to molecular (LUMO/HOMO) orbitals, coupling of transmitted electrons to molecular vibrational degrees of freedom of molecules may occur (Secker et al., 2011). Stochastic nature of vibrations gives rise to strong enhancement of noise which has been described usually as enhanced shot noise (see review (Galperin et al., 2007) as well as recent modeling (Novotný et al., 2004) and references therein).

Some of the above-discussed noise sources have been recently used in molecular imaging and identification (Schaffert et al., 2012; Tsutsui et al., 2010). Molecular identification (Tsutsui et al., 2010) has been made via observation and quantification of inelastic electron transmission induced noise in single organic molecules. The noise imaging/identification simplified for the cases when low-frequency noise has form of two-level fluctuations (RTN). In general, stochastic variation of the conductance between tip and substrate could happen as a consequence of molecular conformal changes or some kind of rotary, vertical or horizontal translational motion of atoms.

The experimental system employed for noise imaging should use scanning tunneling microscope with short (typically 10 ns) time resolution and working at low temperatures (a few K). As we discussed above, the RTN noise emerging with tip approach to the molecule, may originate from fluctuations of the molecule bonds. Presence of tip itself may, in addition, create close by some equal/lower in energy state of the molecule separated by relatively small barrier. The voltage dependence of the current fluctuations (dispersion) has been studied for the high enough biases, when electrons are injected into the LUMO orbital in order to identify molecules.

Case study 10.1: Single-Molecule Identification through Noise

Inelastic tunneling or scattering process could give rise to peaks in derivative of the conductance vs. bias (IETS). Makusu Tsutsui et al., (Tsutsui et al., 2010) shows that inelastic process related extra noise could be used for effective molecules indentification, in addition to IETS. Single-molecule identification has been suggested to carry out via detection of the increased current dispersion at specific biases close to energies of characteristic molecular vibrational modes as shown in Fig. 10.2.

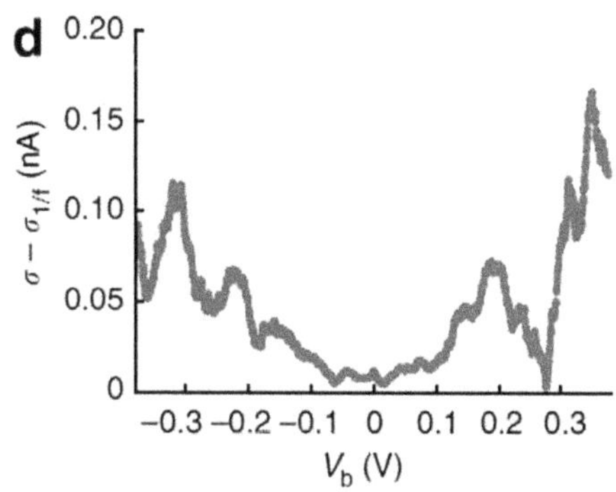

Figure 10.2 Bias dependence of conductance fluctuations (relative dispersion) obtained in Au(HDT)Au single-molecule junctions at 4.2 K. Adapted with permission from Tsutsui et al. (2010).

The following mechanism has been suggested to explain the noise excess. As long as inelastic electron tunneling is accompanied by extra heat dissipation, even with the main dissipation channel directed toward the electrodes, local heating could result in an increase of temperature and correspondingly in increase of thermal noise of the junction. However, within this simplified approach one (based just on thermal noise relation) one estimates effective temperature for some of the noise peaks to reach 1600 K with bath temperature of 4 K. Thermal noise origin of heating contradicts therefore stability of the junction under study. So, the extra noise should be mainly linked with statistical nature of phonon assisted inelastic tunneling process.

The above case study corresponds to the well-established contact between electrodes and the molecules. However, as we mentioned previously, before each full contact is established, interaction of STM tip with molecule could induce configurational instabilities of the molecules which could, in turn, result in strong enough RTN containing another complementary and valuable information on the molecule bonds. Analysis of the bonds hopping rate, the noise amplitude and the relative occupation of the involved states could be used to map spatially individual molecules. While the standard STM maps provide information on the stable electronic states, RTN noise maps transiently populated states. Such situation is considered in the next case study.

Case study 10.2: Noise imaging of molecules

The noise maps of an individually adsorbed molecule (Schaffert et al., 2012) revealed electron-driven hindered rotation between the equilibrium and two metastable positions. In order to carry out real time analysis of RTN for each particular point special electronic scheme has been designed. The figure below depicts an analog electronics circuit used in real time RTN mapping of molecules adapted with permission from Schaffert et al. (2012).

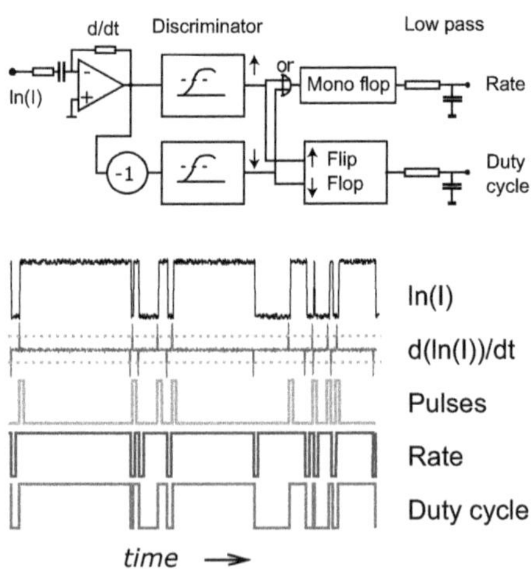

Molecular switching events have been detected by time differentiation of the logarithm of tunneling current. Such analysis ensures detection of the current noise on relative rather than on the absolute scale and allows real time and high precision measurements of switching rate and duty time of RTN.

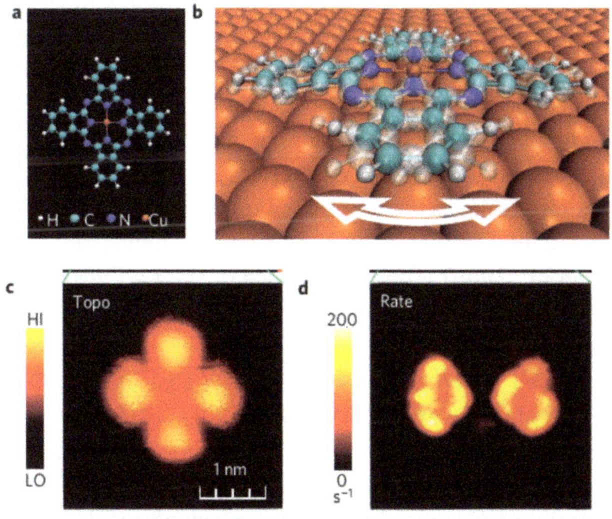

Figure 10.3 Schematic representation of frustrated rotation of CuPc molecule and experimental images of obtained by topography and from real time analysis of RTN rate. Adapted with permission from Schaffert et al. (2012).

In order to get optimized RTN information, electrons have to be injected into the molecules. This is confirmed by observation that the best contrast in RTN rate/duty cycle images has been obtained at biases close to (but not exactly) to the energies of activation of LUMO/HOMO conductance channels. At the same time, the best (i.e., less noisy) topography images have been obtained when transmission to LUMO/HOMO channels is maximum (Schaffert et al., 2012)

10.2 Organic Resistive Switching Memories

Electronic and spintronic devices considered so far change their resistance in fully reversible fashion in the whole range where external bias electric field is varied. In some special cases, however, when oxide and/or organic (polymer) layer form the device barrier, above some critical applied electric field irreversible resistive switching may occur. The main physical mechanism beyond resistive random access memories (RRAMs) memories is the formation of a conductive (ohmic) filament during its operations followed by partial oxidation during reset operation with low conduction through a multi-phonon trap-assisted tunneling process. The RRAM memories could be described as sort of memristic memories (Strukov et al., 2008) where resistance of the two terminal device depends on the charge between two electrodes.

The dependence of the hysteretic resistance on the electric bias has been used to create oxide-based (Puglisi et al., 2013) and polymer (Vandamme et al., 2011)-based RRAMs (Pan et al., 2014). Such memories are among important candidates for fast, reliable and high-density non-volatile memories.

Vandamme and co-authors investigated physical mechanisms of irreversible resistance switching by using low-frequency noise measurements (Vandamme et al., 2011). Their organic RRAM memories have been constructed of a thin semiconducting organic film sandwiched between two metal electrodes, one of which is aluminum covered by a native or thicker oxide (Vandamme et al., 2011). It was observed that the reliable high resistance (H) and low resistance (L) states show pure $1/f$ noise with magnitude following the resistance changes. In both L and H states, the spatial distribution of the current seems to be rather non-uniform with only a fraction of the device carrying the current. Based on LFN analysis it was concluded that the resistive switching takes place at some spots of the Al/Al_2O_3/ polymer interface (Vandamme et al., 2011).

Some recent theoretical models predict reach novel behavior of memristors when spin degree of freedom is considered (Pershin and Di Ventra, 2008). Future noise studies are expected serve as a valuable informative tool to address novel physics in the systems with memristive behavior based on the both charge and spin degrees of freedom. Such devices should allow for a more convenient control of the state of organic and non-organic RRAM nanostructures than those based on purely ionic (vacancies) transport.

10.3 Low-Frequency Noise in Organic Electronics

Before considering noise in organic spintronics, we resume experimental studies of the LFN in non-magnetic organic electronics: both organic materials and devices where thin organic barrier bridges two metallic electrodes. Noise studies in organic spintronics are in their infancy and the first steps will be discussed later in the second part of this chapter. Note that LFN in graphene-based devices is considered separately (see Chapter 12). The present paragraph concentrates mainly on the $1/f$ noise and its bias dependence.

Low-frequency noise in current in n-Si/alkyl chain ($C_{18}H_{37}$) monolayer/Al tunnel junctions shows power spectrum proportional to $1/f^{\gamma}$ (with γ about 1–1.2) (Clément et al., 2007). Additionally, the bias dependence of the LFN has been found to be non-monotonous with increase close to 0.4 V (see Fig. 10.4). However, the characteristic bias corresponding to this noise maxima was observed to vary from sample to sample. This effect has been attributed to an energy-dependent trapping-detrapping process.

Theoretical model describing resonant increase of LFN vs. bias (shown in Fig. 10.4) considers effective two-level tunneling system with an asymmetric double well potential. Each of minima are separated by tunneling barrier (Δ_i) and have difference in their minima of about $2\epsilon_i$. As long as model focuses on the LFN, only traps with long relaxation times for which $\Delta_i \ll \epsilon_i$ are considered. By using usual approximations regarding the uniform distribution in relaxation times in the restricted frequency range, frequency-dependent noise power spectrum (in current) has the following form (Clément et al., 2007):

$$S_I = (E_{max}/A)(\sigma)^2 e^2/(C_{TJ})^2 f, \qquad (10.1)$$

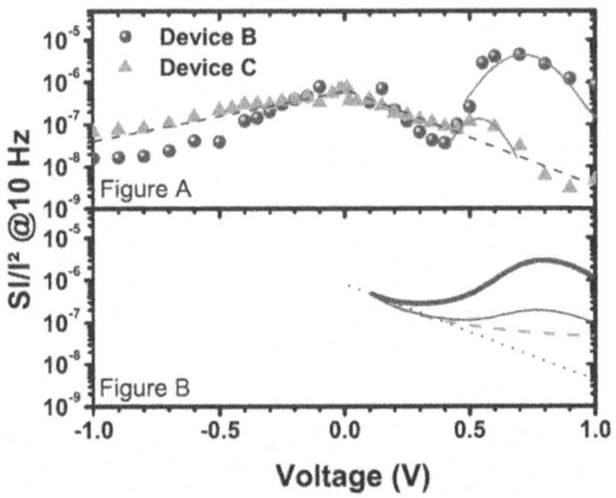

Figure 10.4 Normalized noise power spectrum as a function of bias: experiment and model. Adapted with permission from Clément et al. (2007).

with σ being conductance over empty impurity states, A and C_{TJ} respectively area and capacitance of the tunnel junction and $E_{\max}$ the maximum value of the defect energy.

The above simplified model cannot explain "bump" in the bias-dependent noise around 0.4 V if uniform relaxation time distribution is used (see dashed line, Fig. 10.4). One needs, therefore, to add an additional Gaussian energy-localized distribution of defects (thin solid line) in order to approach the model to the experiment. As we shall discuss in the next paragraph, such "bumps" in the bias dependence of LFN in molecular tunnel junctions could have qualitatively different and simpler explanation associated with effects of vibrational heating on bending of interfacial molecules.

Another observation of non-monotonic bias dependence of the normalized LFN in tunnel junctions with molecular barriers formed by self-assembled monolayers (SAM) (Kim et al., 2010b) has also been explained as a consequence of tunneling thorough localized states. Figure 10.5 shows sample layout and schematically the noise measurement setup.

The typical noise spectra shown in Fig. 10.5b indicates dominant $1/f$ contribution superimposed over $1/f^2$ with later component attributed to generation-recombination processes. Analyzing the noise power for fixed frequency (10 Hz) reveals a maximum in noise for biases around 0.4 V.

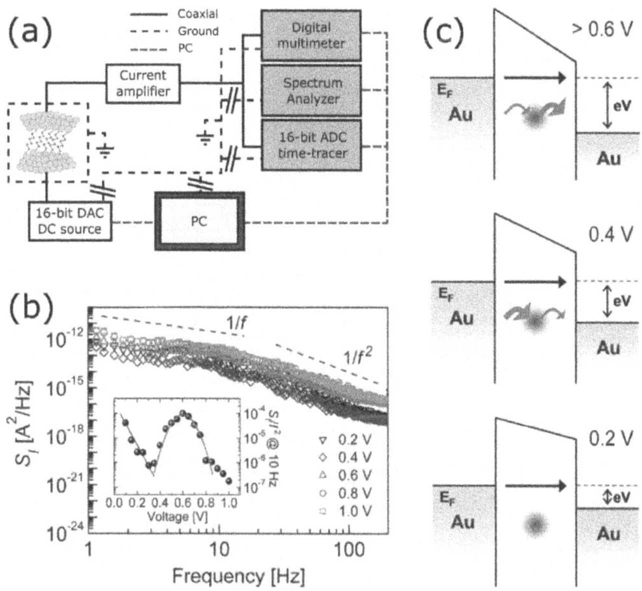

Figure 10.5 (a) Sketch of the noise measurement setup. (b) Noise spectrum at different applied biases observed for metal-organic molecule-metal junctions. (c) Model of tunneling assisted by localized states. Adapted with permission from Ref. Kim et al. (2010b).

Besides the tunneling through the localized states originated from non-intentionally introduced defects in the junction (Fig. 10.5c), the authors (Kim et al., 2010b) suggest alternative explanation of the origin of the localized states as originated from penetration of the gold particles during the metallization. Unfortunately, the bias dependence of noise for the opposite bias polarity, which could clarify better the origin of these defect states, was not analyzed.

10.4 Low-Frequency Noise Related with Molecules Bending

Modeling the LFN anomalies discussed in the previous paragraph considered mainly electronic contributions, without influence of nucleus dynamics (phonons). Composed of light organic elements (such as carbon) molecular barriers have multiple low energy phonon modes facilitating excitation of nucleus phonon modes by inelastic electron tunneling processes. Atomic and molecular bond vibrations could, therefore, be rather relevant for understanding noise in organic electronics and spintronics. *Influence of phonon assisted inelastic tunneling on shot noise will be considered separately.*

Evidently, mainly those bonds vibrations, which contribute to opening/closure of the conduction channels, should be considered when modeling the LFN. While for the single-molecule scenario, such vibration are expected to provide RTN-type contribution, situation could be different in the case of macroscopic junctions where multiple local conductance "hot spots" in the barrier may appear. The previous chapters have already discussed other examples of so-called "hot spots" appearing due to the barrier inhomogeneities in magnetic tunnel junctions with inorganic barriers. In those cases, such RTN has been activated by high local current densities under the external bias and could be originated from the local spin torque. Progressive downscaling of organic electronic components may result in appearance of similar effects with barrier thickness variation, but due to formation and breaking of the chemical bonds under external bias. Contribution of these effects, in general, should depend on relation between barrier thickness and molecule size.

Indeed, in the single-molecule devices (with molecules contacted by scanning tunneling microscope or break junction) with limited influence of the barrier thickness variation, RTN has been found to be a dominant noise source (see for example (Brumme et al., 2011; Ulstrup et al., 2012). Understanding of the noise sources in macroscopic junctions with millions of nearly parallel noise sources, but more apt for the real application, implies an ability to address before at least qualitatively origin of the similar noise sources for the single-molecule junctions.

Brumme and coauthors (Brumme et al., 2011) investigated experimentally and theoretically dynamics of a molecular junction with PTCDA molecule sandwiched between scanning tunneling microscope tip and an Ag(111) surface.

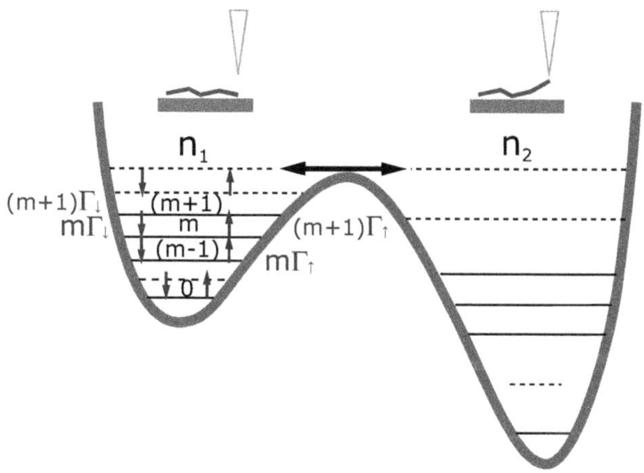

Figure 10.6 Sketch of double-well potential proposed in Ref. Brumme et al. (2011) to describe different vibrational heating effects up (full contact) and down (surface contact) states of PTCDA molecule in the STM. Adapted with permission from Brumme et al. (2012a).

Experimentally, the rates for the up/down transition are sensitive to the applied bias as well as the tip-surface separation (Brumme et al., 2011). The surface to tip switching rate was found to show non-monotonic variation with applied bias, which has been described by assuming vibrational heating and presence of two Lorentzian-type anomalies in the DOS situated below and above Fermi level with only one taken into account at a time (Brumme et al., 2011).

The later extension of the vibrational heating RTN model (Brumme et al., 2012a) considers only a single electronic level pinned to the Fermi energy of the substrate in order to obtain resonant reduction of the electron transfer rate when the applied bias matches one of phonon energies (Fig. see Fig. 10.6). Sharp (few orders of the magnitude) drop of the electron transfer rate (and correspondingly drop in RTN characteristic time) could be reflected in an enhancement of zero frequency noise for biases matching phonon energies.

Interface related vibrational heating could alternatively explain anomalous bias dependence of the noise in macroscopic O-MTJs.

Indeed, due to the samples growth conditions, on the average, molecules could be expected to be attached to the bottom (surface) layer much better than to the upper electrode with usually relatively enhanced roughness. This means that in the first approximation, the bias-dependent dynamical bi-stabilities in macroscopic O-MTJs could be modeled by simple averaging of the results obtained with varying molecule-tip separation in the relevant system (Brumme et al., 2011).

Vibration heating could be related with other kind of noise, which could be called "electron heat noise." Such noise describes fluctuation in the heat transferred by tunneling over molecular barriers electrons (in this sense being different from the heat due to phonons). These fluctuations, being additional to the fluctuations in transmitted electron charge, have been modeled within minimal model with single molecular level coupled to two metallic leads (Zhan et al., 2011). Theoretical results show, in difference to the electron charge noise, power spectrum density of the heat noise changes with frequency even in the zero-temperature limit showing different asymptotic increasing with frequency behaviors in the low- and high-frequency regions. These studies underline the need in detailed investigation of the coupled heat and electron transport across molecular junctions in order to deeper understand rich dynamic behavior excited beyond linear response limit.

Interestingly, the molecular vibrations could induce non equilibrium effects through coupling to electronic structure of electrodes. For example, investigation of the non-equilibrium electron-vibration coupling and conductance fluctuations in C_{60} (Néel et al., 2011; Ulstrup et al., 2012) conclude that full non-equilibrium electronic structure needs to be accounted for in order to understand the observed RTN. More specifically, if was identified (Ulstrup et al., 2012) an existence of strong coupling between the molecular dynamics and the C_{60} electronic structure close to contact formation. Taking into consideration a vibrational heating authors construct a model successfully explaining the experimentally polarity-dependent RTN fluctuations (Néel et al., 2011) in terms of strong variation of LUMO resonance for positive bias. For positive bias the voltage drop over the device is shifted from the C_{60}-tip interface to the C_{60}-substrate interface even with a small C_{60} displacement. Such redistribution of the voltage drop does not occur for the negative bias, explaining the observed asymmetric RTN.

10.5 Phonons Excitation Modified Shot Noise in Molecular Electronics

Molecular dynamics (vibrations) have been predicted to affect not only low-frequency noise through RTN (previous paragraph) but also the shot noise. Electronic contribution to the shot noise in the *tunneling regime* through the molecules and quantum dots is considered in separate paragraphs below.

We start with considering *quantum conductance regime*. Experimental demonstration of how vibration mode scattering could influence electronic shot noise has been presented for Au atomic chains in the high transmission regime (Kumar et al., 2012). Due to the vibration excitations, a negative additional shot noise contribution may appear below a certain transmission because two electrons injected at different energies could compete for the same outgoing state after the phonon emission (see upper left sketch in Fig. 10.7).

Additional change in the shot noise could, however, be dependent on the asymmetry of the level coupling to the electrodes. One-level (LUMO) bridge between two metal contacts with level coupled to a vibrational degree of freedom predicts (see Fig. 10.7) that increase in Fano factor with opening of the inelastic channel could take place only for the resonant tunneling with symmetric coupling to the leads (Galperin et al., 2006) and is negative for the asymmetric coupling (see bottom sketch and figures in Fig. 10.7).

Theory of phonon-assisted current noise in molecular junctions experimentally relevant case of a D_2 molecule predicts a significant inelastic contribution to the current noise which produces negative/positive additional contribution to the shot noise for low/high transmission (Haupt et al., 2009). These results are in line with experimentally obtained by Kumar et al. (Kumar et al., 2012) on atomic gold chains. Presence of phonon heating (i.e., of non-equilibrium phonons due to mass mismatch between molecule and contacts) only enhances the effect.

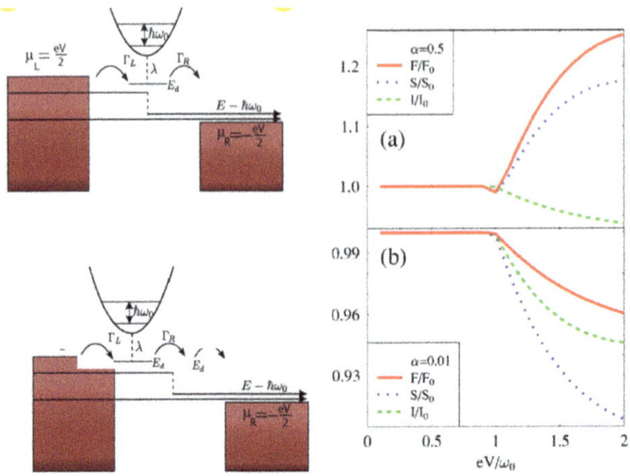

Figure 10.7 Influence of molecular vibrations assisted tunneling on shot noise. Fano factors (solid line), zero frequency noise (dotted line), and current (dashed line) for symmetric (a, left upper sketch) and asymmetric (b, left lower sketch) coupling cases. Right parts (a,b) have been adapted with permission from Ref. Galperin et al. (2006).

On the other hand, numerical simulations of phonon-dependent shot noise (Avriller and Frederiksen, 2012) provided results which only agree with experiments in the ballistic regime when transmission exceeds 0.95. The sign change in the shot noise correction reported in Ref. Kumar et al. (2012) was not reproduced.

In the opposite (low transmission) limit the zero-frequency shot noise for phonon-mediated resonant tunneling with strong coupling to an internal phonons provides super-Poissonian current noise due to electron-phonon coupling induced selective unidirectional cascades of single-electron transitions (Shen et al., 2007).

Full counting statistics of vibrational assisted electronic conduction through the barrier consisting of donor (D) and acceptor (A) states coupled harmonic oscillator (HO) shows (Agarwalla et al., 2015) that in sequential tunneling regime, when vibrational states participate in the conductance, Fano is always super-Poissonian, reaching rather high values (see Fig. 10.8).

Interestingly, the phonon excitation related additional shot noise of nanoscale thermoelectric molecular junctions could affect their thermoelectric efficiency leading to stochasticity in the junction's performance (Esposito et al., 2015).

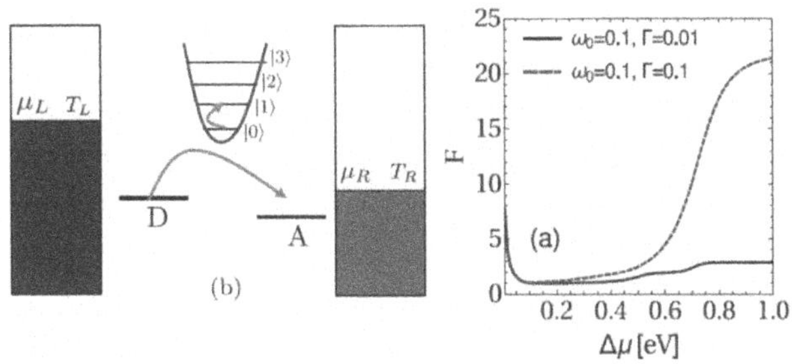

Figure 10.8 Left: Sketch of donor (D) to aceptor (A) harmonic vibron mode mediated electron transport (DA-HO model). Right: Bias-dependent Fano factors for DA-HO model. Adapted with permission from Ref. Agarwalla et al. (2015).

10.6 Electron Transport and Noise in Organic Spintronics

10.6.1 Introduction to Organic Spintronics

There exists a great controversy regarding the magnetoresistance effects observed in spin valve devices containing organic semiconductor layers (such as Alq_3) of thicknesses around 100–200 nm. Some authors associate the observed MR effects to spin injection and transport in the organic layer, while others propose that tunneling takes place through locally thin regions of the layer. The literature also suggests an additional point for disagreement, as studies present both positive (Liu et al., 2009) and negative (Dediu et al., 2002) MR ratios.

Several explanations for the physical origin of the MR sign in junctions with Alq_3 have been proposed. Barraud et al. (Barraud et al., 2010) studied nanoindentation-based LSMO/Alq_3/Co junctions. In this study, the authors aim to explain the fact that both negative (Dediu et al., 2008) and positive (Vinzelberg et al., 2008) MR is observed for large area junctions with thick organic spacers. The authors suggest that the formation of spin-hybridization-induced polarized states in the first molecular layer at the electrode interface completely changes the magnetoresistance of organic spin valves. A conclusion is reached that the sign of the MR at small measurement voltages is determined by the sign of the product of the polarization of the different FM/OSC interfaces, i.e., P(Alq_3/Co) and P(LSMO/Alq_3). Another interpretation of the origin of the MR devices with thick organic spacers is given by Wang et al. (Wang et al., 2012a). The authors study a semi-spin valve device with only one FM electrode with Alq_3 spacers. They find that the electrical conductivity of the organic film at room temperature is affected by the stray fields of the domains present in a magnetically unsaturated ferromagnet. The "fringe" fields from the ferromagnet provide a local magnetic field that changes the electronic transport properties through the hyperfine field, showing that the MR effects in the device originate from electrical transport through the inhomogeneous stray fields from the ferromagnet.

In this part we will compare the different statistics in the transport through thin perylene-3,4,9,10-tetracarboxylic dianhydride (PTCDA) layers in the tunneling regime (1.2–5 nm) or thick (200 nm) tris(8-hydroxyquinoline) aluminum (Alq_3) spacers in the sequential tunneling regime, with the hope of shedding light on the mechanisms which govern the transport through them. The results show qualitatively different electron tunneling statistics when the organic spacer in these devices is in the tunneling regime or well above it. The results regarding OMTJs with PTCDA spacers have been published in Ref. Cascales et al. (2014).

From the point of view of technological applications, the key limiting factor for the operation for any spintronic device is its signal to noise ratio. Therefore, the investigation of noise sources in organic tunnel junctions and spin valves is of fundamental and technological interest, as the noise ultimately determines their practical applications. As we discussed in Section 2, shot noise (SN) measurements have been used systematically to characterize the electronic transport mechanisms in inorganic spintronics (Almeida et al., 2008; Arakawa et al., 2011; Guerrero et al., 2006b, 2007; Nowak et al., 1998; Scola et al., 2007a; Sekiguchi et al., 2010; Tanaka et al., 2014). The precise knowledge of shot noise can provide valuable information on electron correlations near the interfaces with organic barriers, especially in the regime of direct tunneling. In fact, the role of interfaces remains one of the central issues in organic spintronics (Keevers et al., 2013).

The normalized shot noise or Fano factor in MTJs with inorganic barriers has been reported to be equal to unity (Guerrero et al., 2006b, 2007) for high-quality barriers (direct tunneling) or less than unity (Nowak et al., 1999) for non-homogeneous barriers (sequential tunneling). The Fano factor in the O-MTJs with PTCDA barriers varies between 1.5 and 2, which points to spin-dependent bunching taking place when electrons directly tunnel through the barrier. The main findings are explained in terms of a model which includes tunneling through a two-level (or multilevel) system, originated from interfacial bonds of the PTCDA molecules. The results suggest that interfaces play an important role in the control of shot noise when tunneling through organic barriers. The normalized shot noise values in the devices with thick Alq_3 layers is clearly suppressed below the Poissonian value, indicating that the tunneling processes are sequential, with the number of "jumps" depending on the temperature.

10.6.2 Organic Magnetic Tunnel Junctions in the Electron Tunneling Regime

10.6.2.1 Sample growth

The growth of the junctions described in this paragraph was carried out in the group of Minn-Tsong Lin (Li et al., 2011). The layer sequence in the PTCDA organic MTJs studied is

$$\text{NiFe(25 nm)/CoFe(15 nm)/Al}_x\text{O(0.6 nm)/PTCDA(1.2–5 nm)/}\\ \text{Al}_x\text{O(0.6 nm)/CoFe(30 nm).}$$

The structure was deposited onto a glass substrate, and the junctions were prepared in a high-vacuum environment with a base pressure lower than 10^{-8} mbar. The metallic layers were deposited by sputtering with an Ar working pressure of 5×10^{-3} mbar. Organic semiconductor (OSC) PTCDA layers were grown by thermal evaporation at 10^{-8} mbar, with a deposition rate of 0.1 nm/s. Thin AlO_x layers were grown between the PTCDA layer and both ferromagnetic layers by partially oxidizing Al in oxygen plasma for 5 s. Figure 10.9a presents a schematic view of the investigated O-MTJs. As can be seen, the ferromagnetic electrodes are deposited

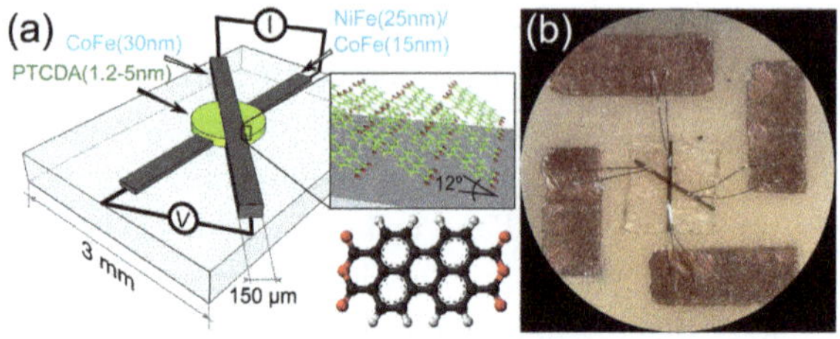

Figure 10.9 (a) Diagram of the sample structure and the PTCDA molecule. (b) View through the microscope of a sample on a glass substrate contacted to Cu pads.

as lines of width of 150 μm. The free layer is a NiFe/CoFe line, parallel to which the external magnetic field is applied. The pinned layer is made of CoFe, and it forms an angle of 45° with the free layer, which results in the pinned layer having a higher effective coercive field than the free layer. The glass substrate is then glued onto a circuit board and the electrodes of the sample are wire-bonded to millimeter sized Cu pads, as shown in Fig. 10.9b. The circuit board is mounted on the sample holder, and the Cu pads are connected to the measurement circuit with gold wires.

Previous studies on this kind of samples revealed a TMR of 12% at room temperature, and an exponential dependence of the resistance with the PTCDA film thickness (Li et al., 2011), indicative of quantum tunneling. The recent work by Hong et al. (2014) presents the characterization of the OSC/FM interfaces in these samples through XPS measurements. The junctions exhibit room temperature, MR ratios of around 12% when the partially oxidized alumina buffer layer is included between the OSC and the FM layers. The authors conclude that the PTCDA molecules lie essentially flat regardless of the layer they are deposited onto (AlO_x/Co or Co). A tilt angle of $12° \pm 3°$ (see Fig. 10.9a) is estimated for the average stacking orientation of the molecules with respect to the contact plane (either AlO_x or Co). The C 1s and O 1s core-level XPS spectra indicate that a hybridized layer is formed at the OSC/FM interface. The addition of a thin AlO_x buffer layer prevents the electronic interaction (hybridized layer) between the Co and PTCDA layers, preserving an effective spin injection into the organic PTCDA spacer thanks to which an MR response is observed.

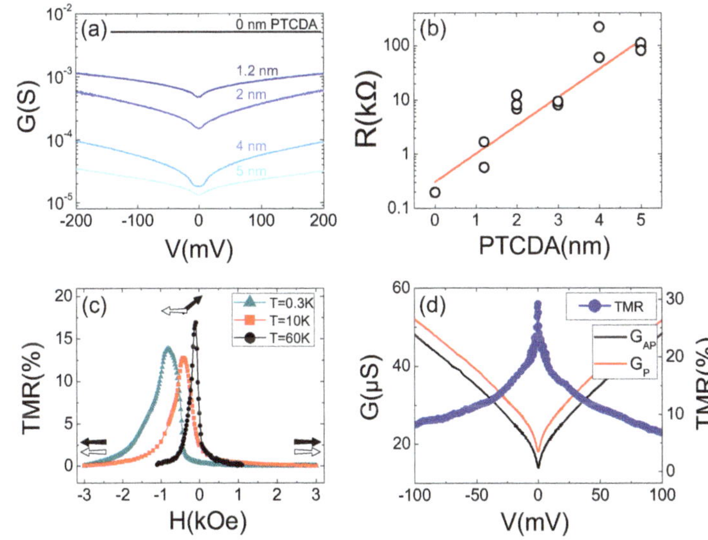

Figure 10.10 Magnetotransport characterization of MTJs with PTCDA barriers. (a) Bias dependence of the differential conductance of the studied O-MTJs as a function of the PTCDA thickness. (b) Experimental dependence of the resistance on the PTCDA thickness. (c) TMR curves at different temperatures for a sample with 2 nm of PTCDA. (d) Dependence of the TMR and differential conductance on the bias voltage in the P and AP states for a 4 nm PTCDA O-MTJ. Adapted with permission from Ref. Cascales et al. (2014).

The conductance in the O-MTJ samples decreases exponentially with the barrier thickness. As can be seen in Fig. 10.10(a), when only the AlO_x buffer layer is grown (no PTCDA) the behavior of the junction is metallic-like, since its resistance decreases when the temperature is lowered, while the opposite is true in general for junctions with PTCDA. Adding an AlO_x buffer layer improves the interface between the FM electrodes and the organic layer (Dediu et al., 2008). Figure 10.10b plots, for all the measured samples, the exponential increase of the resistance with the PTCDA thickness. This indicates that the insulating PTCDA layer indeed acts as a barrier in the single step tunneling regime (Schoonus et al., 2009). Figure 10.10c shows tunneling magnetoresistance (TMR) curves for three different temperatures of a 2 nm PTCDA O-MTJ, where the parallel (P) and antiparallel (AP) magnetic alignment of the

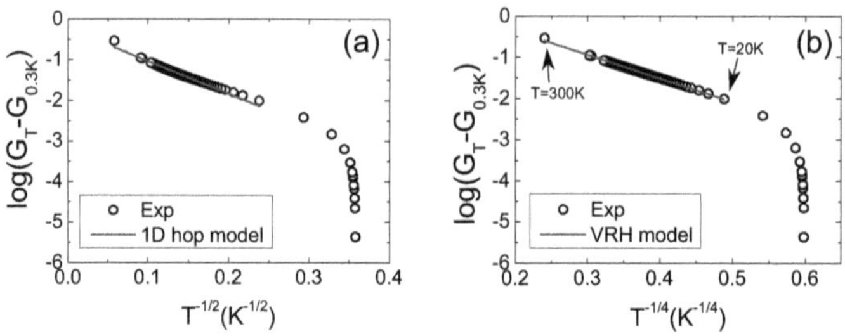

Figure 10.11 Comparison between the dependence of the resistance with the temperature in a 2 nm PTCDA O-MTJ and a (a) one-dimensional and a (b) variable range hopping model. Adapted from Ref. Cascales (2015).

electrodes are indicated by arrows. The TMR decreases with the applied bias, and is reduced considerably when the bias reaches 100 mV (see Fig. 10.10d). Figure 10.10d also shows the differential conductance in the P and AP states as a function of the bias voltage at $T = 0.3$ K for a 4 nm PTCDA O-MTJ.

To provide further proof that direct tunneling takes place, we have compared the temperature dependence of the resistance in the samples with one-dimensional ($G(T) \propto e^{-AT^{-1/2}}$) and variable range ($G(T) \propto e^{-AT^{-1/4}}$) hopping models (Mott, 1969), as is done in Refs. (Raman et al., 2009; Shim et al., 2008). A is a constant related to the bulk-related inverse localization radius. Figure 10.11 presents a fit of both models to the temperature dependence of the resistance of a 2 nm PTCDA O-MTJ. We have plotted $\log(G(T) - G(T = 0.3\text{ K}))$ (or $\log(\Delta G)$) as a function of $1/T^{1/2}$ or $1/T^{1/4}$. If the data obey any of the laws, the graphs should present linear dependence of $\log(\Delta G)$ for some interval of temperature. We indeed find a linear behavior in both Figs. 10.11a,b between room temperature and 20 K. As can be seen, the variable hopping model presents a better agreement with the results. Therefore, from the dependence of the conductance with the temperature we conclude that a three-dimensional hopping transport may take place at high temperatures for around 30% of the tunneling electrons. For temperatures below 20 K the transport turns into direct tunneling.

We found that the magnetic tunnel junctions with PTCDA barriers were more robust than conventional inorganic MTJs, and typically did not experience dielectric breakdown as readily. Out

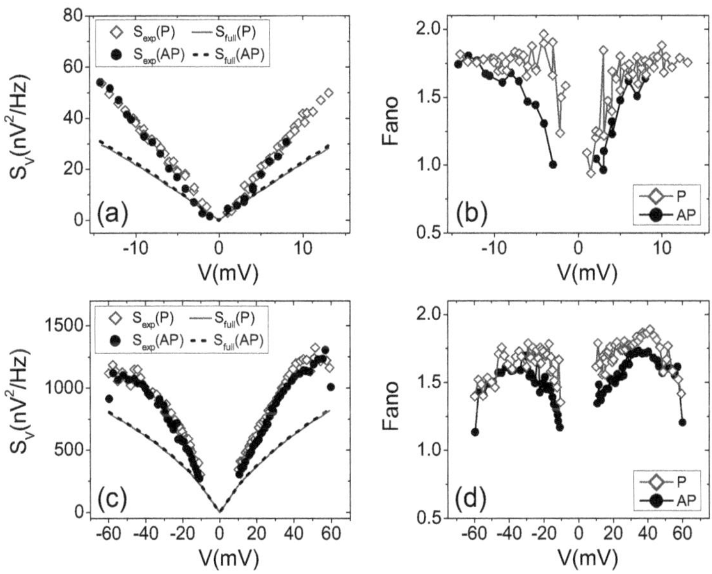

Figure 10.12 Voltage dependence of the shot noise at $T = 0.3$ K in the P and AP states of the (a) experimental (dots) and expected full shot noise (lines). (b) Fano factor for a 2 nm PTCDA O-MJT. (c) and (d) present similar results for a 5 nm PTCDA O-MJT. Adapted with permission from Ref. Cascales et al. (2014).

of 14 samples studied, only 3 have degraded during multiple bias sweeps up to 500 mV.

10.6.2.2 Super-Poissonian shot noise

The main experimental findings systematically show that the shot noise in O-MTJs is enhanced above the classic value ($F = 1$). The Fano factor was obtained as explained in Section 3.2.

Let us denote the Poissonian value of shot noise by S_{full} and the experimental shot noise obtained from the spectra by S_{exp}. The experimental SN and S_{full} and the Fano factor at $T = 0.3$ K for the 2 nm PTCDA junction from Fig. 10.10c are shown in Figs. 10.12a,b for the P and AP states. The dependence of S_{exp} and S_{full} (Fig. 10.12c) and the Fano factor (Fig. 10.12d) with the bias is also shown for another sample with a 5 nm-thick PTCDA barrier. As can be seen, the Fano factor ranges from $F = 1$ at low voltages to $F \simeq 2$ at higher

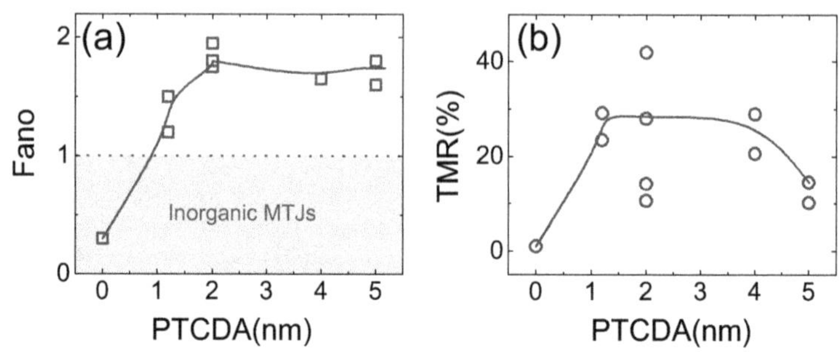

Figure 10.13 (a) Dependence of the maximum Fano factor in the P state on the PTCDA thickness at $T = 0.3$ K. The shadowed region corresponds to the range of Fano values for MTJs with inorganic barriers. (b) Zero bias TMR at vs. PTCDA thickness at $T = 0.3$ K. Adapted with permission from Ref. Cascales et al. (2014).

voltages (Cascales et al., 2014). All the O-MTJ samples measured displayed a qualitatively similar variation of the Fano factor with the bias voltage. The shot noise could be obtained for voltages up to a few tens of mV only. The maximum voltage for which the shot noise is measured corresponds to the energy at which the $1/f$ noise becomes dominant and obscures the frequency-independent part of the noise spectrum. Even though the spectra could be obtained up to 100 kHz, filtering due to the capacitance of the samples (dependent on the PTCDA thickness) allowed shot noise measurements only between 1–10 kHz. The appearance of $1/f$ noise restricted SN measurements in all the studied samples, especially in the AP state.

Figure 10.13a presents the average saturation value of the Fano factor in the P state of the samples for which frequency-independent spectra. Figure 10.13b also shows the variation of TMR with the PTCDA thickness. Control junctions, with only a 1.2 nm AlO_x layer and no PTCDA show TMR values below 1%, and a metallic-like electron transport. (Fig. 10.10a). This points to diffusive electron transport, for which the theory (Blanter and Büttiker, 2000a) predicts the Fano factor equal to 1/3, which agrees quite well with Fig. 10.13a. Therefore, the control measurements prove that the super-Poissonian SN observed in the samples is due to the PTCDA barriers. The O-MTJs with PTCDA between

1.2 and 5 nm show relatively high TMR and super-Poissonian tunneling statistics with the Fano factor approaching 2, which is indicative of co-tunneling or tunneling with bunching. Eight O-MTJs of different barrier thicknesses, from three sample sets, have shown qualitatively similar SN values (Fig. 10.13a).

10.6.2.3 Model and discussion of the results

A number of electron tunneling mechanisms (Kondo effect (Yamauchi et al., 2011a), co-tunneling (Okazaki et al., 2013; Onac et al., 2006), and others (Kießlich et al., 2007; Lambert et al., 2007)) are capable of producing super-Poissonian shot noise. However, they are mostly relevant for small quantum dots. The observed SN has been accounted for in terms of the approach developed by Belzig (Belzig, 2005), extended to include spin-dependent transport.

The following theoretical model has been developed in the groups of Jozef Barnaś and Vitalii Dugaev (Szczepański et al., 2016). It has been shown that the observed SN could be accounted for in terms of the approach developed by Belzig (2005). The corresponding model is based on tunneling through a two-level system (or multi-level system in a more general case), with remarkably different tunneling rates through the two levels. A diagram presenting the main features of the model is shown in Fig. 10.14. Physically, the origin of these two levels with different couplings could be bonding and antibonding interface states. Thus, one level is either below the Fermi level of the drain electrode or it is weakly coupled to the electrodes, while the second level is in the tunneling window (between the Fermi levels of the source and drain

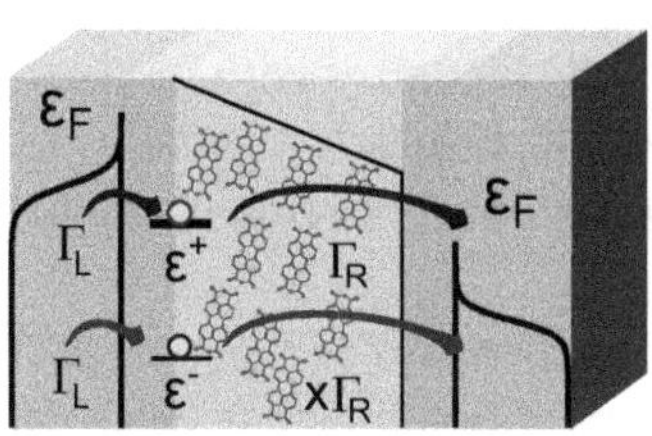

Figure 10.14 Sketch of the theoretical two-level model used to calculate shot noise. Electrons tunnel from one lead to another via two energy levels, leading to super-Poissonian shot noise.

electrodes) and is strongly coupled to the electrodes. Essentially, *the new feature of the model is the assumption that the coupling of the molecular level to the electrodes is spin dependent.*

To describe the transport through the two-level system, the authors use the relevant master equation (expressed in matrix form) to find the probability that the molecule is occupied the upper or lower energy states. The rates in the matrix are parameterized by the bare tunneling rates $\Gamma^{\pm}_{L\sigma}$ and $\Gamma^{\pm}_{R\sigma}$ to the upper (+) and lower (−) level from left (L) and right (R), which capture the basic features of the assumed model and level positions. The temperature-dependent factor x corresponds to the tunneling from the lower level to thermally activated holes, and we assume that $x \ll 1$. Following Refs. (Bagrets and Nazarov, 2003; Belzig, 2005), the noise and Fano factor are calculated using the full counting statistics method.

The following factors are introduced the factors: $\alpha^{\pm} = \Gamma^{\pm}_{R\uparrow}/\Gamma^{\pm}_{L\uparrow}$ to describe left-right asymmetry, $\beta_{L(R)} = \Gamma^{\pm}_{L(R)\uparrow}/\Gamma^{\pm}_{L(R)\downarrow}$ to describe spin asymmetry on the left and right sides (note this spin asymmetry is equal for both levels), and $x_{L(R)} = \Gamma^{+}_{L(R)\uparrow}/\Gamma^{-}_{L(R)\uparrow}$ to account for asymmetry in the bare tunneling to the two levels. In the limit of independent of spin tunneling rates, $\Gamma^{\pm}_{L\uparrow} = \Gamma^{\pm}_{L\downarrow} = \Gamma_{\mathrm{L}}$ and $\Gamma^{\pm}_{R\uparrow} = \Gamma^{\pm}_{R\downarrow} = \Gamma_{\mathrm{R}}$, one obtains the result of Belzig, $F = 3$. In a general case, the Fano factor depends on the parameters introduced above. The dependence of F on α^{+} for P and AP configurations is shown in Figs. 10.15a,b, respectively, for $\beta_{\mathrm{R}} = 4$ and indicated values of β_{L}. Note, the Fano factor is super-Poissonian, but it can vary in a relatively broad range, roughly from $F \approx 1.5$ to $F \approx 3.5$ for the assumed parameters. The dependence of the Fano factor on x_{R} is presented in Figs. 10.15c,d for the P and AP states, for $x_{\mathrm{L}} = 1$, $\beta_{\mathrm{R}} = 4$ and β_{L}. The Fano factor strongly increases with increasing x_{R} and decreases when x_{R} becomes small, although it remains larger than 1. This is because at small values of x_{R}, the role of the parameter x becomes reduced.

The generating function $S(\chi)$ (a sort of partition function) can be presented as a sum of independent Poissonian processes transferring *ne* charge (Belzig, 2005). This means that the occurrence of a process with large n during one cycle is possible because tunneling to the left lead from the lower level is strongly suppressed by the

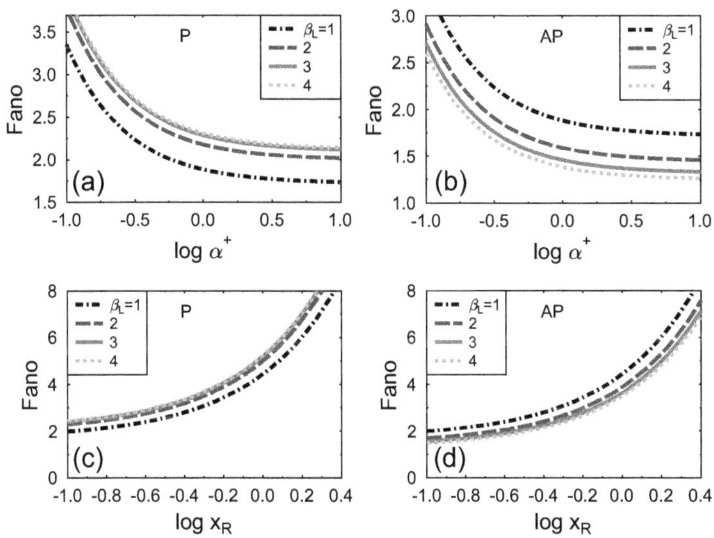

Figure 10.15 Fano factor in the P (a) and AP (b) configurations as a function of α^+ for $x_R = 0.3$, $\alpha^- = 0.2$, $\beta_R = 4$, and indicated values of β_L. The Fano factor in the P (c) and AP (d) configuration as a function of the parameter x_R for $\alpha^+ = \alpha^- = 1$, $x_L = 1$, $\beta_R = 4$, and indicated values of β_L. Detailed theoretical analysis of the model is provided in Ref. Szczepański et al. (2016).

temperature factor (small x) or small coupling of this level to the leads. *In other words, several electrons can be quickly transferred through the upper level until the cycle is stopped by an electron at the lower level. This leads to a super-Poissonian process, and to a Fano larger than 1.* The description can be generalized to a multilevel system.

For appropriate parameters one can find a reasonable agreement with the experiment (Figs. 10.16b,c). The fit of the experimental data to the theory is done in the following way. With the experimental Fano factor in the P and AP states (obtained when the Fano factors in Fig. 10.12 saturate), and the TMR ratio of each tunnel junction, we have three conditions and three unknown parameters. We have fixed x_R, which is related to the parameter x, at $x_R = 0.3$ which seems to give the best agreement between theory and experiment. Also, we assume the spin asymmetry is the same for tunneling right and left,

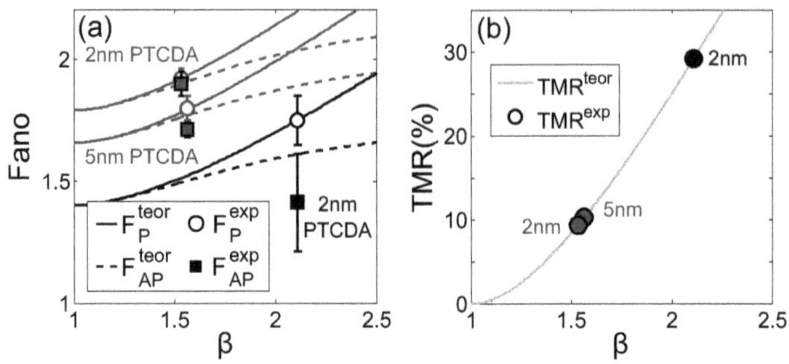

Figure 10.16 Fit of the theory to experimental values of the (a) Fano factor and (b) the TMR for the samples with 2 and 5 nm of PTCDA. The Fano points are obtained from the average saturation value in the Fano factor vs. bias plots, and the error bars indicate the dispersion of the average. Adapted with permission from Ref. Cascales et al. (2014).

we set $\beta_L = \beta_R = \beta$. Thus, we have three parameters to determine from a fit between experiment and theory: α^+, α^- and β.

The parameters chosen are those that make all three functions F(P), F(AP) and TMR the closest to the experimental value. The graphs in Fig. 10.16 show the experimental points of F(P), F(AP) and TMR compared to the theoretical curve. For simplicity, all parameters except β were fixed in order to show a line plot and not a more complicated graph (surface plot, etc.). The obtained parameters are presented in the following table.

Table 10.1 Fitting parameters obtained for the samples in Fig. 2a–c respectively.

t (nm)	F_P	F_{AP}	TMR (%)	α^+	α^-	β
2	1.75 ± 0.1	1.41 ± 0.2	29.1	0.49	0.33	2.11
2	1.92 ± 0.04	1.9 ± 0.05	9.3	0.77	1.01	1.53
5	1.8 ± 0.05	1.71 ± 0.03	10.2	0.92	1.01	1.57

Since the tunneling probabilities are different for spin up and down electrons, the factor F also depends on the magnetic (P and AP) configuration, which agrees with the experimental Fano and TMR values. The most important property of the model (Fig. 10.14) is a strong asymmetry in tunneling rates through the two levels, leading to super-Poissonian noise. Note, for each sample the TMR and Fano factors have been fitted with the same parameter β, which justifies validity of the approach.

Therefore, the model essentially accounts, qualitatively, for the experimental data (Figs. 10.12 and 10.13). In real junctions, we have a large number of molecular chains, which slightly differ from one another. Moreover, they interact with the thin AlO_x buffer layers. All this can lead to some splitting of the levels and may result in tunneling through a quasi-continuum band (multilevel system).

Physically, the two or more levels with different couplings, which are responsible for the observed super-Poissonian shot noise, can have its origin in localized states arising from interfacial bonds between the PTCDA molecules and the AlO_x buffer layers. The following arguments suggest that the localized states in the model have an interfacial nature: (i) the exponential dependence of the tunneling resistance on PTCDA thickness (Fig. 10.10a) including the metallic character of the conductance when only the AlO_x buffer layer is present (see Fig. 10.10a); (ii) a lateral size of the junctions larger than a micron, for which the influence of Coulomb blockade is minimized. The physical origin of the interfacial states could be a charge neutrality level (Vázquez et al., 2004), or gap states (Yogev et al., 2013), which appear due to the alignment of the energy levels at metal/organic interfaces (Braun et al., 2009). The bias dependence of the interfacial density of states could explain the suppression of the Fano factor at large voltages (Fig. 10.12d).

A qualitatively similar to above model which considers correlated tunneling through two-level system (provided by magnetic molecule) has been considered theoretically (Xue et al., 2015). By using third electrode attached to the molecule, both conductance and shot noise could be tuned by a gate voltage. the obtained negative differential conductance and super-Poissonian shot noise depend on sensitively on the spin polarization of the source and drain electrodes as well as on the gate voltage. Dependence on the relative alignment of the electrodes has not been considered.

Finally, we mention also some other observations on the enhanced shot noise due to tunneling through localized states. Among them is report (Garzon et al., 2007) on observation the enhanced spin-dependent shot noise in magnetic tunnel barriers, suggesting transport through localized states within the barrier. Safonov et al. (2003) also discussed enhanced shot noise in non-magnetic tunnel barriers due to two interacting localized states and presented a classical model that can be used to calculate the Fano factor for arbitrary tunnel barriers.

10.6.3 Sequential Tunneling through Thick Organic Layers Organic Spintronics: Conductance and Noise

As we discussed earlier, there are a number of open questions concerning the MR responses detected in spin valve devices with organic semiconductor spacers of thicknesses well above the tunneling regime. In the previous section we concluded (thanks to shot noise measurements) that direct tunneling can take place through thin (1.2–5 nm) organic tunnel barriers, which influence the tunneling statistics so these O-MTJs present super-Poissonian shot noise.

We will now go above the tunneling regime by considering the shot noise of organic spin valve devices with 200 nm thick, Alq_3 organic spacers. The study of the tunneling statistics in these devices, through shot noise measurements, could to resolve some of the standing questions about electronic transport mechanisms which take place in thick organic layers.

10.6.3.1 Growth and sample characteristics

The growth of the junctions with thick Alq_3 organic spacers (for which conductance and noise measurements (Martinez et al., 2016) will be described below) were carried out by the group of Dediu (Dediu et al., 2008, 2002).

The layer sequence of the organic spin valves with Alq_3 is

$$La_{0.7}Sr_{0.3}MnO_3(20\ \text{nm})/Alq_3(200\ \text{nm})/Al_2O_3(2\ \text{nm})/Co(20\ \text{nm}).$$

The junctions with Alq_3 barriers were grown by the following procedure (Prezioso et al., 2011). The bottom ferromagnetic electrode of the devices is a 20 nm thick, $1 \times 5\ \text{mm}^2$ strip of

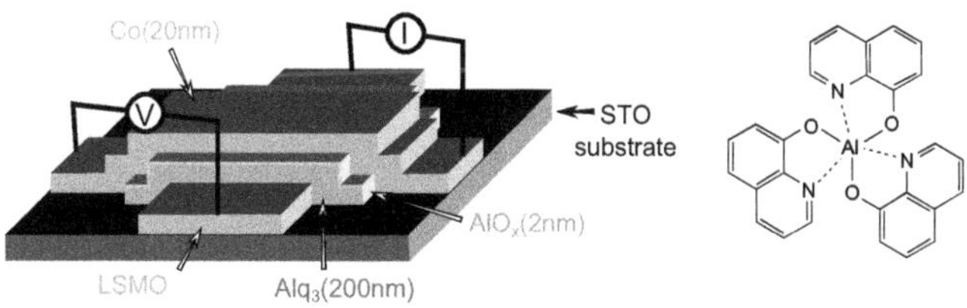

Figure 10.17 Diagram of the layered structure of an Alq_3 device and the Alq_3 molecule.

$La_{0.7}Sr_{0.3}MnO_3$ (LSMO). This film is grown on a matching $SrTiO_3$ (STO) 5×10 mm^2 substrate grown by pulsed plasma deposition (PPD) in a 4×10^{-2} mbar oxygen atmosphere, with the substrate kept at 880°C. The sample is then exposed to air and moved to the organic semiconductor (OSC) deposition chamber (base pressure $2 \cdot 10^{-8}$ mbar). Before depositing the OSC layer, the is heated to 250°C for 20 min to recover its surface properties. Afterwards, an Alq_3 layer 200 nm thick is evaporated on top of the LSMO at a rate of 0.05 Å/s at room temperature. The sample is exposed to air yet again and taken back to the PPD chamber where a 2 nm-thick AlO_x tunnel barrier is deposited at an oxygen pressure of 2.5×10^{-2}. Lastly, the sample is exposed to air and moved to a metal deposition chamber, where a 20 nm-thick Co strip is evaporated with an electron gun at a 5×10^{-8} mbar base pressure.

A previous works studied MR effects in devices with 250 nm-thick Alq_3 layers, at temperatures between 100 K and 300 K (Prezioso et al., 2011). The best result yielded an MR of 22% at −0.1 V and 100 K (see Fig. 10.18a). The authors (Prezioso et al., 2011) also found that the magnetoresistance quickly decreased when the voltage was increased up to 1 V. What is perhaps more surprising is the discovery of an electrical hysteresis cycle of the resistance of the sample (see Fig. 10.18b). The authors (Prezioso et al., 2011) find that they can switch the device to a low or high resistance state increasing the voltage to a maximum positive or negative value. The MR response in the high resistance state is found to disappear.

Conductance measurements (Martinez et al., 2016) reveal MR during first measurements at $T = 92$ K as shown in Fig. 10.19a.

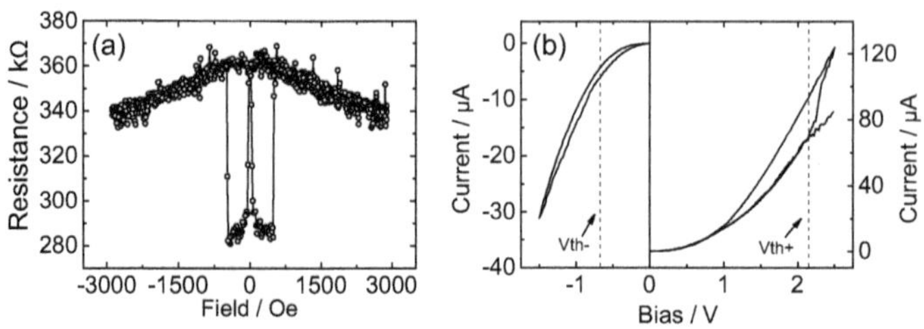

Figure 10.18 (a) Magnetoresistance curve of an organic spin valve with a 250 nm-thick Alq_3 spacer. (b) I–V curve at 100 K showing the electrical hysteresis of the same device as in (a). Adapted with permission from Ref. Prezioso et al. (2011).

The curve does not present the clear transitions from earlier results, and it was found that the sample was in an intermediate (between low and high) resistance state, since high resistance proved to be around 30 kΩ. Figure 10.19b plots the first electrical hysteresis cycle tried on the sample in (a). As can be seen, the sample resistance of the sample is changed from low resistance (lower branch) to a high resistance (upper branch).

In the process of extracting shot noise results (Martinez et al., 2016), the device was cooled down to 4 K to eliminate the influence of charge traps (appearing as random telegraph fluctuations). It

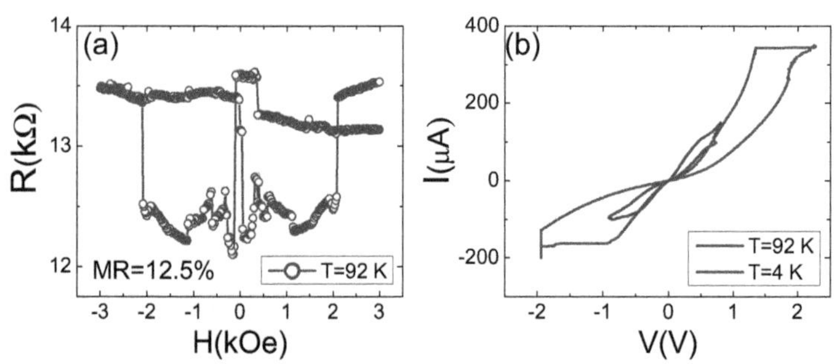

Figure 10.19 (a) Magnetoresistance curve of an organic spin valve with a 200 nm-thick Alq_3 spacer. (b) I–V curve at 92 K and 4 K showing the electrical hysteresis of the device in (a). The switch becomes increasingly difficult as the temperature is lowered below 10 K. Adapted with permission from Martinez et al. (2016).

was found that the critical current values needed for switching the resistance of the spin valves became increasingly higher as the temperature dropped (see Fig. 10.19b). The MR response was also lost and not recovered afterwards, although the resistance of the sample did not degrade. Losing the MR effect at low temperatures supports the MR origin presented by Wang et al. (2012a) since: (i) the resistance of the sample did not degrade, hence the loss of the MR response cannot be attributed to dielectrical breakdown; (ii) decreasing the temperature makes the ferromagnetic electrode magnetically harder, changing the size and coercive field of the magnetic domains.

10.6.3.2 Suppressed shot noise in the variable-range hopping regime

Let us now discuss the shot noise results obtained for the Alq_3-based spin valves (Martinez et al., 2016). We have found that these samples with thick molecular barriers present a different type of electronic transport, as is apparent from the analysis of the Fano factor and conductance at different temperatures.

Figure10.20a presents the main observations regarding shot noise in these organic spin valve devices. As can be seen, the normalized shot noise or Fano factor is sub-Poissonian and independent of the voltage, but is heavily influenced by the temperature. A lower temperature yields a higher Fano factor, which could mean that the transport is dominated by variable range hopping. The fact that the shot noise is sub-Poissonian shows that multi-step tunneling takes place and the strong influence of the temperature agrees with the variable range hopping hypothesis, since an increase in Fano when the temperature is lowered points to an increase in the characteristic hopping length. Indeed, we show that in Fig. 10.20b the dependence of the conductance (minus $\sigma_{T\to 0}$) with $1/\sqrt{T}$ of a sample with a 200 nm-thick Alq_3 barrier, obeys the variable range hopping law $\sigma = \sigma_0 \exp(-(T_0/T)^{\frac{1}{d+1}})$ for $d = 1$ (one dimension). A diagram illustrating the temperature-dependent, multistep hopping is shown in Fig. 10.20c.

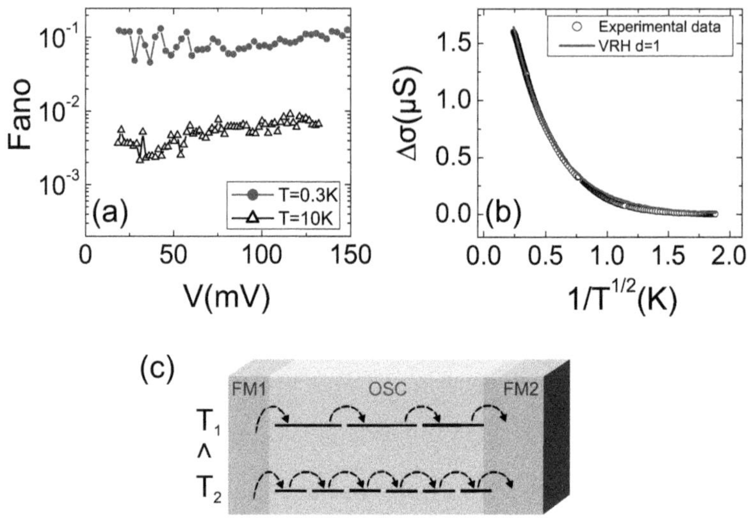

Figure 10.20 Shot noise and conductance in variable hopping regime. (a) The Fano factor is independent of the voltage but it decreases when the temperature is lowered. The noise is always sub-Poissonian ($F < 1$). (b) Dependence of the conductance with $1/\sqrt{T}$ of a sample with a 200 nm-thick Alq_3 barrier, fit by a variable range hopping law for $d = 1$. (c) Schematic representation of temperature-dependent, multistep hopping through the organic barrier. Adapted with permission from Martinez et al. (2016).

Therefore, one may conclude that the temperature dependence of the shot noise and conductance in O-MTJs with thick Alq_3 barriers (Fig.10.20) has signatures of Mott variable range hopping (VRH) (Jonscher, 2004). It has been previously found that sequential (multistep) tunneling decreases the shot noise below Poissonian values (Kuznetsov et al., 2000). When an electron conduction through the barrier occurs by a VRH process, it has been predicted that the expected Fano $F = L_0/L$ is not universal, with L_0 being some characteristic scaling length and L the length of the sample (which is the barrier thickness here) (Kuznetsov et al., 2000). In the random hopping network the characteristic length L_0 represents a typical distance between the most resistive hops. If we consider L_0 to be hopping length (Kuznetsov et al., 2000) $L_0 \approx r_c \sim (\frac{1}{T})^{\frac{1}{1+d}}$, with d being the dimensionality. The Fano factor increases about 7–10 fold when the temperature decreases from 10 K to 0.3 K (Fig.10.20a), which points to a quasi one-dimensional VRH transport through the Alq_3 barrier at low temperatures.

10.7 Possibility of Phonon Noise Spectroscopy in Organic Spintronic Devices

While the interplay between the electronic states and the mechanical motion (electron shuttle) and between spin and electron transport (spintronics) are well established, the only physical mechanism which yet remains obscured within electron-phonon-spin triangle is a possible interplay between spin and phonons.

In that direction, it would be of great importance to complement previously explored conductance spectroscopy with monitoring also voltage fluctuations as a function of applied bias, magnetic state and temperature in order to detect influence of molecular phonons excitation on the LFN. Such experiments are preferably to be carried out on macroscopic O-MTJs where efficient heat transport away from the molecules is guaranteed and magnetic state of electrodes could be well controlled. Following predictions (Brumme et al., 2012a) one could expect resonant enhancement of the low-frequency noise at specific biases to occur due to bending mode-related vibrational heating (if energy-dependent density of states in the proximity of Fermi energy exists). Separate experiments on the possible variation of the noise with magnetic state could help to resolve possible presence of recently predicted vibron-assisted spin relaxation at a metal/organic interfaces (Droghetti et al., 2015). Alternatively, phonon modes and noise could be influenced also by magnetostriction of the ferromagnetic electrodes.

Chapter 11

Electron Transport and Noise in Small Dots Connected to Ferromagnetic Leads

When a small object (further quantum dot or QD) is connected through weakly transparent barriers to macroscopic ferromagnetic leads, electron interaction or confinement effects emerge. The spin dependent transport and noise then become qualitatively different from those discussed for the double-barrier magnetic tunnel junctions. The experimental fabrication of such a device is a great challenge because the need of contacting small (often nm size) objects to the extended ferromagnetic electrodes. Evidently, the nm size ferromagnetic leads terminations would most probably be paramagnetic at room temperature. Studies of electron transport and noise in QD-based spintronic devices are currently mostly of academic character.

In general, depending on the QD dimensions, electron levels in QD could have quasi-continuum (larger dots) or discrete (small dots) spectrum. However, as we shall see, depending on QD capacitance, even in quasi-continuum spectra case, a single-electron spin could be confined at discrete levels originated from classic (electron-electron) interaction on the dot. Quantum size effects in FM/QD/FM devices with discrete electron number will then

Noise in Spintronics: From Understanding to Manipulation
Farkhad G. Aliev and Juan Pedro Cascales

ISBN 978-981-4774-34-5 (Hardcover), 978-1-315-11088-2 (eBook)
www.panstanford.com

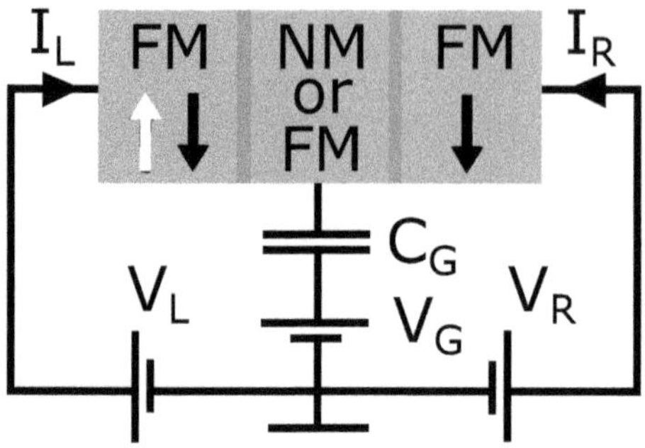

Figure 11.1 Sketch of a small (quantum) dot connected to FM leads.

represent an ultimate degree of miniaturization in semiconductor (or in some cases molecules)-based spintronics (see Fig. 11.1).

The chapter is organized as follows. We start with analyzing electron transport and noise in FM/QD/FM structures in the Coulomb blockade regime as long as they have been in the focus of the research so far. We will discuss spin accumulation and spin flip-related contribution to the shot noise. Then we shall consider influence of Kondo effect on electron transport and shot noise in quantum dots. Finally, we shall introduce shot noise in FM/QD/FM devices in the presence of co-tunneling.

In order to simplify the comparison with (so far scare) experimental data, we skip the consideration of the $1/f$ noise contribution and mainly analyze shot noise (typically measured for the frequency range above few MHz). We start with a brief introduction to the main physical mechanisms responsible for electron transport through quantum dots.

11.1 Brief Introduction in Coulomb Blockade

We first briefly review Coulomb blockade (CB) phenomena in QDs connected to normal metal leads. We already mentioned that CB dominates electron transport in relatively large dots when one may omit electron confinement effect. If electron charge Q is transferred from the source to the drain, the change of the energy of the

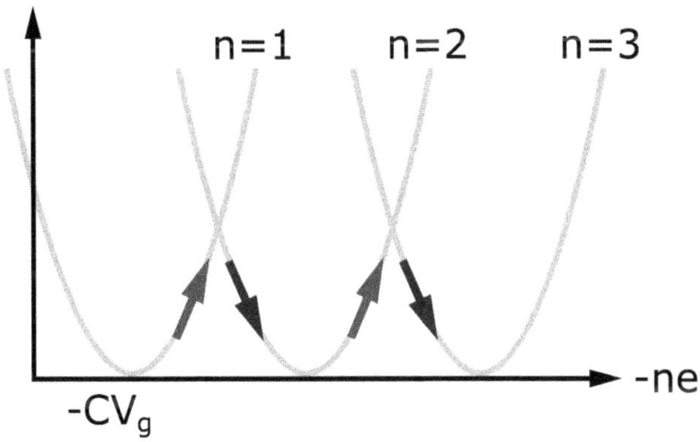

Figure 11.2 Illustration of the Coulomb staircase in a quantum dot.

system is

$$\Delta E = QV_{\text{g}} + Q^2/2C, \tag{11.1}$$

being a sum of the energies needed to overcome gate potential V_{g} and energy cost to charge the QD (C is capacitance of the QD).

If the transferred charge is quantified in a single-electron charge, i.e., has units (ne), then the energy dependence on the charge for different n electron occupation of the QD leads to overlap of parabolic dependences, which is sometimes called *Coulomb staircase* (see Fig. 11.2).

The difference in energy of the occupation by consecutive electron charges is then

$$\Delta E(n+1) - \Delta E(n) = -eV_{\text{g}} + (n - 1/2)e^2/C. \tag{11.2}$$

In the condition of $\Delta E(n+1) - \Delta E(n) = 0$ or in other words, when the applied gate voltage is $V_{\text{g}} = (n-1/2)e/C$, one could expect extra electron charge added/removed from QD due to resonant transmission through QD levels. This leads to conductance peaks in otherwise blocked for electron transport QD.

We note that the above discussed CB is entirely classic effect *visible* at temperatures $k_{\text{B}}T < Q^2/2C$ only. As long as physical origin of electron levels in the QD could also be quantum, the total QD energy has to be considered in the general case, summing the Coulomb interaction and quantum electron confinement energies.

11.2 Theoretical Considerations of Shot Noise in FM/QD/FM Structures

Depending on the transparency of the barriers connecting QD to FM leads, there could be two possibilities:

(a) transport in the Kondo regime
(b) single-electron transport regime with (periodic in gate voltage) CB

The Kondo effect appears only for sufficiently transparent barriers facilitating a singlet state formation due to large enough coupling between electrons on the QD and those in the ferromagnetic contacts.

We start considering the case which calculates numerically transport through non-magnetic QD connected to FM leads. The first theoretical study of shot noise in spintronic devices (Bułka et al., 1999) calculated the influence of spin accumulations on shot noise FM/QD/FM system in sequential tunneling limit (with co-tunneling or Kondo effects omitted) with both charge and spin fluctuations possible. It was predicted that besides the charge, also spin fluctuations, especially relevant in the low-frequency limit, could also have significant effect on the shot noise. This study underlines the specifics of the FM/QD/FM systems in comparison with NM/QD/NM structures where spin fluctuations have practically no influence on the shot noise.

Frequency dependence of the shot noise is determined by (i) spin fluctuation and (ii) charge fluctuations times. It was found that spin accumulations could induce negative differential resistance (NDR) regimes on the IV curves, providing super-Poissonian shot noise, similarly to known to happen in the resonant tunneling diodes with NDR (Bułka et al., 1999).

Shot noise in FM/QD/FM has been obtained analytically (Bułka, 2000) in the extreme regime when spacer particle (QD) is small enough to have only one electron level available for tunneling process. This simplification allows getting a better insight into spin-dependent tunneling and solving the problem analytically. New effects related with Coulomb interactions between electrons with different spins have been predicted. More specifically, depending on the QD parameters, due to the CB effect, the device could operate as a diode or resonant diode. The noise power spectrum shows a continuous transition from sub-Poissonian to super-Poissonian regimes controlled by magnetic asymmetry between spin up and spin down channels (Bułka, 2000).

Electron transport and noise in magnetic structures in the presence of the Coulomb interaction has been reviewed by Martinek et al. (2003). Electron transport in FM/QD/FM structures has been described in a 2D space of charge and spin states by calculating spin and charge accumulation and fluctuations, along with the TMR. For a broad range of applied biases, in the regime when current periodically varies with the applied bias, due to spin accumulation the current noise has been found to be reduced in the AP state in comparison with the P configuration. In the QB regime sequential tunneling is suppressed. However, close to resonances in the conductance, sequential and co tunneling (see also below) processes could have similar probabilities. In these conditions, close to resonances, the shot noise could become super-Poissonian, in agreement with previous calculations (Bułka, 2000; Bułka et al., 1999)

In the conditions when electron states in QD are strongly coupled to electrodes, spin flips could dominate spin dependent transport and noise in FM/QD/FM devices (Brito et al., 2001; Mishchenko, 2003). In the absence of external magnetic field or some magnetic inclusions providing exchange spitting in QD, spin up/down channels are degenerate.

Within this approach, Brito et al. (2001) treat tunneling in a double-barrier system from fully spin-polarized emitter (left electrode) through QD with broken spin up/down degeneracy and to the normal metal collector (right electrode, Fig. 11.3). In the limiting range of parameters (close to symmetric for the injected spin-polarized current barrier) spin flips have been found to suppress Fano factor down to 0.5 (see Fig. 11.3). Such result logically describes situation when spin flips transform resonant ($F = 1$) tunneling into sequential and respectively suppress the shot noise.

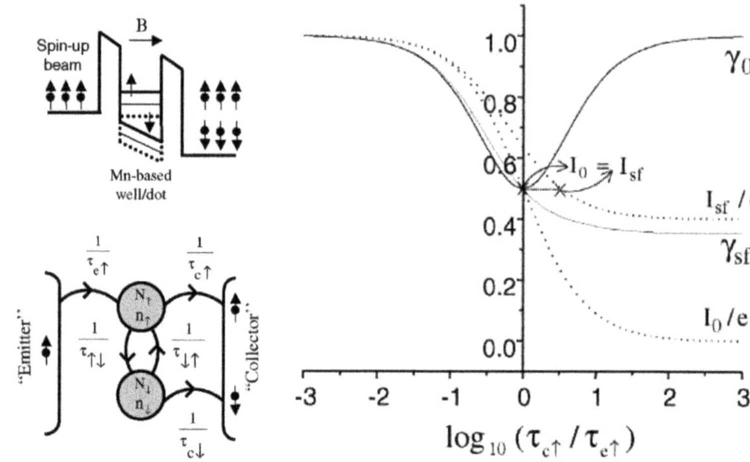

Figure 11.3 Model and results on influence of spin flips on SN in FM/QD/FM system. Left parts present sketch of the double-barrier tunneling system and the corresponding sequential model. The right graph shows calculated Fano factor (solid lines) and average current (dashed lines) for different spin flip times. Adapted with permission from Ref. Brito et al. (2001).

11.3 Electron Transport and Noise in FM/QD/FM in the Presence of Kondo Effect

Coulomb Blockade regime is characterized by weak coupling of the localized on QD electrons to the leads, which leads to finite conductance only for specific biases when an extra electron "*pushes put*" the localized one (at $T \to 0$). Situation dramatically changes with increasing of coupling (overlap) between localized on QD and extended in the leads states. In the strong coupling limit qualitatively new ground state emerges at sufficiently low temperatures which could be described in terms of Kondo effect for which electron spin of the extended (leads) states screens localized electron spin of the dot (Fig. 11.4). This happens with odd number of electrons on the dot only, resulting in formation of a Kondo resonance in the density of states close to the Fermi level (Martinek et al., 2003). Kondo effect in QDs has been observed experimentally through

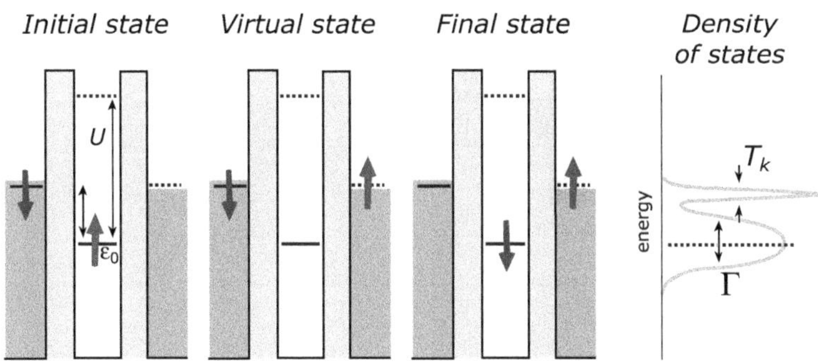

Figure 11.4 Sketch explaining Kondo screening and spin flip-related transmission through QDs. The right part describes schematically resulting density of states with narrow (with width of the order of Kondo temperature T_k) peak emerging al low temperatures close to Fermi energy (so-called Kondo or Abrikosov-Suhl resonance).

logarithmic increase of the conductance below characteristic Kondo temperature and splitting of zero bias anomaly at sufficiently high (exceeding thermal energy) magnetic fields (Goldhaber-Gordon, 1999).

In the absence of ferromagnetic leads shot noise for QD in Kondo regime has been investigated experimentally by Yamauchi et al. (Yamauchi et al., 2011a). They observed gradual increase of shot noise below Kondo temperature (see Fig. 11.5) with Fano factor approaching values $F = 2$. This behavior signals the presence of two-particle scattering process as Kondo effect evolves.

As we shall see below, the Kondo effect with ferromagnetic leads presents a few unique and non-trivial features, not existing in NM/QD/NM devices. More specifically, the zero bias anomaly becomes split in the P state, while the logarithmic enhancement of the conductance (sign of Kondo effect) is observed for the AP alignment only (Martinek et al., 2003). Tunneling magnetoresistance for FM/QD/FM devices in the Kondo regime also shows some peculiar behavior which we discuss below.

Indeed, as long as the Kondo effect is accompanied by spin flips, higher conductance could be expected in the AP configuration and resonant Kondo transmission should be blocked in the P state

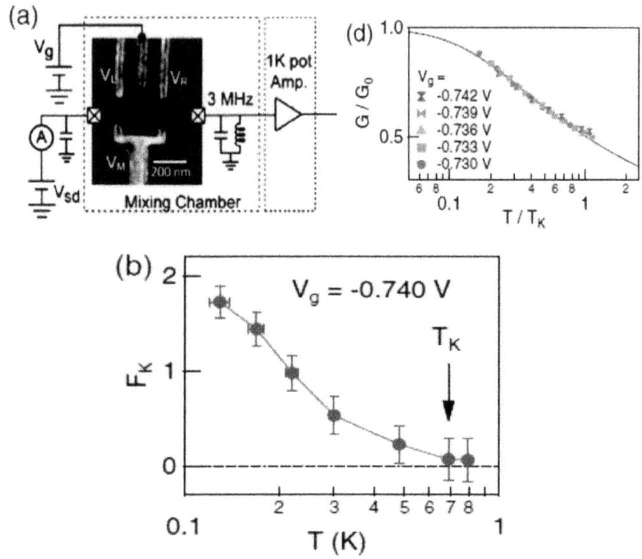

Figure 11.5 Shot noise measurement in quantum dot. Upper left part shows experimental configuration. The upper right part presents temperature dependent conductance measured at different applied biases. Close to logarithmic conductance increase indicates presence of Kondo-type correlations. The lower part shows experimentally measured shot noise as a function of temperature. Adapted with permission from Ref. Yamauchi et al. (2011b).

(Martinek et al., 2003). This leads to the negative TMR values at low biases, contrary to what is typically observed in spintronic devices. On the other hand, for the applied biases exceeding characteristic Kondo temperatures (i.e., $V > k_B T_K$) normal sequential tunneling through QD is restored. In these conditions FM/QD/FM loses coherency of the spin transmission and becomes electronically similar to DMTJs with strong spin relaxation and correspondingly a positive TMR (see Fig. 11.6).

Experimental confirmation of the above-discussed theoretical predictions for the electron and spin transport through QD connected to ferromagnetic leads in the presence of Kondo effect has been realized by Pasupathy et al. (Pasupathy et al., 2004) for $Ni_1/C_{60}/Ni_2$ system.

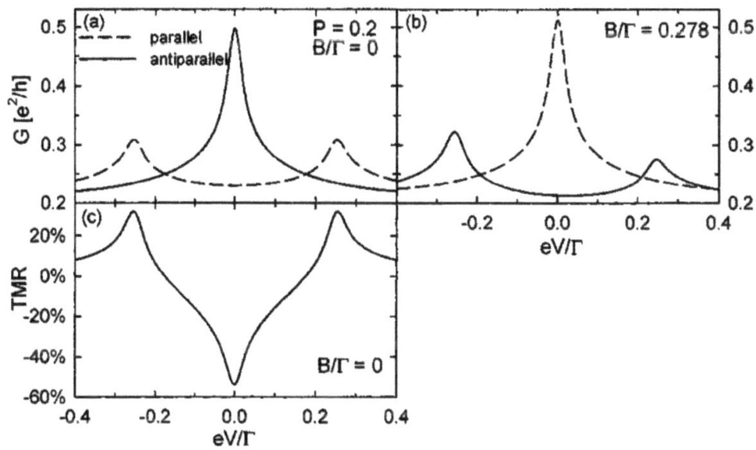

Figure 11.6 Influence of Kondo effects on conductance as a function of magnetic state (upper two graphs with data for different couplings) and TMR (lower graph) on FM1 QD FM2 device. Adapted with permission from Ref. Martinek et al. (2003).

11.4 Shot Noise in FM/QD/FM in the Presence of Kondo Effect

Below we concentrate only on physical models describing different aspects of shot noise in FM/QD/FM in the presence of Kondo effect because there have been no related experimental studies yet.

Various aspects of the electron transport and shot noise in QD coupled to FM leads in the Kondo regime have been considered by López and Sánchez (2003). First of all, Kondo temperature is suppressed with increasing magnetization of ferromagnetic contacts in parallel alignment. The Kondo state, however, is found to be remarkably stable to the leads magnetization values in the antiparallel configuration providing, in agreement with Martinek et al. (2003), a negative zero bias TMR.

In the absence of spin flips shot noise is supressed in the P state in the small bias limit due to resonant transmission through unsplit Kondo resonance (see Fig. 11.7a). As long as spin flips split the Kondo resonance and/or bias voltage exceeds Kondo

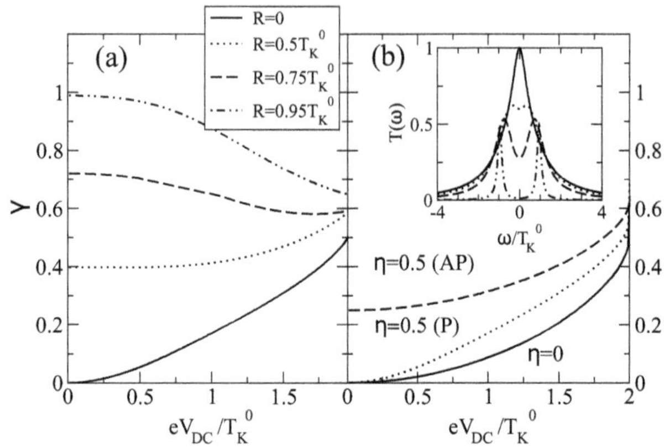

Figure 11.7 Shot noise for Kondo Effect in FM/QD/FM device. Part (a) Fano factor vs. bias voltage for different reflections R for the parallel alignment with symmetric barriers. Part (b) shows voltage dependence of the shot noise for the unpolarized, parallel and antiparallel case at $R = 0$. Inset shows energy dependent transmission probability for the same values of R as in part (a). Adapted with permission from Ref. López and Sánchez (2003).

temperature (i.e., $eV_{DC} > k_B T_k$), the Fano factor trends to specific for sequential tunneling through symmetric barriers value $F = 1/2$. For the maximum spin flip rate, the Fano factor in the P state tends to 1. This is because over one of two barriers becomes shorted by the spin-up channel while the second barrier is blocked. Such situation corresponds to electron transport through system with effectively single barrier, characterized by a Poissonian shot noise.

Finally, let us mention that in the absence of Kondo effect the physics of electron and spin transport becomes more trivial: spin-flips just make the P and the AP configurations to be degenerate reducing in this way effect effective spin polarization of the ferromagnetic leads and correspondingly influence of their alignment on the spin transport (De Souza et al., 2002). Introduction of spin flips reduces TMR and increase both AP-current and AP-noise weakly affecting both the P-current and P-noise (De Souza et al., 2002).

11.5 Shot Noise in FM/QD/FM in the Presence of Co-Tunneling

We have already pointed out that CB with ($U \to \infty$) blocks single-electron transport through the QD. In these conditions higher order processes, particularly co-tunneling (CT) could dominate the electron transport in FM/QD/FM structures (Dong Bing, 2007). Figure 11.8 schematically sketches CT involving spin-flips.

In general terms, CT represents tunneling events related with simultaneous tunneling of two or more electrons. Under conditions of suppressed single-electron tunneling, CT may lead to super-Poissonian shot noise. Initially, electron CT attracted considerable attention mainly as a unexpected source of limitation to the accuracy of functioning of single-electron devices (Likharev, 1999). The important relevance of CT processes for the research on quantum dots is related with its capability to enhance deliberately tunnel coupling between single or interacting quantum dots and metallic ferromagnetic) leads.

Depending on the applied bias, CT regime could be classified into elastic and inelastic processes (see for discussion Ref. (Okazaki et al., 2013)). While the elastic processes are related with tunneling over the QD ground state, inelastic ones involve dynamic charge fluctuation between the ground and excited states.

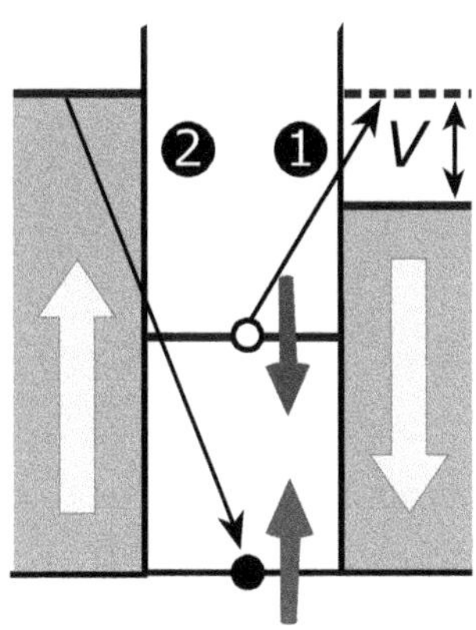

Figure 11.8 Diagram of spin flip co-tunneling in a FM/QD/FM device.

The set of sketches in Fig. 11.9 resumes different channels which could contribute to inelastic CT through QD (Kung et al., 2012). In general, two CT channels: electron and hole-like contribute to the electron transport. While the electron-like channel consists of tunneling from the source into the dot followed by the tunneling from dot to the drain, for the hole-like channel the sequence of tunneling events is inverted with tunneling from the dot to the drain followed by tunneling from the source into the dot. Virtual intermediate states represent respectively $(N+1)$ and $(N-1)$ electron states.

After the CT process occurred, depending on energy of the excited state in which dots is left, QD could relax to the lower (ground) state either by phonon or photon emission. Elastic tunnel could also help relaxing the dot close to the ground state by CT (without charge transfer) or completing tunneling process to convert it to the sequential tunneling (parts *d*, *e*, *f*, *g* in Fig. 11.9).

Generally, super-Poissonian statistics could originate from number of different bunching mechanisms. Among them is dynamical channel blockade (DCB), suggested initially to occur in a multilevel

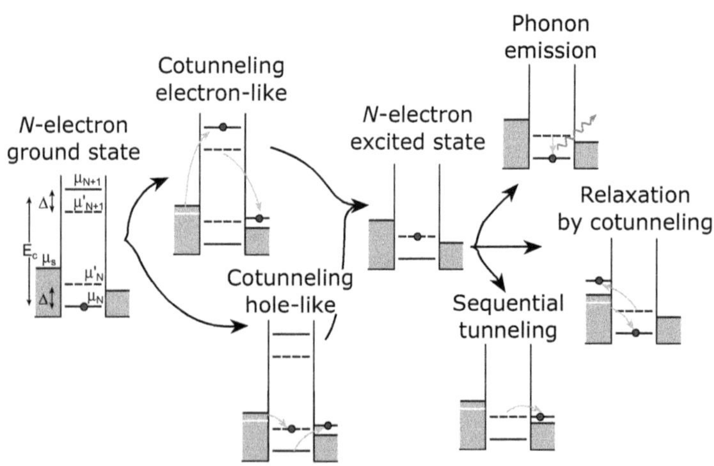

Figure 11.9 Schematics of possible inelastic co-tunneling processes, redrawn from Ref. Kung et al. (2012). Part (a) sketches N-electron ground state and introduces nomenclature. Chemical potential of dot ($\mu_{\rm N}$; while of source $\mu_{\rm S}$ and of the drain is $\mu_{\rm D}$). The source-drain voltage is labeled as Δ while Coulomb energy is represented by $E_{\rm c}$.

QD (Belzig, 2005) and which was discussed in contest of super-Poissonian SN observed in organic magnetic tunnel junctions (Cascales et al., 2014).

An alternative mechanism to explain super-Poissonian shot noise in FM/QD/FM systems is based on the above-introduced spin accumulation in the QD (Bułka et al., 1999). In the third scenario, so-called dynamical spin blockade (DSB), the spin minority channel with a slow tunneling rate modulates the fast transport of the spin majority (Cottet et al., 2004a,b; Dong et al., 2009; Lindebaum et al., 2009). Spin blockade has been observed experimentally in double QD by Pioro-Ladrière et al. (2003).

Finally, we note that mixed effects of the above-mentioned different super-Poissonian statistics models developed for FM/QD/FM systems have been investigated theoretically in Ref. Wang et al. (2011a).

Experimental studies of spin-related shot (or $1/f$) noise in electron transport through QDs are scare. For example, dynamic spin blockade has been reported indirectly by analyzing changes of the spectrum of QD- subject to external magnetic field (Rogge et al., 2004, 2006). Indeed, measurements of spin dependent shot noise in QD connected to FM leads present great experimental challenge. Ubbelohde et al. (2013) presented measurements of electron shot noise in the spin blockade regime (Fig. 11.10).

Super-Poissonian noise has been experimentally seen in the biased region corresponding to the conductance peaks connecting CB regions and interpreted in terms of the dynamical charge channel blockade mechanism with underlying competition between different spin channels (Ubbelohde et al., 2013).

11.6 Multiterminal and Double Quantum Dots Connected to Ferromagnetic Leads

Quantum dots connected to more than two ferromagnetic leads present yet more versatile spintronic devices. Question on the possible influence of correlations between (partially polarized) spin currents with QD functioning as splitter has been first addressed by

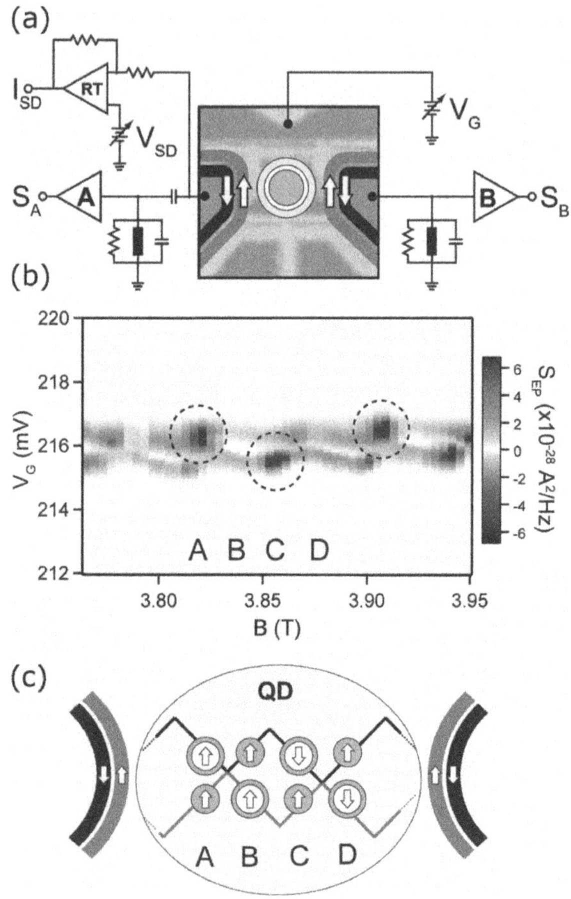

Figure 11.10 Experimental detection of spin blockade. Part (a) shows schematic diagram of the setup. Part (b) shows excess noise with red color indicating super-Poissonian noise. Part (c) shows simplified energy diagram of the multilevels in a magnetic field with ground state marked in red and excited state in black. (A–D) mark four different configurations of the multilevel QD. Adapted with permission from Ref. Ubbelohde et al. (2013).

Cottet et al. (Cottet et al., 2004b). The dot operates as beam splitter (see Fig. 11.11) spin currents injected from contact 2 are split by QD to contacts 1,3 and only energy level of the dot is tuned to the condition of resonant tunneling at V_0 has to be taken into account. The model addresses shot noise in the total current I_2 as well as

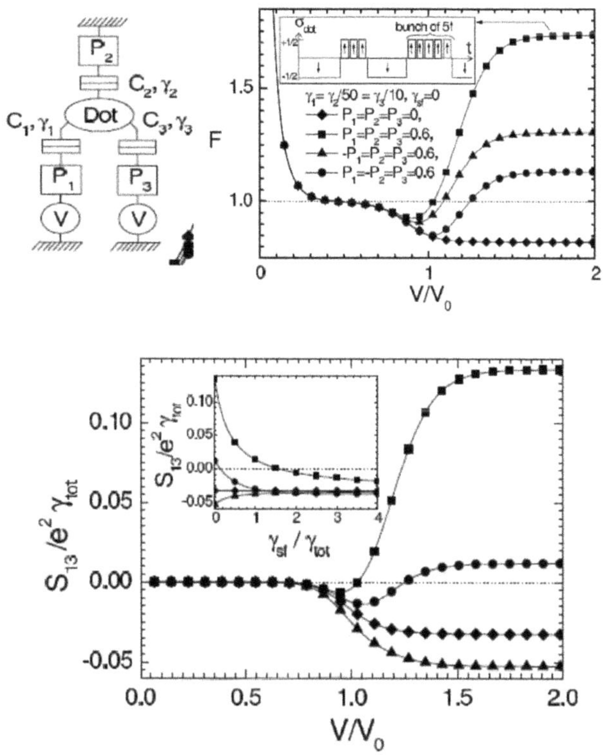

Figure 11.11 Sketch of the QD connected to three FM leads. Bias dependence of the shot noise calculated for different magnetization orientations. Adapted with permission from Ref. Cottet et al. (2004b).

cross correlation between currents I_1 and I_3 in the leads which apply bias to the dot and investigates SN as a function of sign of collinear orientation of the magnetization of the FM leads ($P_{1,2,3}$, see Fig. 11.11).

When all FM leads are oriented in the same direction, the spin-dependent bunching of the tunneling events occurs due to dynamic spin blockade with most of the time electron transport through QD being blocked in the conditions $P_1 = P_2 = P_3$. We remind, however, that the model treats partially polarized spin currents emitted/received by ferromagnetic contacts. In these non-ideal ferromagnet conditions, there should occur rare events (with probability inversely proportional to difference $P_{1,2,3} - 1$) related

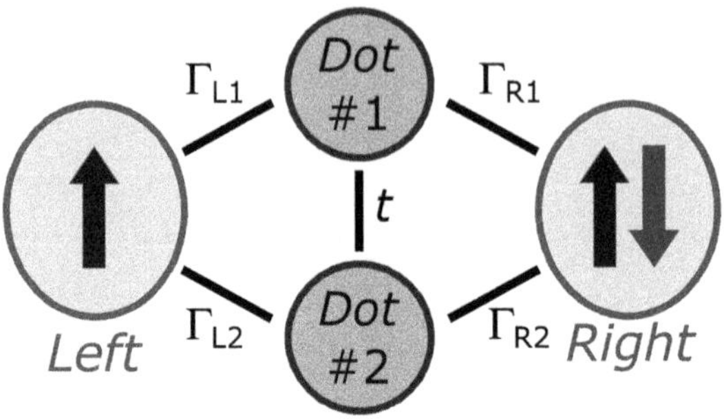

Figure 11.12 Sketch of the symmetrically biased double QD connected with strength $\Gamma_{L,R}$ to two left (L) and right (R) ferromagnetic leads. Redrawn following model in Ref. Weymann (2008).

with minority spin tunneling, say from lead (1) to QD. As long as such minority spin channel tunneling event is not blocked, it will produce spikes in the current through the QD. As the electron transport through QD remains blocked for the long time periods, the current could flow only in short time windows (corresponding to unblocked by spin down channel). Resulting DSB produces therefore a bunching of tunneling events, which explain the super-Poissonian Fano factor (see Fig. 11.11).

Interestingly, super-Poissonian SN (in the total current though the dot) could be realized also in the conditions of negative cross correlation in current between leads $S_{1,3}$. This occurs, for example, with configuration ($-P_1 = P_2 = P_3$). Here the cross-correlations $S_{1,3}$ are negative (Fig. 11.11, bottom) in spite of the super-Poissonian shot noise ($F > 1$) because the correlated (bunched) electrons are mostly up electrons flowing through the lead (3) to the drain (2). In other words, positive current cross correlations in the drain contact does not require the injection of correlated electron pairs.

More exotic four-terminal configuration with QD connected to ferromagnetic leads gives direct access to the spin-flip scattering rate by measuring cross-correlations between currents in terminals with opposite spin orientations (Belzig and Zareyan, 2004).

Additional control over spin currents and shot noise could be obtained with coherent double quantum dots (DQD) weakly coupled to external ferromagnetic leads (Weymann, 2008) (see Fig. 11.12). These considerations have been limited to the full weak-coupling regime including sequential, co-tunneling, and co-tunneling-assisted sequential processes. Among recently considered novel geometries are isolated DQD structures where coherent coupling between two antiparallel electron spins has been demonstrated (Bertrand et al., 2015) opening way to novel types of spin qbits.

11.7 Concluding Remarks and Open Problems

We have seen in this chapter that electron transport between two (or more) ferromagnetic contacts connected to quantum dot (optionally with gate) exhibits very rich new physics due to quantization effects of the confined electrons and the possibility to gain better control over spin transport between ferromagnetic terminals in comparison with double-barrier MTJs. In spite of a tremendous (mainly theoretical) progress in the recent years, the level of understanding of the electron and spin transport in FM/QD/FM devices is yet not comparable to one reached in similar structure but with non-magnetic leads. The main problem is related with experimental challenge of controlled spin injection into the quantum dot.

Shot noise represents inherent current fluctuations of nano devices. The most versatile device allow control over both electron and spin transport and shot noise similar to FM/QD/FM spin transistor with electric (gate) control of the energy of QD levels (Hauptmann J. R., and E.). In the long term perspective, FM/QD/FM devices with QD being small nanomagnets could form a new versatile class of spintronic logic devices with the possibility of having electric (gate or spin torque driven) control of the electronic and magnetic state of the dot.

Chapter 12

Charge and Spin Transport and Noise in Two-Dimensional Materials

It is textbook knowledge that the free electron density of states (DOS) in two dimensions is energy independent, being dramatically different from DOS in 3D (square root of energy) and in the 1D (where DOS is one over square root of energy for each electron channel). The 2D systems in the presence of external magnetic field are known to support dissipationless chiral edge currents which lead to so-called quantum Hall effect (Yennie, 1987). Charge and spin edge currents flow separately in quantum spin Hall effect in 2D topological insulators (TI) (Maekawa, 2012) providing opportunities for next generation spintronic devices. Peculiarities of the spin transport in 2D materials such as graphene or transition metal dichalcogenide monolayers (TMDC) are anomalously long mean free path and spin relaxation times due to negligible spin orbit coupling (Duan et al., 2015; Peres, 2010; Schmidt et al., 2015). In order to construct spintronic devices one should find a ways to modify in a controlled way locally Rashba SOI (Manchon, 2015), for example by using heavy ad-atoms or placing graphene on MoS_2 (Gmitra and Fabian, 2015). This chapter briefly summarizes the current state of the art of 2D materials as an essential part of

Noise in Spintronics: From Understanding to Manipulation
Farkhad G. Aliev and Juan Pedro Cascales

ISBN 978-981-4774-34-5 (Hardcover), 978-1-315-11088-2 (eBook)
www.panstanford.com

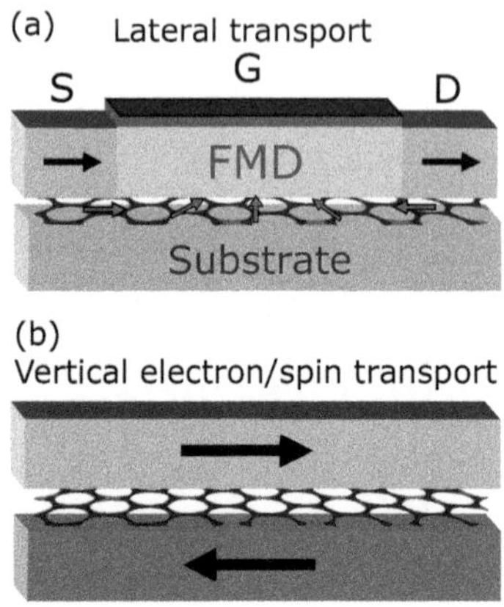

Figure 12.1 (a) Sketch of the spin FET based on a graphene (circles with bonds) with gate being a ferromagnetic dielectric (FMD). The source (S) and drain (D) are ferromagnetic metals with controlled alignment of their magnetic moments. The ferromagnetic dielectric has magnetization alignment along the x axis (see filled circle with a dot). The moving along graphene channel electron spin precesses due to the exchange interaction with the gate which results in the device conductance suppression. The bias electric field may alter the exchange interaction and therefore spin precession changing the probability for spin to be collected by the drain. (b) Sketch of a magnetic tunnel junction with a graphene layer as barrier.

emerging area of low-dimensional spintronics. Scarce information on the low frequency noise and modeling of spin-related noise in these systems will also be summarized.

In general, two geometrically different spintronic structures have been considered where charge and spin transport through 2D materials have been studied either in lateral (Semenov et al., 2007) or vertical (Cobas et al., 2012) geometries (see Fig. 12.1). For vertical transport the TMR of few percent only was found to be consistent with single-layer graphene spin filtering (Karpan et al., 2007) in the conditions of disordered interface. The advantage of the lateral geometry is easier fabrication and versatile gate control of charge and spin transport but with the evident drawback related with the difficulty of scaling down the devices. Up to date, the experimentally reported spin FETs (Yan, 2016) showed below 10% variation of spin-related current at temperatures below liquid nitrogen (50 K).

12.1 Noise in Graphene

12.1.1 An Introduction to Graphene

Graphene is the first truly two-dimensional crystalline material to have been obtained experimentally. It consists on a single layer of carbon arranged in a hexagonal or honeycomb lattice (Fig. 12.3a), with the carbon atoms separated by 0.142 nm. The first observation of such a material was remarkable since the existence of 2D crystals was seriously questioned due to the Mermin–Wagner theorem (Mermin, 1968), which states that a 2D crystal at a non-zero temperature loses its long-range order and melts due to thermal fluctuations. In reality, free-standing graphene is not perfectly flat but presents some rippling which may explain its thermodynamic stability (Fasolino et al., 2007). Graphene has led to the discovery of a whole class of single layer (or 2D) materials, which include Boron-Nitride (BN) and molybdenum-disulphide (MoS_2).

There are varied reasons for the enormous scientific interest in graphene. Firstly, graphene holds quite a lot of technological promise, as was proven in 2004 by the Novoselov and Geim, where graphene was tested as the building block of a field effect transistor (Novoselov et al., 2004). Current silicon-based electronics are reaching their limit of miniaturization, which is on the order of 50 nm for an electric channel, while a graphene strip of width of only a few nanometers may be used as a transistor.

Furthermore, electrons in graphene behave as relativistic particles, making graphene an ideal candidate for the test of quantum-field theoretical models which have been developed in high energy physics. Additionally, electrons in graphene may be viewed as massless *charged* fermions in a 2D space, which differs from massless elementary particles like photons or neutrinos which happen to be electrically neutral.

Graphene was theoretically described as far back as 1947 (Wallace, 1947), where P. R. Wallace already predicted the electronic structure and noted the linear dispersion relation. The derivation of the wave equation for excitations was carried out in 1956 by J. W. McClure (McClure, 1956), and the similarity to the Dirac equation was observed by Semenoff (1984) and DiVincenzo and Mele (1984) in 1984.

As we mentioned earlier, even though it was well known that graphite consists of stacked one-atom-thick layers, obtaining a single sheet of graphene at a finite temperature was not thought possible, let alone isolate one for electrical measurements could be performed. Therefore, Konstantin Novoselov, Andre Geim et al. (Novoselov et al., 2004) surprised the physics community in 2004 by isolating a graphene sheet, transferring it to a substrate and carrying out its electrical characterization (Novoselov, 2005). A further milestone in the study of graphene was the experimental observation of an unusual quantum Hall effect (Novoselov et al., 2005; Zhang et al., 2005). Scientific publications have since then increased dramatically and graphene has become a most prolific branch of condensed matter physics.

Therefore, and as is mentioned throughout this manuscript, the investigation of noise sources in graphene is of both scientific and practical significance. There exist numerous reasons why the study of $1/f$ noise in graphene are particularly important, most of them related to the properties of graphene and its future applications. Graphene will most like by technologically useful for applications which take advantage of its high electron mobility, thermal conductivity and the fact that carrier concentration may be tuned electrically by a gate voltage. Some examples could be the fabrication of transparent electrodes, optoelectronic devices, ultrafast transistors and chemical or biological sensors. High frequency devices for communication have also been proposed as well as an increase in the functionality of amplifiers and phase detectors used for communications and signal processing (Meric et al., 2011; Schwierz, 2010; Yang et al., 2011). For all these mentioned applications, the characterization of electronic noise in graphene, and specially $1/f$ noise is of both fundamental and technological importance.

12.1.1.1 Obtaining graphene

Ever since its discovery, several ways have been devised for obtaining graphene. Depending on the method, graphene flakes

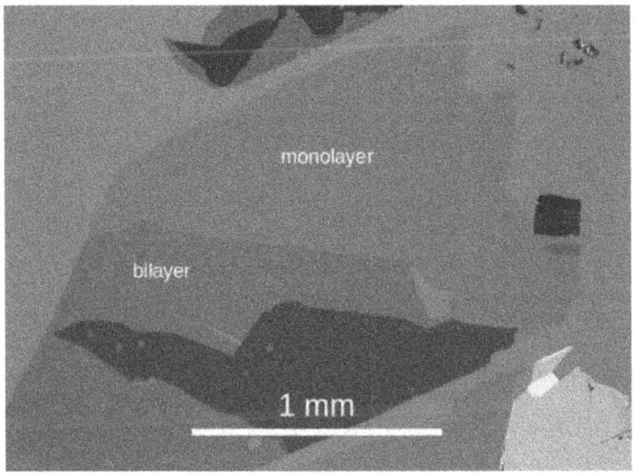

Figure 12.2 An optical image of a graphene flake, obtained from the exfoliation of graphite. Adapted with permission from Peres (2010).

of different purity and size may be obtained. The first graphene samples to be observed were fabricated by *micromechanical cleaving* of graphite (Novoselov et al., 2004). In this approach, possibly the best known, the graphene sheets are separated from the crystalline graphite either by means of an adhesive tape or by rubbing small graphite crystals against each other. Single-layer graphene sheets of sizes around $10 \times 100\ \mu m^2$ may be obtained by using this method.

The single-layer graphene sheets may be optically identified (Blake et al., 2007) by interference-induced color shifts (which can be compared with an atomic force microscopy profile) as shown in Fig. 12.2.

Raman spectroscopy may be additionally employed to not only determine the number of layers of each flake, but the arrangement between layers (Malard et al., 2009). Other methods include (Eletskii et al., 2011) the liquid phase exfoliation of graphite, graphite oxidation, synthesis of graphene by chemical vapor deposition, electric arc graphene production, thermal decomposition of silicon carbide and epitaxial growth of graphene on a metal surface.

12.1.1.2 Crystal structure of graphene

The carbon atoms in graphene assemble in a honeycomb lattice due to their sp^2 hybridization, which can be thought of as two

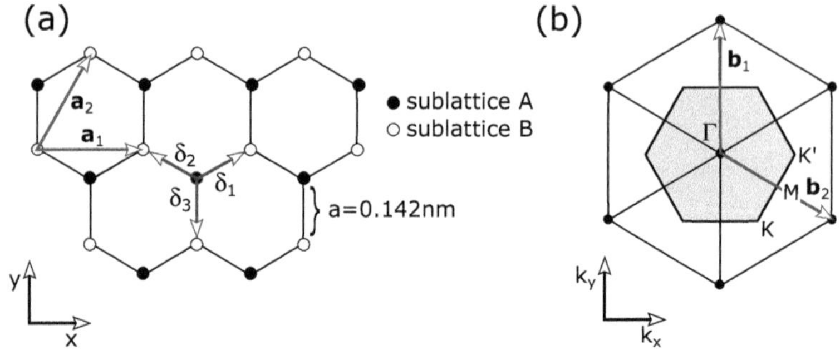

Figure 12.3 (a) Graphene's honeycomb lattice. The vectors a_1 and a_2 are basis vectors of the triangular lattice and δ_1, δ_2, and δ_3 connect nearest-neighbor carbon atoms separated by $a = 0.142$ nm. (b) Reciprocal lattice of the triangular lattice, with primitive lattice vectors a_1^* and a_2^*. The shaded region represents the first Brillouin zone, with its center Γ and the two inequivalent corners K and K' and the three inequivalent crystallographic points M, M', and M''.

triangular sublattices A and B, or a triangular Bravais lattice with a two-atom basis (see Fig. 12.3a). The distance between nearest-neighbor carbon atoms is 0.142 nm, the average distance between single (C–C) and double (C=C) covalent σ bonds (as in the case of benzene). The reciprocal lattice of the triangular Bravais lattice is shown in Fig. 12.3b and is spanned by the reciprocal lattice vectors a_1^* and a_2^*.

One may note the six corners of the first Brillouin zone (BZ), consisting of the inequivalent points K and K'. These crystallographic points play a fundamental role in the electronic properties of graphene because the low-energy excitations are centered around K and K'.

12.1.1.3 Electronic band structure of graphene

As we mentioned above, the first calculation of energy bands was carried out as early as 1947 (Wallace, 1947). Electrons in graphene can be described by a tight-binding Hamiltonian which includes nearest and next-nearest neighbor hopping. Electronic transport in graphene is mostly determined by the energy spectrum around the reciprocal space points K and K', where the dispersion relation is

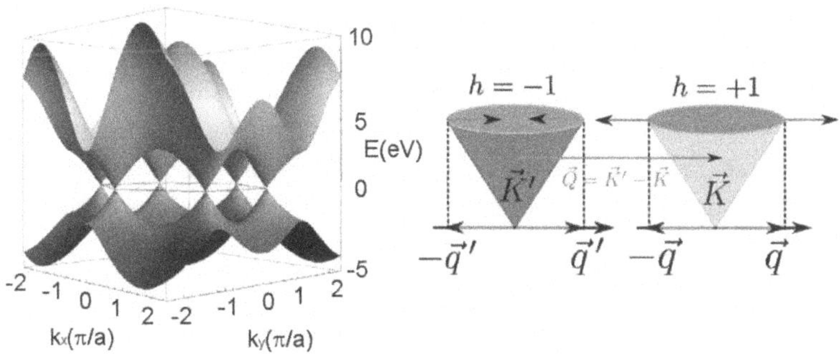

Figure 12.4 Tight-binding band structure of graphene. The low-energy band structure near the points K and K' in can be approximated by a two cones with linear dispersion relations. Adapted with permission from Peres (2010).

conical and given by $E(k) = \pm v_F \hbar k$, where k is the momentum measured relatively to either K or K' and the Fermi velocity $v_F = 3ta_0/(2\hbar)$ (Peres, 2010). The tight-binding calculated band structure of graphene and the conical structure around the K and K' points is depicted in Fig. 12.4.

Several experimental studies have obtained consistent values of $v_F \simeq 1.1 \times 10^6$ m/s (Novoselov et al., 2005). The band structure in graphene obtained by the angle-resolved photoemission spectroscopy (ARPES)(Bostwick et al., 2006), shown in Fig. 12.5.

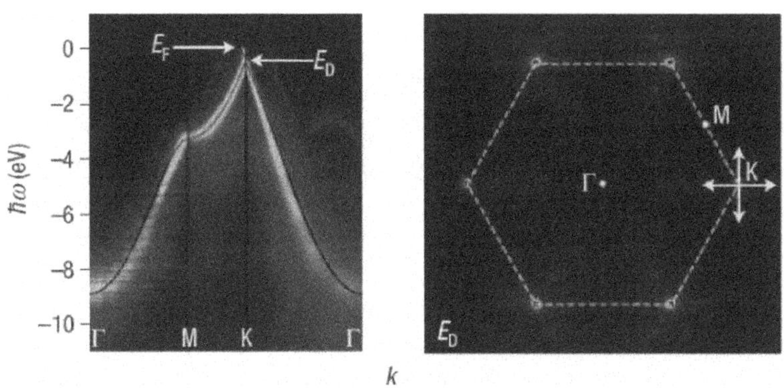

Figure 12.5 ARPES band structure of graphene. Adapted with permission from Bostwick et al. (2006).

The energy spectrum around the reciprocal space points K and K', i.e., the low-energy band structure of graphene, may be approximated by cones located centered at the points (Fig. 12.4a). The energy dispersion in these cones is linear and the electron dynamics can be treated as "relativistic," where the Fermi velocity v_F of graphene substitutes for the speed of light c. Therefore, undoped graphene is a zero gap semiconductor and the conduction and valence bands meet at Fermi energy at a point called *the Dirac point.*

12.1.2 Transport Properties of Graphene around the Dirac Point

As mentioned, the Fermi energy of undoped graphene coincides the Dirac point, where the material has a vanishing density of states. This could intuitively suggest that graphene has zero conductivity at zero energy. What is found is, however, that graphene shows a finite conductivity at the Dirac point. If graphene is used in a field-effect transistor configuration, then the density of states or the conductance can be studied as a function of the gate voltage V_g.

A limiting factor for the conductivity (and the electronic mobility) of graphene is disorder. Some sources of disorder in graphene can be adsorbed atoms (hydrogen) or molecules (hydrocarbons), defects such as folded regions wrinkles, vacancies, and topological defects such as of Stone–Wales type, specially at the edges (Huang et al., 2009).

12.1.3 1/*f* Noise in Graphene

The sensitivity of many types of sensors is limited by their noise level at low frequencies. Also, even though $1/f$ noise contributes to the spectrum mainly at low frequencies, it influences high frequency oscillators as phase noise and ultimately determines the practical implementation of devices for communications. In the case of graphene, downscaling any material system for nanometer-sized devices can additionally increase the amplitude of $1/f$ noise levels and hamper its practical applications (Balandin, 2013). Graphene is a particular system in which to study $1/f$ noise because it is purely two-dimensional, presents a zero-energy band gap, a linear energy dispersion for electrons and holes and metallic-type conductance. Since it is only a surface, conduction electrons may be exposed, for example, to charged impurity traps resulting in a high fluctuation of carrier numbers. Also, because of its zero-energy bandgap metal, mobility fluctuations may heavily influence the $1/f$ noise.

The first measurements of $1/f$ noise in graphene appeared in 2007 (Chen et al., 2007; Lin and Avouris, 2008), soon followed by several studies of $1/f$ in graphene under different configurations and biasing (for a complete view of $1/f$ noise in graphene read (Balandin, 2013)). All reports agree that the fluctuations observed in graphene below 100 kHz show typical $1/f$ behavior, with a low amplitude. Interestingly, the magnitude of the $1/f$ fluctuations in graphene is similar to the typical noise levels in regular Si n-channel MOSFETs and in Si MOSFETs (Rumyantsev et al., 2010). Random telegraph noise (or generation-recombination noise) occasionally appears due to edge defects of the material (Shao et al., 2009). The characterization of $1/f$ noise in graphene samples of different areas has determined that the $1/f$ fluctuations arise from the graphene itself, and not from the metal contacts (Rumyantsev et al., 2010).

The typical dependence of the normalized $1/f$ noise on the gate voltage in graphene has been shown to be of two different shapes (Kaverzin et al., 2012). As shown in Fig. 12.6, the dependence of the normalized noise (black curve) presents an (a) M-type or a (b) V-type behavior. The difference was found to depend on the presence of water in the M-type samples, which is a long-range scatterer and increased the noise by an order of magnitude without affecting the resistance.

$1/f$ may also be used to carry out band spectroscopy of the material, even in the case when band structure information is smeared out in the conductance. The authors in (Xu et al., 2010b) present $1/f$ noise peaks at energies which are correlated with the DOS of the samples, which show quantum confinement effects along the widths of the samples.

The layered structure of graphite also allows to determine whether $1/f$ fluctuations come from the volume or the surface of an n-layer sample, if the dependence of the $1/f$ noise level is studied against the layer thickness. Unlike metal or semiconductor films, the thickness t of a few-layer graphene sample may be varied by

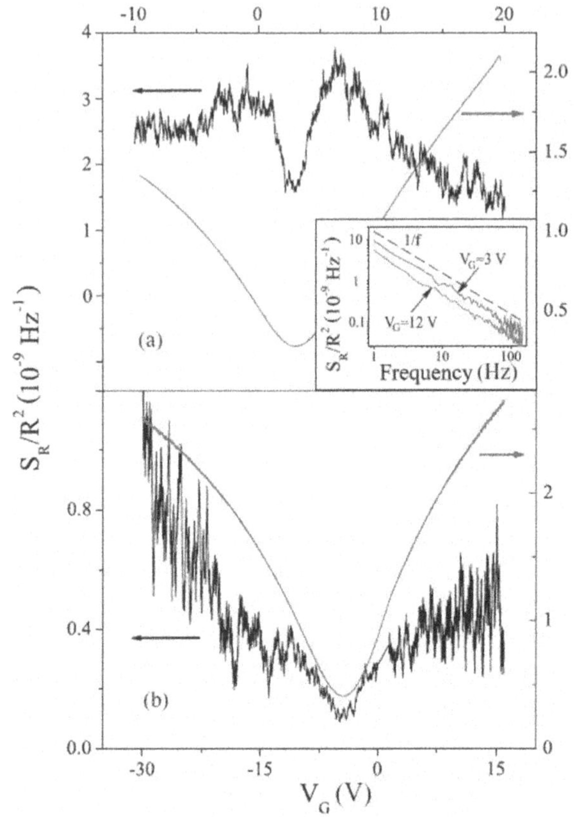

Figure 12.6 Normalized noise spectral density S_R/R^2 at f = 30 Hz (black line) and conductivity σ (red line) as a function of V_g. Two behaviors can be seen: (a) M-type and (b) V-type. Adapted with permission from Kaverzin et al. (2012).

a single atomic layer down to a two dimensional structure (single-layer graphene). If the noise arises from fluctuations in a volume, then the normalized noise should scale as $S_I/I^2 \propto 1/t$, while if the current flows in parallel through all layers, then $S_I/I^2 \propto 1/t^2$ since the current is proportional to the thickness. It was found that $1/f$ noise in a bilayer of graphene was strongly reduced compared to a single layer sample (Lin and Avouris, 2008). A comparison between the normalized $1/f$ noise in a single and double layer of graphene is shown in Fig. 12.7. It has since been found in few-layer graphene

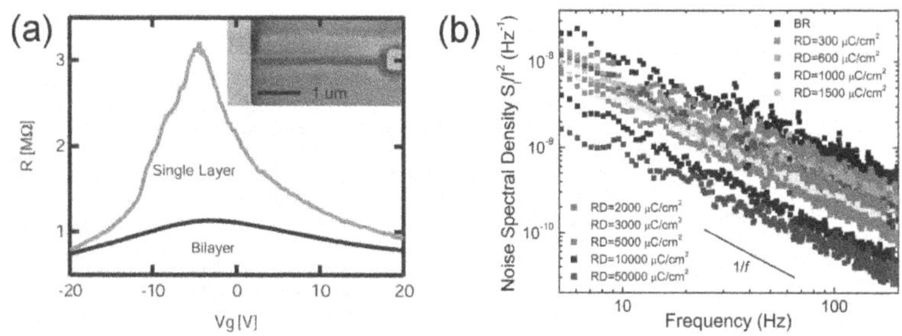

Figure 12.7 (a) Resistance of one single-layer and one bilayer graphene nanoribbon devices measured as a function of gate voltage of identical channel layout. Adapted with permission from Liu et al. (2013). (b) Normalized current noise spectral density, S_I/I^2, showing a monotonical decrease of $1/f$ the increasing irradiation dose. Adapted with permission from Zahid Hossain et al. (2013).

samples that volume noise dominates for thicknesses above seven layers (2.5 nm) while a surface contribution dominates below this thickness (Liu et al., 2013). *Therefore, $1/f$ noise may be substantially suppressed if the sample has more than one layer.*

Another approach to reducing the $1/f$ noise level in graphene was presented in (Zahid Hossain et al., 2013), where graphene devices were irradiated by an electron-beam. The authors found that $1/f$ noise decreases with increasing concentration of defects induced by irradiation, which lowers the mobility while still retaining values useful for applications (Fig. 12.6b).

12.1.4 Shot Noise in Graphene

Only a handful of experimental studies regarding graphene have been reported. They mainly agree that transport through graphene is ballistic and is similar to that in diffusive metals. The Fano factor for graphene strips of large aspect ratio W/L is predicted to be equal to 1/3 at the Dirac point, where transport takes place via electronic evanescent waves (Tworzydło et al., 2006) and the conductivity is minimal.

The first report of shot noise measurements on a number of graphene devices with large W/L was carried out by DiCarlo et al.

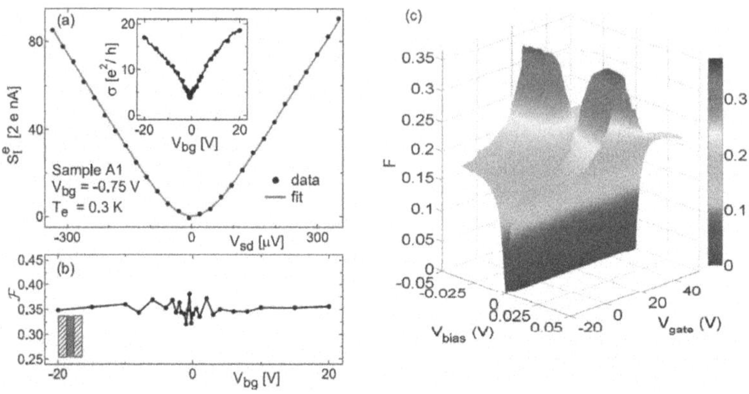

Figure 12.8 (a) Shot noise as a function of V_{sd} around the charge-neutrality point, with a theoretical fit of shot noise yielding a Fano factor of $F = 0.349$. (b) Each Fano is plotted vs. the gate voltage, reaching electron and hole densities of around 1.4×10^{12} cm^{-2}. Adapted with permission from DiCarlo et al. (2008). (c) Mapping of the Fano factor as a function of gate voltage and bias voltage at $T = 8.5$ K. Adapted with permission from Danneau et al. (2008).

(2008). The Fano factor is calculated by sweeping the source-drain voltage V_{sd} for different gate voltages. The authors find that for three single-layer graphene samples, the Fano factor values between 0.35 and 0.37 are found for both electrons and holes which shows no dependence on the electronic sheet density. An example of the shot noise vs. the source-drain voltage and the dependence of the Fano factor with the gate voltage can be seen in Fig. 12.8a,b, respectively. Single-layer p-n junctions were characterized and revealed similar results as well. Finally, the Fano factor in a multilayer sample was found to evolve with V_{gate} from 0.33 at the charge-neutrality point to 0.25. These results may be compared to the theoretical result for ballistic single-layer graphene ($F = 1/3$) (Tworzydło et al., 2006), with the addition of coupled finite density leads finds which slightly increases the Fano factor from 1/3 (Schomerus, 2007), in agreement with the experiment.

A simultaneous report (Danneau et al., 2008) of graphene strips with large W/L, reproduce the value of $F = 1/3$ although it varies with the gate voltage so at very large carrier density, the Fano factor tends to zero (expected for a ballistic system). This is shown in

Fig. 12.8c. The authors changed the aspect ratio to a large enough *L* and the Fano factor at the Dirac point was reduced in agreement with (Tworzydło et al., 2006). If the aspect ratio is changed further by increasing *L*, a strong shot noise suppression occurs, which can be explained by with a variable range hopping effect, more specifically by inelastic hopping conduction in the quasi-1D limit due to contaminants and rough sample edges (Danneau et al., 2010).

Further studies in bilayer graphene samples corroborate the value of $F = 1/3$ with a decay down to $F = 0.26$ at high bias (Fay et al., 2011). This decrease is attributed to the coupling between electrons and optical phonons.

Up to date there have been no reports on experimental studies of spin dependent shot noise in graphene or transition metal dichalcogenides (TMDCs). One way to create conditions to observe and control spin-dependent shot noise in graphene in the presence of Rashba SOI could be by depositing magnetic superlattice. It has been predicted that in such a structure the shot noise could be controlled by the number of barriers, strength of SOI and by the external magnetic field (Sattari, 2015).

12.2 Low-Frequency Noise in Single/Few-Layer Transition Metal Dichalcogenides

Ultrathin (with thickness down to few monolayers) TMDCs films from alternative/complementary to graphene class of two (quasi-two) dimensional materials with important for applications physical properties. For example, single layer MoS_2 has been widely investigated as core material for applications ranging from low-power digital electronics (Wang et al., 2012b), light detection (Lee et al., 2012) to chemical sensing applications (Perkins et al., 2013) and very recently in spintronics (Zibouche et al., 2014). Being easy exfoliated down to atomically thick layers, the main characteristics of TMDCs include finite direct band gaps in the optical range (close to that of Silicon) or a rather large SO coupling with giant spin splitting in the valence bands of 400 meV (Zibouche et al., 2014) which perfectly complement graphene making possible to design future spintronic applications from 2D-TDMCs (Xie and Cui, 2016).

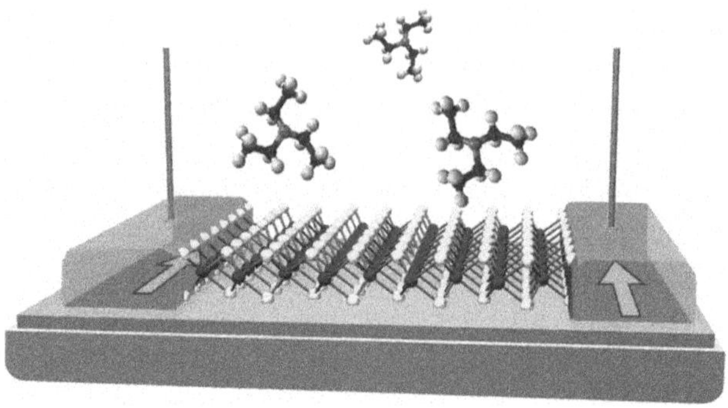

Figure 12.9 Sketch of spintronic sensor based on black phosphorus. Adapted with permission from Perkins et al. (2013).

Requirements for low $1/f$ noise levels is a great challenge for 2D materials forming core of FET, which are the core of currently developed FETs. Being deposited over insulators, their conductance channels are substantially exposed to the influence of fluctuations in charge of the traps localized in the gate dielectric layers and inside substrates.

So far, low-frequency noise studies mainly aimed to address the problem of relative importance of charge number fluctuation (mainly coming from interfaces) and mobility fluctuations, affecting already conducting within TMDCs materials electron channels. Besides, due to imperfection of metal deposition on TMDCs, part of the LFN could come from the contributions of contacts.

Besides, spintronics with few layer TMDCs currently emerges as separate area of the research on TMDCs 2D materials (Dankert et al., 2014; Kamalakar et al., 2015). Figure 12.9 shows a schematic example of such future spintronic sensors with sensitivity possibly tuned by the magnetic state of the device.

So far, noise studies in devices formed by TMDCs have been carried out with non-magnetic contacts only, below we briefly summarize the main activity related with studies of LFN in TMDC-based FETs.

The first extensive study of low-frequency electronic noise in single layer MoS_2 FETs by Sangwan et al. (2013) showed that mobility fluctuations (MF) model present a dominant source of $1/f$ noise at room temperature. Gate voltage was used in order to vary number of the carriers in MoS_2 FETs. At low temperatures (below 10 K) generation-recombination noise emerges either due to traps in the underlying SiO_2 substrate or midgap states in single layer MoS_2. The lowest values of the Hooge parameter were found to be comparable to other materials with surface dominated transport such as Carbon nanotubes.

Similar conclusions have been obtained by (Na et al., 2014) for few-layer (11–40 nm)-thick MoS_2, where the presence of Al_2O_3 passivation layer induced an additional conduction channel characterized by generation-recombination noise. Without passivation, the Hooge mobility fluctuation model corresponds to mainly bulk conduction described satisfactorily by drain current fluctuations.

LFN studies on MoS_2 devices (Renteria et al., 2014) contradicts the mobility fluctuation and claims that both the channel and contacts contribute to the overall $1/f$ noise level of the as fabricated and aged transistors. In contrast to graphene devices (Balandin, 2013) and the above-mentioned papers (Na et al., 2014; Sangwan et al., 2013), the intrinsic LFN characteristics have been described by the carrier number fluctuations (NF) model (Simoen and Claeys, 1999).

The main criteria which has been used far to distinguish between NF and MF models is gate voltage dependence of the inverse Hooge parameter $(\alpha)^1$ vs. $(V_g - V_{th})^n$ where V_g and V_{th} are gate and insulator-metal transition threshold voltages and n-parameter discriminating between models. It was shown that one should expect $n = 1$ for MF and $n = 2$ for NF models correspondingly (Simoen and Claeys, 1999). Interestingly, transition between NF and MF regimes could be tuned using photodoping (Martinez et al., 2017).

12.3 Outlook: Spin-Related Noise in 2D Electronics

Although up to date understanding and control over charge noise has been the main challenge for 2D material-based FETs and related devices, very recent developments in the field pay growing attention on getting precise control over spin injection and detection in 2D structures. Indeed, from one side, due to negligible spin orbit coupling (Duan et al., 2015; Peres, 2010; Schmidt et al., 2015) spin transport in 2D materials is characterized by very long spin relaxation times. From the other side, recent theories and experiments suggest local control and modification of spin orbit interaction in or outside contact area (Avsar et al., 2014; Gmitra and Fabian, 2015; Manchon, 2015; Wang et al., 2015).

The origin of SOI induced current fluctuations is a quantum mechanical effect. In the case of lateral devices, SOI produces the splitting of the electron trajectories. Great advantage to study shot noise in spin FETs such as schematically shown in the upper part of Fig. 12.1 is the capability to tune the SOI using a gate. If entangled electron pairs are injected into 2D layer split two branches, the local SOI in one of the branches could be investigated thought shot noise (Egues et al., 2002). Indeed, following this prediction the Rashba phase between original entangled and (spin rotated) triplet states could be discriminated if noise measurements are carried out on the above beam splitter geometry. It has been shown that SOI determines the Fano factor, providing a direct measure of the Rashba constant through shot noise (Egues et al., 2002).

An alternative way to investigate the SOI would be through magnetic state dependent shot noise in MTJs with 2D layers serving as barriers (see the lower part in Fig. 12.1). If SOI relaxes the spin accumulation in the AP state, one would observe excess shot (or in the presence of interface disorder, normalized $1/f$) noise in the antiparallel alignment, similarly to previously seen in MTJs with AOx (Guerrero et al., 2006b) or MgO barriers (Aliev et al., 2007).

Chapter 13

Magnetoresistive Sensors: Operation Principles and Noise

The main operation principle of magnetic sensors based on GMR or TMR effects is related to their capability to transform linearly (or abruptly) the external magnetic field value into the voltage. These capabilities gave raise to one of the most important and direct applications of spintronic devices: magnetic fields sensing. Technically sophisticated (based on a few elements directed along three axes) sensors are even capable of tracking both absolute value and direction of the local magnetic field.

Among the most important characteristics of the GMR/TMR sensors is their detectivity related to the external field which sensor is capable to resolve. Evidently, this parameter depends directly on the intrinsic sensor noise. That is why sensor noise plays a fundamental role in the minimum field a given sensor is capable to detect. Corresponding parameters known as *noise floor* accompanied by sensor sensitivity determine the signal output and is among the most important figures of merit of a sensors to be discussed in details further below.

The present chapter discusses the main types and operation principles of GMR/TMR sensors, their applicability on a wide range of

Noise in Spintronics: From Understanding to Manipulation
Farkhad G. Aliev and Juan Pedro Cascales

ISBN 978-981-4774-34-5 (Hardcover), 978-1-315-11088-2 (eBook)
www.panstanford.com

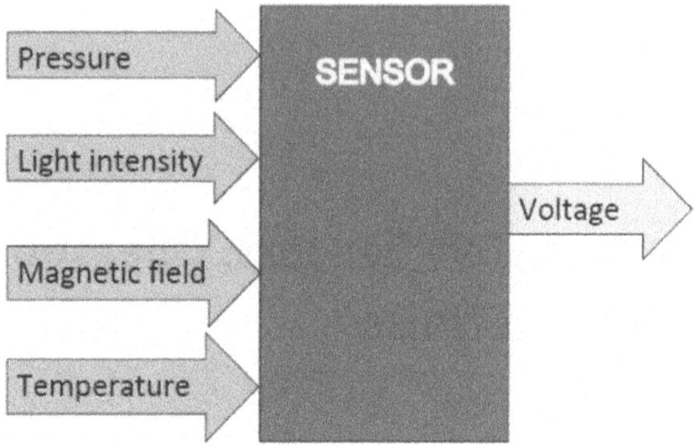

Figure 13.1 Sketch describing general idea of analog sensor.

areas from biomedical and automotive market to information storage and processing. We shall discuss methods to minimize and optimize the GMR/TMR field sensors' characteristics paying attention to their noise. Performance and noise optimization in AMR sensors will be discussed only in brief, because from physical principles of operation AMR sensors stay somewhat apart from spintronic devices. Following excellent reviews are recommended for readers who wish deepen the knowledge of the topic: (Ana V. Silva and Cardoso, 2015; Cardoso et al., 2014; Freitas et al., 2007; Jr. et al., 2009; Lei et al., 2011; Reig et al., 2009; Valadeiro et al., 2015).

13.1 General Introduction to Magnetic Field Sensors

In general terms, sensor is a device that coverts some physical quantity (light intensity, magnetic field, temperature, pressure, etc.) into other easy to read by electronic device, typically voltage (see Fig. 13.1).

There are two main "figures of merit" of sensors: accuracy and precision. Bad sensor accuracy implies that the mean is not correct but the error bar could be not small, whereas a good precision means that the error bar is small but the mean is correct (see sketches in Fig. 13.2). It is task of absolute sensors (example absolute

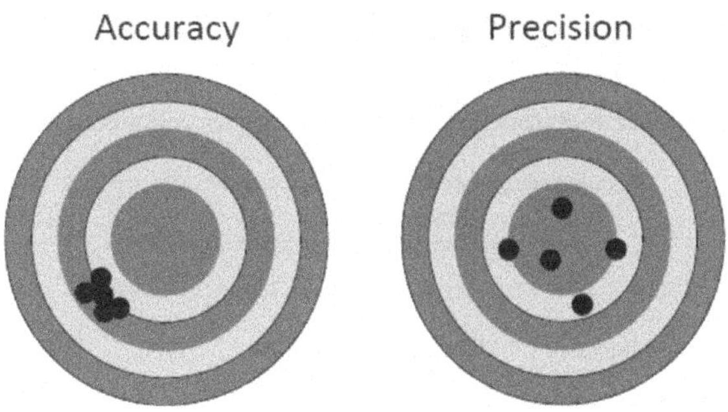

Figure 13.2 Sketch describing accuracy (left) and precision (right) of a sensor.

thermometers, or absolute light power meters, etc.) to provide high accuracy of the measured physical quantities, mainly at the cost of the measurement time. On the contrary, most of the sensors are designed to provide quick high-precision measurements at the cost of accuracy.

The sensitivity of a magnetic field sensor is determined by its transfer curve, i.e., dependence of the voltage output on the applied magnetic field. The formal definition for such a quantity is

$$\mathrm{SENS}(H) = dV_{\mathrm{out}}(H)/dH. \quad (13.1)$$

There is a direct relation between the sensitivity of the sensor and its accuracy. Figure 13.3a shows that in case of a magnetic field sensor, high precision would mean larger error bar in the field measurements. Indeed, the higher SENS is, more sensible (and

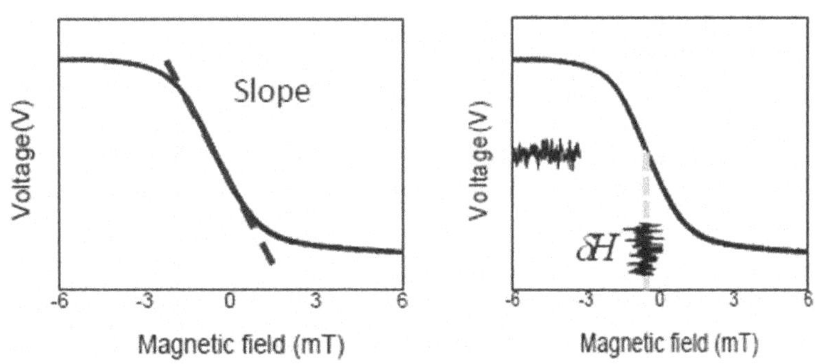

Figure 13.3 Sketch explaining sensor sensitivity (left) and relation between sensor voltage noise and its sensitivity (right).

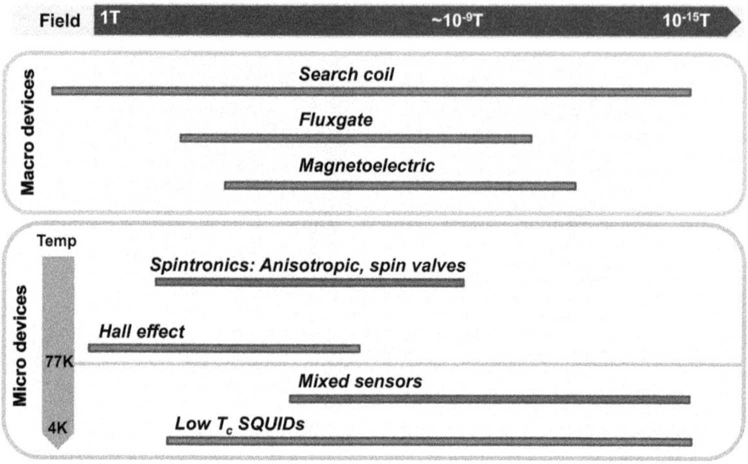

Figure 13.4 Summary of different types of magnetic field sensors.

correspondingly noisier; see Fig. 13.3b) the measured voltage signal from the sensor is.

From equation 13.1, one also observes that the sensitivity in general may change with the magnetic field $S = S(H)$. Generally, S depends on the state of the spintronic device and, more importantly, on the anisotropy of magnetically free layer. The characteristics of the free layer will determine responses to the important questions on the sensor factors such as the following:

(i) Is the sensor hysteretic?
(ii) Is the transfer curve linear (is SENS constant with $V_{\text{out}} \propto H$) in range of field of interest?
(iii) What is the field range where my sensor is an active device?

To finalize the introduction part, Fig. 13.4 summarizes different types of magnetic field sensors, referring their applicability range. From the above summary one could see that among the most sensitive GMR/TMR-based field sensors are so-called mixed sensors based on hybrid structures made of superconducting field transducer transforming the external flux to high-density local current which acts on GMR sensor through its Oersted field (Pannetier et al., 2004).

Case study 13.1: What is the transfer curve of sensor based on a pick up coil?

A pick up coil is a sensor that works at a given frequency f and is able to sense AC magnetic fields. It is based on the induction law, so the amplitude of the current induced by an AC magnetic field of H in a coil of area A and with N turns is

$$V_{\text{out}} = 2\pi f A \mu_0 N H.$$

So the AC sensor sensitivity will be

$$\text{SENS} = 2\pi f A \mu_0 N$$

In the considered case, the sensitivity is independent on the applied magnetic field. Moreover, the sensor is non-hysteretic and the linear region includes all range of the fields and there is no saturation effect.

Case study 13.2: Sensitivity of a GMR sensor

Suppose that in some GMR sensor the transfer curve is $V \propto I_0 H\,|\cos(\theta)|$, θ being the angle between magnetizations of the two layers and I_0-sensing current. Considering that one of the magnetized layers (the hard layer) is totally pinned and the second (free layer) has a strong uniaxial anisotropy and the field to aligned with the hard magnetization axis. In that conditions magnetization of free layer may rotate coherently as function of the magnetic field, as shown in the figure below.

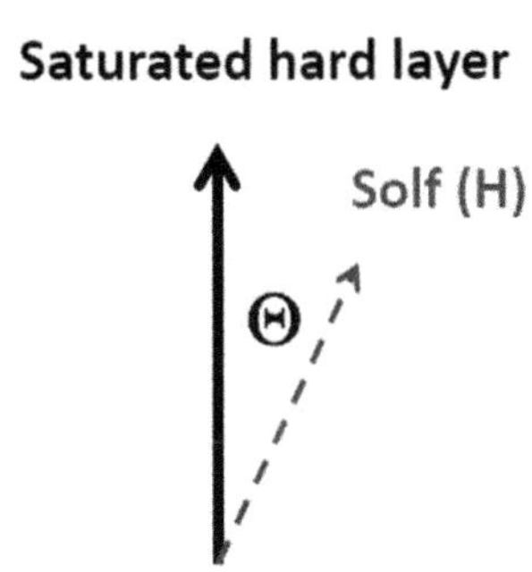

The sensitivity of such a system (in terms of angle θ between fixed and free layers) is given by expression: SENS(θ) $\propto I_0|\sin(\theta)|$. So, the sensor is less sensitive when external field is aligned with hard axis direction. The sensor (in theory) is non-hysteretic as we have introduced ideal conditions for the hard axis. In real systems, this does not happen due to the presence of pinning centers and magnetic topologic anomalies, for example domain walls or vortices or other artifacts. For instance, when the hard layer magnetization could be altered by the applied field or the Censor response to the external field is non-linear. Indeed, the saturation of the sensor happens at fields given by the saturation magnetization of the free layer and the anisotropy constant. Therefore, above some threshold field linear dependence of the soft magnetization layer on the magnetic field will not be valid anymore.

Note also that as one increases the input sensor power (in our example controlled by the sensing current $\propto I_0$), the output signal per unit of field also increases. So in theory the I_0 parameter could also be adjusted to reach the detection of very tiny magnetic fields. In other words, here we have a tradeoff between the electric power fed to the sensor and the small magnetic field that it is able to detect. This tradeoff could be conserved to a certain level as we will see in the next point.

13.2 Sensor Sensitivity in Terms of Detected Power

In the previous chapters, we described the main noise sources in spintronic devices. They could be summarized as (i) thermal noise, (ii) shot noise, (iii) so-called $1/f$ noise and (iv) other colored noise (RTN, brown, etc.).

So far, the noise description has been done using the power spectral density. In sensing applications, this quantity is very convenient. However, taking into account that in most of the sensors applications aim on detection of signals below some threshold power by averaging signal over some period of time, use of power spectral density requires an additional explanation.

Parameter known as noise equivalent power (NEP) is sensor figure of merit defined as signal power (W) incident on the detector required to produce signal to noise ratio equal to one ($S/N = 1$) in a one hertz output bandwidth (which is equivalent to half a second of integration time). NEP parameter has units of $\mathrm{W}/\sqrt{(\mathrm{Hz})}$ and is mainly of importance for sensors such as bolometers, which detect signal power. One can calculate directly NEP from the noise spectral density (for example expressed in units of $\mathrm{V}/\sqrt{\mathrm{Hz}}$) dividing it by the response function (V/W). Evidently, more sensitive detectors have smaller NEP.

Case study 13.3: Noise equivalent power

Consider detector with $NEP = 10^{-15}\,\mathrm{W}/\sqrt{\mathrm{Hz}}$. It means that it should be possible to detect a signal with one femtowatt power with SNR $= 1$ having averaged over half second. Averaging for 5000 s (reducing here bandwidth factor of 10^4) improves the SNR factor of $\sqrt{10^4} = 100$, i.e., as the square root of the averaging time.

Another alternative approach discussed below defines noise floor in terms of total measured by the electronic device noise power. The spectral density definition is an amount of power in signal per unit of frequency. As we already mentioned, in some sensing application, however, our concern could be to detect a signal with a given power. The question is, how do we translate between power density to useless signal, i.e., noise? This translation is performed by using following relationship:

$$P = \int_0^{\infty} S(f)df, \tag{13.2}$$

where P is the noise power (see the definition of SNR above). This quantity by definition is divergent, as the thermal noise does not depend on frequency; thus, its associated total power could be infinite. Nevertheless, real systems have a finite bandwidth that limits this integral. Thus if we consider a linear system defined by a transfer function $H(f)$, one can rewrite this equation as

$$P = \int_0^{\infty} H(f)S(f)df. \tag{13.3}$$

So, now the noise power remains finite and can be modified by changing the transfer function associated with the system. Indeed in the usual cases and for the white noise sources we can define a equivalent noise bandwidth (EQNB) as the brick wall filter needed to obtain the same noise power for a given $H(f)$. This could be expressed as

$$P = \int_0^\infty H(f)S(f)df = S_0\,[EQNB]. \tag{13.4}$$

The quantity EQNB is generally tabulated as function of the number of poles in the transfer function and allows to calculate the noise power given the nominal noise cutoff (f_{-3dB}) for the white noise power density (see for more details http://analog.intgckts.com/equivalent-noise-bandwidth/).

The above described calculations allow us to determine the SNR for a given sensor and if desired for given chain of detection (amplifiers+filters). However, this quantity depends on the field to detect so we have to fix a target field to be detected in order to give at least an order of magnitude of the SNR. In order to keep the description of a sensor as general as possible, i.e., application independent, usually we define the noise floor of a sensor as

$$\mathrm{NF}(f) = \frac{S(f)}{\mathrm{SENS}(H)}. \tag{13.5}$$

given in units of field squared per unit of frequency. This quantity allows straightforward calculation of the SNR given a magnetic field H_0 to detect as

$$SNR = \frac{[H_0]^2_{\mathrm{rms}}}{\mathrm{NF}(f)\,[EQNW]}. \tag{13.6}$$

Note that SNR has no dimensions. The dimensions of NF are Tesla^2/Hz. Therefore magnetic field to detect (H_0) should be defined as the power of the magnetic signal.

Assuming a white noise floor, the NF allows a classification of sensor as function of its performance. For instance, low temperature SQUIDs are in the range of pT^2/Hz and are currently the most performant sensor. Spintronic sensors perform at room temperature in the range of nT^2/Hz and mixed sensors (see next section)

Case study 13.4: Tuning noise floor by bias current

Using the spin valve introduced in the previous sections one may define the noise floor $\mathrm{NF}(f)$ for the three different noise sources. First, suppose that the output signal per unit of field with $\mathrm{SENS}(H, I_0, \theta)/H \propto I_0 \sin(\theta)$. For the maximum sensitivity point ($\theta = 90°$) we can then calculate the $\mathrm{NF}(f)$ as

$$\mathrm{NF}(f) = \frac{S(f, I_0)}{I_0^2}.$$

Introducing specific noise sources, for example the thermal noise, we obtain

$$\mathrm{NF}(f) = \frac{4k_B T R}{I_0^2}.$$

So, the tradeoff power performance described in the previous point is conserved, i.e., more bias current is applied, lower is the noise floor.

In the case of shot noise we obtain a similar result:

$$\mathrm{NF}(f) = \frac{2eI_0}{I_0^2}$$

However, in the case of $1/f$ noise with $S(f) \propto$ Const $(I_0)^2/f$ and in general other resistance fluctuation with noise power proportional square of the bias current, the $\mathrm{NF}(f)$ results in

$$\mathrm{NF}(f) = \frac{\mathrm{Const}\, I_0^2}{f I_0^2},$$

meaning that one cannot improve the performance of the sensor by just by increasing the power delivered to the sensor through bias. This aspect limits the performance of spintronic noise sensors. In spite of its high sensitivity, the magnetic noise (see Chapter 4) scales as the derivative of the magnetoresistive response on the field, i.e., the sensitivity. So the noise floor is limited in the best case by the magnetic thermal fluctuations that are translated into resistance fluctuation. In this chapter, we will describe the recent advances performed to overcome this intrinsic limitation in magnetic field sensing based on spintronic devices.

in the range or pT^2/Hz and are limited at low frequency by the magnetic $1/f$ noise.

Further, we shall concentrate mainly on the operation principles and noise of spintronic-based field sensors.

13.3 AMR Sensors

As we have seen in the previous chapter, typical MR curve of ferromagnetic material presenting AMR is nearly symmetric respect the external magnetic field. Presence of AMR asymmetry mainly depends on the DW pinning strength. The straightest way to make AMR sensor linear field sensitive for the low magnetic field range is modifying its transfer curve by using so-called "barber poles" (see Fig. 13.5a).

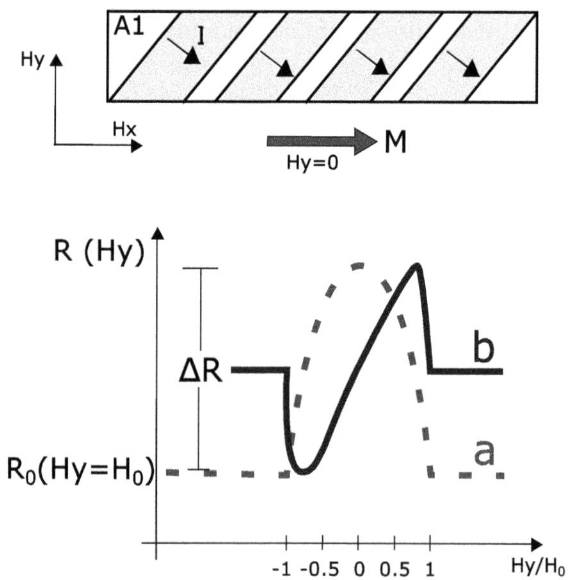

Figure 13.5 Upper part: sketch presenting AMR sensor A based on so-called "barber poles" with arrows indicating how current direction is changed. The lower part schematically compares AMR before (a) and after (b) "barber poles" are introduced.

This could done by coverage of the soft ferromagnetic film with aluminum bars under 45° inclination to the stripe axis. As long as resistivity of Aluminum is roughly 5 times lower than one of Permalloy, the barber poles change the direction of current forcing it to be about 45° to the strip H_x axis. In these conditions, H_y component of the magnetic field will change the resistance of the stripe almost linearly with the magnetic field. Figure 13.5b sketches changes in the transfer curve with the barber poles. In order to achieve predictable (hysteresis free) operation of the sensor some small auxiliary stabilizing H_x field may be applied by the permanent magnet.

Even with well-linearized characteristics, the AMR-resistor by itself is not well suited as an MR sensor as long as its zero field resistance is finite, magnetoresistance is small (below 2–3%) and could be comparable with possible small temperature drifts. Using a Wheatstone bridge (so-called AMR-Bridge, see Fig. 13.6) compensates the above-mentioned AMR disadvantage.

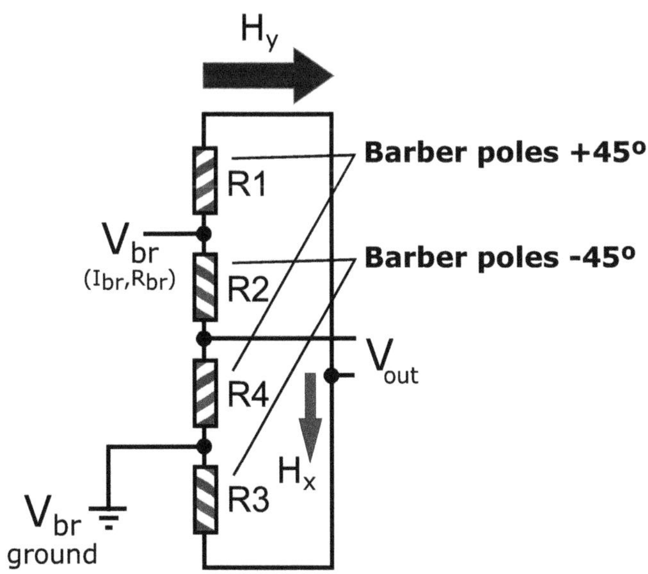

Figure 13.6 Sketch describing the operation of AMR-Bridge sensor (see text for details).

Additionally, the AMR-Bridge doubles the output differential signal in comparison with AMR resistor. Special tools such as laser trimming are used to balance the bridge and make bridge being extremely sensitive field detector. Figure 13.6 shows typical AMR-Bridge circuit.

13.4 GMR Field Sensors

Operation principle of GMR and TMR devices depends on particular applications. For the MRAM applications, the free and pinned layers easy axes are set to be parallel in order to maximize the difference in resistance between two magnetic states (0 or 1). Situation is different for magnetic field sensors where two layers are set to be orthogonal, usually by applying external magnetic field during deposition, providing an induced magnetic anisotropy. Besides, as we have seen above, the field sensor output must be linearized as a function of external magnetic field.

In order to achieve linearization and maximum angle (i.e., field) sensitivity one usually fixes magnetization direction of the pinned layer transversally (i.e., with $\theta = \pi/2$) to the unperturbed free layer direction (see sketch in Case Study 13.2). Indeed, as long as angular dependence of magnetoresistance of GMR-SV sensor is given by the following angular dependence (we consider conservative definition, see (Xu and Mai, 1999))

$$\text{GMR} = (1 - \cos(\theta))/2$$

Then maximum angular sensibility

$$\text{MAX}[d[\text{GMR}]/d(\theta) \propto \text{MAX}[d[\text{GMR}]/d(\theta) \propto \sin(\theta)$$

will occur for $\theta = \pi/2$. Following recipes of Ref. Freitas et al. (2007), Fig. 13.7 explains the main procedures to engineer satisfactory magnetic field sensor.

Recent new generations of read heads include at the first stage change from current in plain (CIP) to current perpendicular to the plain (CPP) configurations the metallic spin valves, followed by the CPP configuration for MTJs now actively being used. The CPP geometry reduced spacing between shields allowing reduction of

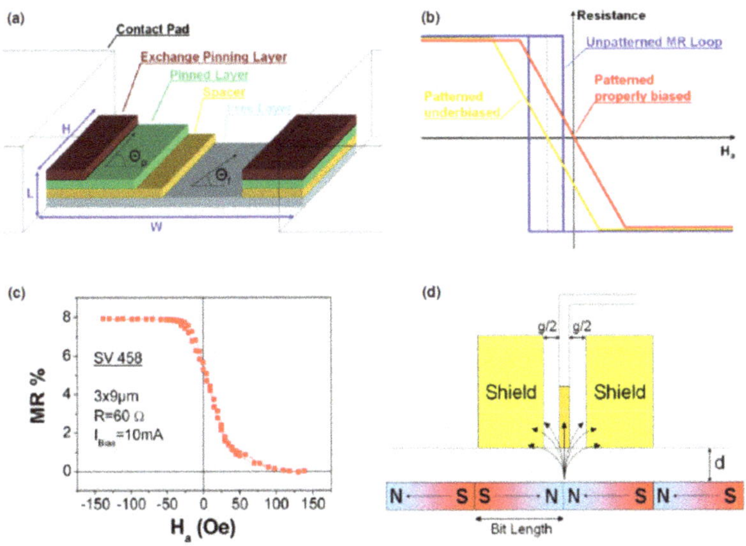

Figure 13.7 (a) Schematics of the unshielded GMR sensor; (b) transfer curve of the under-biased and linearized GMR sensor and in comparison of properly biased and linearized GMR sensor; (c) experimentally measured transfer curve for an unshielded GMR sensor; (d) Sketch of the shielded GMR sensor used in the read head. Adapted with permission from Ref. Freitas et al. (2007).

the bit length (Cardoso et al., 2014; Freitas et al., 2007). Below we briefly review recent advances in the optimization of MTJ-based heads and field sensors.

13.5 TMR Field Sensors

The main advantage of MTJ sensors is related to their reduced dissipation and the possibility to reach higher signal-to-noise ratios than that for GMR sensors. This is achieved due to (about factor of 10) increase in resistances and magnetoresistance (TMR) ratios, but typically accompanied by somewhat higher than in GMR sensors noise levels. Below we present a related case study based on Ref. (Almeida et al., 2006).

Case study 13.5: Optimizing transfer curve for TMR sensor

The sample geometry (see insert in the figure) is similar to one used for SVs with the fixed layer direction determined during deposition and free layer magnetization direction determined by its shape and directed transverse to direction of the sense field. As could be seen in the figure below (adapted with permission from Ref. Almeida et al. (2006)), from the initial tests it becomes clear that the transverse field is hysteretic, which is important drawback for any field sensor. In order to tackle the problem, some additional longitudinal bias field of a few Oersted has been applied. Such field, through suppression of the DW formation, dramatically reduces magnetization hysteresis. In that particular case we discuss the progress in reduction of sensors hysteresis has been achieved at the cost of sensor sensitivity, due to induced changes in the angle between fixed and free layers by finite longitudinal bias field of 15 Oe acting on the equilibrium angle of the hard layer.

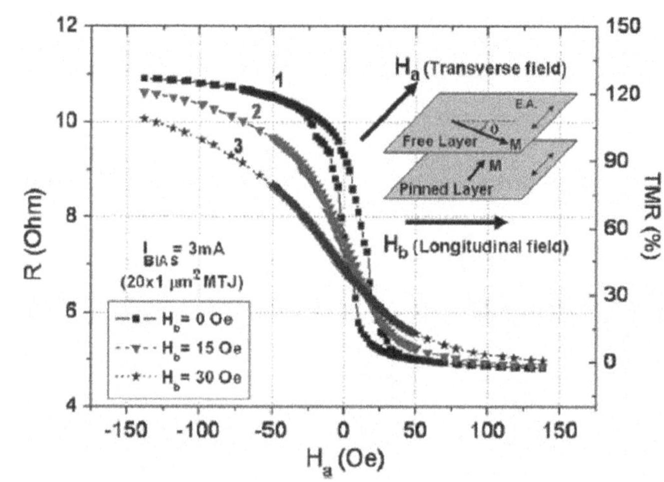

We note that for the read-head applications the MTJs resistance by area product is reduced down to about 1 $\mu\Omega*cm^2$ by using small (typically below 1 nm thick) oxide barrier (recently MgO) in order to compensate resistance increase due to unavoidably small lateral sensor dimensions. Such resistance levels are a condition to provide the read-head resistance in the range of and below 100 Ω, needed to fulfill the compatibility conditions for coupling between sensors and microwave amplifier.

13.6 Low-Frequency Noise in MTJ Field Sensors

Precise vectorial detection of the low-intensity–low-frequency magnetic signals is one of the most challenging current areas where the application of MTJ-based sensors have been having major progress. Brain activity generates signals from pT range and at frequencies of a few tens of Hz (Cardoso et al., 2014; Freitas et al., 2007). Although MgO-based sensors have already shown trends in direction of detection of signals even below $pT/Hz^{1/2}$, this has been achieved at relatively high frequencies where $1/f$ noise becomes

Case study 13.6: Definition of field detectivity for MTJ sensor

As long as $1/f$ noise of MTJs is expressed using the Hooge factor, we can describe noise of the sensor in terms of voltage fluctuation over $\sqrt{\mathrm{Hz}}$ (i.e., square root of the noise power density in voltage in units $V/\sqrt{\mathrm{Hz}}$) based on the relation for $1/f$ noise power presented in previous chapters):

$$\sqrt{S_\mathrm{V}} = V\sqrt{(\alpha_\mathrm{H}/Af)}$$

where V is the applied bias, α_H the Hooge factor, A the junction area, and f the frequency.

In order to characterize the sensor detectivity (in terms of (magnetic field/$\sqrt{\mathrm{Hz}}$) we multiply the above relation by the derivative of the sensor transfer curve: $\Delta H/\Delta V$

Then we may see that magnetic field sensor detectivity D(H) is

$$D(H/\sqrt{\mathrm{Hz}}) = (\Delta H/\Delta V)V\sqrt{(\alpha_\mathrm{H}/Af)} = \frac{\Delta H\sqrt{(\alpha_\mathrm{H}/Af)}}{\mathrm{TMR}},$$

with $\mathrm{TMR} = \Delta V/V$ and with ΔH being linear sensor field range.

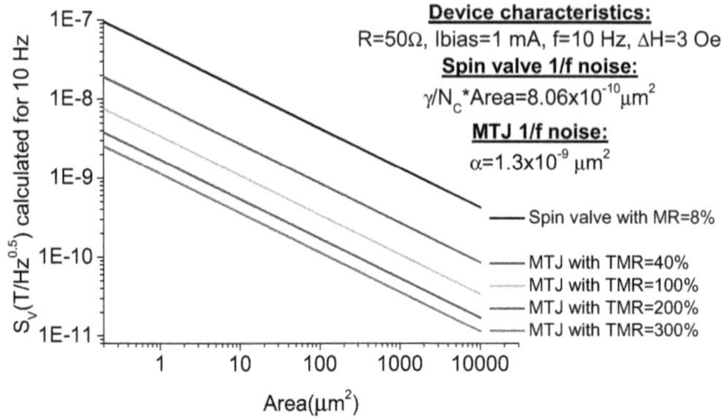

Figure 13.8 Estimated magnetic field detection limits at 10 Hz vs. sensor area and with different magnetoresistance values. Adapted with permission from Ref. Freitas et al. (2007).

negligible. The main problem with application of TMR/GMR sensors in pT field detection is that maximum of the $1/f$ noise occurs exactly in the field range corresponding to switching of the magnetization of the sensing layer (i.e., range designed to detect external field using linearized response, see, for example, Fig. 13.7).

Below we consider so-called field detectivity which represents sensors characteristics, describing its capability to detect external magnetic field.

Figure 13.8 shows magnetic field detection limits calculated for the frequencies around 10 Hz for GMR/TMR sensors with different areas and magnetoresistance values. For this calculation, the Hooge parameter was taken in the magnetically saturated conditions in order to exclude influence of the magnetic noise contribution. We note that except for correction to the TMR dependence with the bias voltage, both the sensor signal and the noise are proportional to the applied current making detectivity independent on the bias current.

13.7 Low-Frequency Noise Arrangements for Field Sensors

As we have seen in the previous paragraph, the noise floor is mainly determined by two factors: the power noise spectrum and the sensitivity of the sensor. The most obvious action to reduce noise floor would be to decrease the intrinsic noise of the sensor. Nevertheless, this is the hard task, as the sensor noise in some cases is determined by the fluctuation-dissipation theorem (thermal noise: electronic and magnetic). So, straightforward solutions are often limited. In that point, we shall review some other alternative techniques which may allow an improvement of the noise floor by mainly two different handles: sensitivity improvement and detection band displacement.

One could improve the sensitivity of a sensor simply by adding an amplifier, so that the number of volts per unit of the field would increase. This solution, however, implies using an amplifier with a noise figure close 0 dB. Otherwise, we would just add noise into the output and make things worse (see Chapter 3). The above-mentioned root could be implemented in a straightforward way by simply using a Wheatstone bridge. An even smarter solution would be to increment the measured by the sensor magnetic signal directly by incorporating some kind of low noise "magnetic amplifier" in the way that that the main noise source would remain to be provided by the sensor itself. The latter method has been implemented by adding flux transformer and flux concentrators. Below we resume the above-mentioned solutions to improve sensors sensitivity.

(a) Wheatstone bridge

A GMR Wheatstone bridge (WB) with concentrator (note that WB have been also discussed in application to AMR sensors but in the conditions when all parts of the WB varied their resistances under external field) is sensing arrangement performed by four resistors and used to sense tiny changes in the resistance of one, two, three and even all the resistors in the arrangement. The simplest configuration is depicted in Fig. 13.9, where only part of the resistors changes under an external disturbance and the rest remain constant "R" corresponding to the value under zero magnetic field. A change in the resistance will yield a ΔV between point D and ground under drive current applied between points A–C.

$$\Delta V = \frac{\Delta R}{4R} V_0 \quad (13.7)$$

This configuration yields an output proportional to the variation of resistance. Besides, the noise output (if all elements are identical in terms of noise), is the same as with one of single resistor.

(b) Magnetic amplification

In order to magnify influence of the local magnetic field acting on the sensor some useful signal amplification (for example by so-called flux concentrators) could be used. In order to create flux concentration one may deposit an additional Permalloy structures on the substrate close to TMR/GMR sensor. Figure 13.9 sketches such modification of the Wheatstone bridge by placing active resistors (R1, R4) in the gap between two flux concentrators. As a consequence, these resistors experience magnetic field much larger than external one, being enhanced in a factor approximately proportional to the ratio of the gap between the flux concentrators (D1) to the length of one of the flux concentrators (D2). Simultaneously, flux concentrators could work as shields for two other resistors (R3 and R4). Design of flux concentrators could be specially adapted to use with TMR/GMR sensors, depending on their sensitivity. For example, if a GMR material saturates at approximately

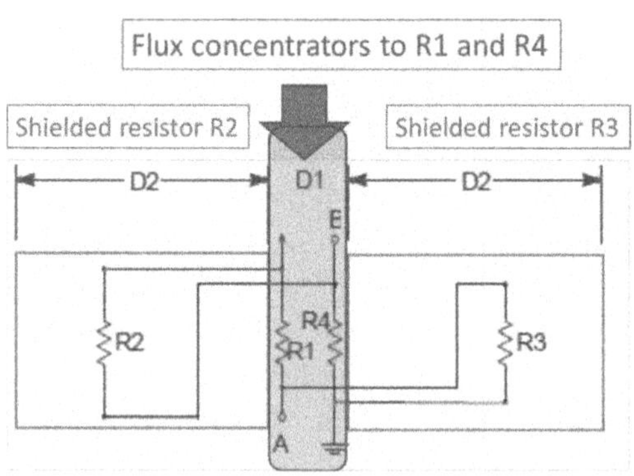

Figure 13.9 Wheatstone bridge sensor based on GMR resistors with flux concentrators.

400 Oe, different concentrators geometries could be used to build sensors saturating at say 30, 50, or 100 Oe.

(c) Chopping techniques for detection band displacement
Jandler and co-authors (Jander et al., 2003) were first to introduce chopping techniques with the intention to reduce the influence of $1/f$ noise in MTJ sensors due to band displacement. The methods proposed included sensitivity modulation via orthogonal fields, parallel field chopping driving sensor in non-linear regime or modulate the permeability of the flux concentrator. Only the parallel chopping showed some advantage at lowest frequencies, while exhibited excess noise at higher frequencies (see Fig. 13.10).

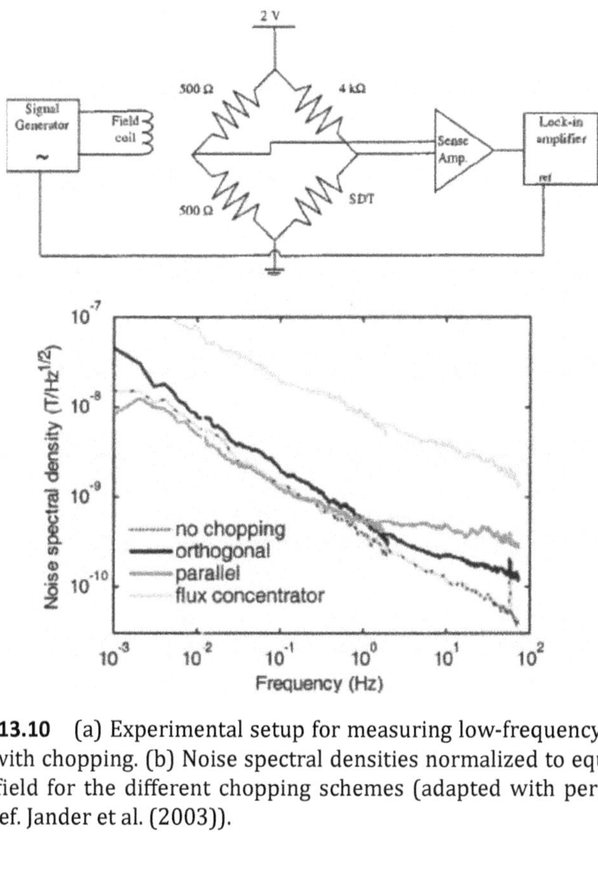

Figure 13.10 (a) Experimental setup for measuring low-frequency sensor noise with chopping. (b) Noise spectral densities normalized to equivalent sense field for the different chopping schemes (adapted with permission from Ref. Jander et al. (2003)).

Later on, Edelstein and co-authors (Edelstein et al., 2006) proposed and demonstrated the validity of new device concept with band displacement suppressing the influence of the $1/f$ noise in GMR sensors through incorporation of the microelectromechanical system (MEMS). In the proposed sensor the flux concentrator is periodically moved with kHz frequency shifting therefore the operation frequency and reducing the $1/f$ noise by one-three orders of the magnitude. It was shown that incorporation on MEMS did not changed intrinsic noise of the spin valves and only consumed microwatt power, providing strong evidence for the validity of novel device concept.

Valadeiro and co-authors (Valadeiro et al., 2015) summarized alternative strategies for pT field detection by using TMR-based sensors with soft pinned sensing layer and when device lateral size would not be an issue. In order to improve sensors sensitivity, two main methods discussed: (i) using large number of sensors connected in series and (ii) design of individual sensor with a sufficiently large sensing area and with integrated magnetic flux guides.

Implementation of the first solution has been realized in the device consisting of 952 MTJs in series with MgO barrier a sensitivity of 29.3%/mT. Detection level of 455 $pT/Hz^{1/2}$ has been demonstrated at chopping frequencies of 100 Hz. On the other hand, integration of magnetic flux guides in a single sensor devices resulted in sensitivity of about 138.3%/mT and a detection level of 576 $pT/Hz^{1/2}$ also with chopping of 100 Hz. These two strategies are possible to implement only when large spatial sensitivity is not required.

We conclude this Chapter with description of the important collaboration effort of NIST and NVE Corporation (Stutzke et al., 2005) which investigated the low-frequency noise in the range between 0.1 Hz to 10 kHz for variety of commercially available at that time (2005) magnetic field sensors such as AMR, GMR and TMR.

It has been found (in agreement with previous discussion) that for the AMR and GMR sensors both electronic and magnetic components contribute to the sensor noise with maximum noise correlated with maximum sensor sensitivity (Stutzke et al., 2005). In TMR field sensors, however, noise was found to be mainly due to resistance fluctuations possibly from defects in AlO_x barrier (note that MTJs with spin filtering by epitaxial MgO barriers were not on the market yet). It has been observed that the best low-field detectivity is of the order of 100 pT/$Hz^{0.5}$ at 1 Hz.

Chapter 14

Outlook and Perspectives

Since the first reports on the noise in spin valves (Hardner et al., 1993a) and magnetic tunnel junctions (Ingvarsson et al., 2000), giant steps have been made toward the understanding of the fundamental mechanism of the noise sources in these newly developed spintronic devices. Although we intended to cover a pedagogically wide range of possible contributions to the noise in spintronic devices and their control, many of the topics discussed in this book, as could not be otherwise, have rather close relation to the background, as well as recent research interests of the authors. Naturally, therefore some important issues related to spin-dependent noise sources could have been left out or some other aspects were kept to give a wider possible coverage of the research by different authors. This last chapter concluded the book with a brief summary of the state of the art in the field through the suggestion of perspective directions in the area which we believe could be in spotlight of the community investigating noise in spintronics in the coming years.

(i) With respect to noise control and manipulation, such possibility has been demonstrated on the example of magnetic control of the shot noise in double-barrier magnetic tunnel

Noise in Spintronics: From Understanding to Manipulation
Farkhad G. Aliev and Juan Pedro Cascales

ISBN 978-981-4774-34-5 (Hardcover), 978-1-315-11088-2 (eBook)
www.panstanford.com

junctions (Cascales et al., 2012). Further extension of the shot noise control through the implementation of the third (gate) leads to the variation in the potential of the central electrode and QWSs and is currently an important challenge.

(ii) Another perspective direction of the voltage-controlled low-frequency magnetic noise suggests exploring voltage influence on the perpendicular magnetic anisotropy. Bias-voltage-induced anisotropy modulation could provide a reduction in the magnetic noise and improvement in the signal-to-noise ratio (Wisniowski et al., 2014).

(iii) Thermal fluctuation-related magnetic noise in perpendicular magnetic spin valves have been suggested to be suppressed by the DC current-induced spin-transfer torques (Bandopadhyay et al., 2011)

(iv) In general, control over DW-related magnetic noise which dominates magnetization inversion area remains a critically important challenge to overcome in order to improve the detectivity of magnetic field sensors. Among the most promising applications are biochip-based GMR and TMR sensor arrays designed for applications in biomedicine, molecular diagnostics of cancer, cardiac, and other diseases (see for review (Wang and Guanxiong, 2008)).

(v) An alternative view on the noise in spintronics would be to use magnetic noise itself as a tool to detect creep in different physical processes such as current-induced domain wall motion in synthetic antiferromagnets (Herranz et al., 2009). Qualitatively similar method of "hearing RTN" has been applied to detect and analyze the spin torque acting on DW motion in magnetic tunnel junctions with perpendicular to the ferromagnetic lead current (Herranz et al., 2011). Critical current densities for the vertical-current-induced domain-wall motion in elliptic MgO-based magnetic tunnel junctions were found to be orders of magnitude smaller than for DW motion in lateral geometries. Interestingly, similar effects have been independently reported through direct transport measurements in specially designed curved MTJs (Chanthbouala et al., 2011).

(vi) Current-induced magnetization reversal in about 100 atom clusters with tunneling current injected from spin polarized tip (Krause et al., 2007) has been demonstrated also by analyzing random telegraph noise. Similar RTN analysis method has been later applied to demonstrate spin torque in artificially constructed magnets, composed of a few exchange-coupled atomic spins adsorbed on a metallic substrate (Khajetoorians et al., 2013).

(vii) Noise studies could probably help reveal internal instabilities in skyrmions. More specifically, Carpentieri and co-authors (Carpentieri et al., 2015) reported recently on the skyrmion instabilities in time domain due to transitions of the skyrmion number when system is DC current biased. These coupled to emission of incoherent spin wave transitions could be observed as extra noise in the time and frequency domains (Carpentieri et al., 2015).

(viii) Low-frequency dynamics of the structural defects inside the barrier and close to the barrier region presents an important source of the nonmagnetic noise yet to be controlled. Reduction of the interface mismatch at ferromagnet–barrier interface has been a first clear step in this direction. Herranz et al. (Herranz et al., 2010b) revealed more than an order decrease of the normalized $1/f$ noise accompanied by a substantial increase of the tunneling magnetoresistance in epitaxial FeV/MgO/FeV MTJs due to reduced FeV/MgO interface mismatch (Herranz et al., 2010b). Vanadium content exceeding 25% suppresses ferromagnetism of the electrodes and TMR.

(ix) In relation to interfaces, the LFN spectroscopy of the buried interfaces in spintronics and electronics in general remains yet a widely unexplored area with great potential (Aliev et al., 2014). The BENS method, recently applied to resolve band edges in topological insulators (Cascales et al., 2015b), has some conceptual similarity to the so-called Tedrow–Meservey technique (Meservey and Tedrow, 1994) intensively used to determine real spin polarization of electrons emitted from ferromagnetic buried interfaces through a thin insulator to thin superconducting films.

(x) Due to high spin polarization of some ferromagnetic oxides (among other manganites), magnetic tunnel junctions with complex oxide electrodes and barriers (Lu et al., 1996) have recently attracted considerable attention. Among the appealing properties are their giant (above 1000%) TMR (usually observed at low temperatures only when the ferromagnetic order becomes weakly established (Werner et al., 2011)) and possibility to electrically tune magnetism and TMR characteristics (Ngo et al., 2014). Low-frequency noise characteristics of such MTJs, however, have not yet been optimized. Recent studies show that LFN in such MTJs could be affected by oxide traps inside the barrier and by the low-temperature structural phase transition (Guerrero et al., 2012).

(xi) Molecular spintronics, including single-molecular devices, are another important field where the noise studies could provide valuable information on the internal molecular structure and dynamics. For example, for MTJs with organic barriers applied bias activate vibrational modes and may produce vibrational heating for those molecule which contribute to conductance channels only through molecular bending. Vibrational heating, considered so far in the single molecular case and without ferromagnetic electrodes (Brumme et al., 2012b), could be an important perturbing factor for molecular spintronics. Alternatively, the noise studies in realistic molecular spintronic devices with rough interfaces could provide information on molecule-leads bonding effectiveness.

Besides, detailed noise studies are needed to address novel physics in organic memristive devices, with both charge and spin degrees of freedom playing an important role.

(xii) Shot noise studies could provide valuable information on transport mechanisms and charge/spin correlations through single-molecules spintronic devices (Bogani and Wernsdorfer, 2008). For example, current and noise oscillations have been predicted to occur as a function of bias voltage due to the violation of spin-selection rules by quantum tunneling (Romeike et al., 2006).

(xiii) There has been recent interest in the investigation of topological insulators and 2D materials where carrier and spin transport over surfaces and edges becomes an important factor for both conductance and conductance fluctuations. We envisage a valuable and informative contribution from the noise studies to understand charge and spin dynamics at topologically different edges of graphene (open, closed, zigzag, or armchair) and similar effects in other 2D materials where the influence of broken bonds could provide strong RTN signals in conductance. General trend of downscaling in spintronics and 2D materials shall increase the relative importance of spin flips on electron and spin transport leading to $1/f$ noise when spin flips are associated with low-frequency photon emission (Jozwiak et al., 2013)

(xiv) While the magnetic field separates degenerate spin (up-down) channels, a similar role of the effective magnetic field could be played by the spin orbit (SO) interaction in spintronics (Žutić et al., 2004). Therefore, the SO interaction is able to generate effectively pure spin currents in the absence of external magnetic field. Only recently, research started to note that SO coupling, being one of the most important channels for both spin current generation and spin relaxation, is a stochastic process. Research has now revealed an important influence of the SO coupling on the shot noise in spintronics (Mirzanian et al., 2015). It has been shown that the spin Hall noise originating from SO process fluctuations (Kamra et al., 2014) is consistent with the fluctuation-dissipation theorem.

(xv) One of the open directions for the future research on noise in spintronics is related to numerical simulations of the noise. Particularly challenging are simulations of the $1/f$ noise. Despite some initial attempts in that direction (Rebei et al., 2005), the field with great application potential remains practically unexplored.

(xvi) Last but not least, shot noise of spin-polarized transport through a single atom could be used to characterize transport channels (Burtzlaff et al., 2015). Effective spin polarization of the current could be evaluated through the Fano factor. The fields of single-atom spintronics and associated noise are among the perspective fields of research.

References

Agarwalla, B. K., Jiang, J.-H., and Segal, D. (2015). Full counting statistics of vibrationally assisted electronic conduction: Transport and fluctuations of thermoelectric efficiency, *Phys. Rev. B* **92**, p. 245418, doi:10.1103/PhysRevB.92.245418, http://link.aps.org/doi/10.1103/PhysRevB.92.245418.

Aliev, F. G., Cascales, J. P., Hallal, A., Chshiev, M., and Andrieu, S. (2014). Band-edge noise spectroscopy of a magnetic tunnel junction, *Phys. Rev. Lett.* **112**, p. 216801, http://link.aps.org/doi/10.1103/PhysRevLett.112.216801.

Aliev, F. G., Guerrero, R., Herranz, D., Villar, R., Greullet, F., Tiusan, C. and Hehn, M. (2007). Very low 1/f noise at room temperature in fully epitaxial Fe/MgO/Fe magnetic tunnel junctions, *Appl. Phys. Lett.* **91**, 23, 232504, http://scitation.aip.org/content/aip/journal/apl/91/23/10.1063/1.2822812.

Almeida, J. M., Ferreira, R., Freitas, P. P., Langer, J., Ocker, B., and Maass, W. (2006). 1/f noise in linearized low resistance MgO magnetic tunnel junctions, *J. Appl. Phys.* **99**, 8, 08.

Almeida, J. M., Wisniowski, P., and Freitas, P. (2008). Low-frequency noise in MgO magnetic tunnel junctions: Hooge's parameter dependence on bias voltage, *Magn., IEEE Trans.* **44**, 11, pp. 2569–2572.

Alvarez-Hérault, J. (2010). *Magnetic Memory Using a Thermally Assisted Spin Polarized Current Writing*, Theses, Université de Grenoble, https://tel.archives-ouvertes.fr/tel-00565255.

Silva, A. V., Leitao, D. C., Valadeiro, J., Amaral, J., Freitas, P. P., Cardoso, S. (2015). Linearization strategies for high sensitivity magnetoresistive sensors, *Eur. Phys. J. Appl. Phys.* **72**, 1, pp. 10601–p1–10601–p20, doi:10.1051/epjap/2015150214, http://dx.doi.org/10.1051/epjap/2015150214.

Arakawa, T., Sekiguchi, K., Nakamura, S., Chida, K., Nishihara, Y., Chiba, D., Kobayashi, K., Fukushima, A., Yuasa, S., and Ono, T. (2011).

Sub-poissonian shot noise in CoFeB/ MgO/ CoFeB-based magnetic tunneling junctions, *Appl. Phys. Lett.* **98**, 20, 202103, http://scitation.aip.org/content/aip/journal/apl/98/20/10.1063/1.3590921.

Arakawa, T., Shiogai, J., Ciorga, M., Utz, M., Schuh, D., Kohda, M., Nitta, J., Bougeard, D., Weiss, D., Ono, T., and Kobayashi, K. (2015). Shot noise induced by nonequilibrium spin accumulation, *Phys. Rev. Lett.* **114**, p. 016601, doi:10.1103/PhysRevLett.114.016601, http://link.aps.org/doi/10.1103/PhysRevLett.114.016601.

Aurelio, D., Torres, L., and Finocchio, G. (2010). Thermal effects on spin-torque-driven switching in high-tunneling-magnetoresistance magnetic tunnel junctions, *J. Appl. Phys.* **108**, 8, 083911, http://scitation.aip.org/content/aip/journal/jap/108/8/10.1063/1.3499641.

Avriller, R., and Frederiksen, T. (2012). Inelastic shot noise characteristics of nanoscale junctions from first principles, *Phys. Rev. B* **86**, p. 155411, doi:10.1103/PhysRevB.86.155411, http://link.aps.org/doi/10.1103/PhysRevB.86.155411.

Avsar, A., Tan, J. Y., Taychatanapat, T., Balakrishnan, J., Koon, G., Yeo, Y., Lahiri, J., Carvalho, A., Rodin, A. S., O'Farrell, E., Eda, G., Castro Neto, A. H., and Özyilmaz, B. (2014). Spin–orbit proximity effect in graphene, *Nat. Commun.* **5**, p. 4875, doi:10.1038/ncomms5875, http://dx.doi.org/10.1038/ncomms5875.

Bagrets, D. A., and Nazarov, Y. V. (2003). Full counting statistics of charge transfer in Coulomb blockade systems, *Phys. Rev. B* **67**, p. 085316, http://link.aps.org/doi/10.1103/PhysRevB.67.085316.

Baibich, M., Broto, J., Fert, A., Van Dau, F., Petroff, F., Etienne, P., Creuzet, G., Friederich, A., and Chazelas, J. (1988). Giant magnetoresistance of (001) Fe/(001)Cr magnetic superlattices, *Phys. Rev. Lett.* **61**, pp. 2472–2475, doi:10.1103/PhysRevLett.61.2472, http://link.aps.org/doi/10.1103/PhysRevLett.61.2472.

Balandin, A. A. (2013). Low-frequency 1/f noise in graphene devices, *Nat. Nanotechnol.* **8**, pp. 549–555, http://www.nature.com/nnano/journal/v8/n8/full/nnano.2013.144.html.

Bandopadhyay, S., Brataas, A., and Bauer, G. E. W. (2011). Feedback control of noise in spin valves by the spin-transfer torque, *Appl. Phys. Lett.* **98**, 8, 083110, doi:http://dx.doi.org/10.1063/1.3556270, http://scitation.aip.org/content/aip/journal/apl/98/8/10.1063/1.3556270.

Barraud, C., Seneor, P., Mattana, R., Fusil, S., Bouzehouane, K., Deranlot, C., Graziosi, P., Hueso, L., Bergenti, I., and Dediu, V. (2010).

Unravelling the role of the interface for spin injection into organic semiconductors, *Nat. Phys.* **6**, pp. 615–620, http://www.nature.com/nphys/journal/v6/n8/abs/nphys1688.html.

Beenakker, C. W. J., and Büttiker, M. (1992a). Suppression of shot noise in metallic diffusive conductors, *Phys. Rev. B* **46**, pp. 1889–1892, doi:10.1103/PhysRevB.46.1889, http://link.aps.org/doi/10.1103/PhysRevB.46.1889.

Beenakker, C. W. J., and Büttiker, M. (1992b). Suppression of shot noise in metallic diffusive conductors, *Phys. Rev. B* **46**, pp. 1889–1892, http://link.aps.org/doi/10.1103/PhysRevB.46.1889.

Beenakker, C. W. J., and Patra, M. (1999). Photon shot noise, *Mod. Phys. Lett. B* **13**, 11, pp. 337–347, doi:10.1142/S0217984999000439, http://www.worldscientific.com/doi/abs/10.1142/S0217984999000439.

Belashchenko, K. D., Velev, J., and Tsymbal, E. Y. (2005). Effect of interface states on spin-dependent tunneling in Fe MgO Fe tunnel junctions, *Phys. Rev. B* **72**, p. 140404, http://link.aps.org/doi/10.1103/PhysRevB.72.140404.

Bell, L. D., and Kaiser, W. J. (1988). Observation of interface band structure by ballistic-electron-emission microscopy, *Phys. Rev. Lett.* **61**, pp. 2368–2371, http://link.aps.org/doi/10.1103/PhysRevLett.61.2368.

Belzig, W. (2005). Full counting statistics of super-poissonian shot noise in multilevel quantum dots, *Phys. Rev. B* **71**, p. 161301, http://link.aps.org/doi/10.1103/PhysRevB.71.161301.

Belzig, W., and Zareyan, M. (2004). Spin-flip noise in a multiterminal spin valve, *Phys. Rev. B* **69**, p. 140407, doi:10.1103/PhysRevB.69.140407, http:// link.aps.org/doi/10.1103/PhysRevB.69.140407.

Berger, L. (1996). Emission of spin waves by a magnetic multilayer traversed by a current, *Phys. Rev. B* **54**, pp. 9353–9358, http://link.aps.org/doi/10.1103/PhysRevB.54.9353.

Berger, L. (2003). Multilayer configuration for experiments of spin precession induced by a dc current, *J. Appl. Phys.* **93**, 10, pp. 7693–7695, http://scitation.aip.org/content/aip/journal/jap/93/10/10.1063/1.1555771.

Berger, R. F., Fennie, C. J., and Neaton, J. B. (2011). Band gap and edge engineering via ferroic distortion and anisotropic strain: The case of $SrTiO_3$, *Phys. Rev. Lett.* **107**, p. 146804, http://link.aps.org/doi/10.1103/PhysRevLett.107.146804.

Bertrand, B., Flentje, H., Takada, S., Yamamoto, M., Tarucha, S., Ludwig, A., Wieck, A. D., Bäuerle, C., and Meunier, T. (2015). Quantum manipulation of two-electron spin states in isolated double quantum dots, *Phys. Rev. Lett.* **115**, p. 096801, doi:10.1103/PhysRevLett.115.096801, http://link.aps.org/doi/10.1103/PhysRevLett.115.096801.

Binasch, G., Grünberg, P., Saurenbach, F., and Zinn, W. (1989). Enhanced magnetoresistance in layered magnetic structures with antiferromagnetic interlayer exchange, *Phys. Rev. B* **39**, pp. 4828–4830, doi: 10.1103/PhysRevB.39.4828, http://link.aps.org/doi/10.1103/PhysRevB.39.4828.

Blake, P., Hill, E. W., Castro Neto, A. H., Novoselov, K. S., Jiang, D., Yang, R., Booth, T. J., and Geim, A. K. (2007). Making graphene visible, *Appl. Phys. Lett.* **91**, 6, p. 063124, doi:http://dx.doi.org/10.1063/1.2768624, http://scitation.aip.org/content/aip/journal/apl/91/6/10.1063/1.2768624.

Blanter, Y., and Büttiker, M. (2000a). Shot noise in mesoscopic conductors, *Phys. Rep.* **336**, pp. 1–166, http://www.sciencedirect.com/science/article/pii/S0370157399001234.

Blanter, Y. M., and Büttiker, M. (2000b). Shot Noise in Mesoscopic Conductors, *Phys. Repor.* **336**.

Blöchl, P. E. (1994). Projector augmented-wave method, *Phys. Rev. B* **50**, pp. 17953–17979, http://link.aps.org/doi/10.1103/PhysRevB.50.17953.

Bogani, L., and Wernsdorfer, W. (2008). Molecular spintronics using single-molecule magnets, *Nat. Mater.* **7**, 202501, doi:10.1038/nmat2133, http://dx.doi.org/10.1038/nmat2133.

Bonell, F., Andrieu, S., Bataille, A. M., Tiusan, C., and Lengaigne, G. (2009a). Consequences of interfacial Fe-O bonding and disorder in epitaxial Fe/MgO/Fe(001) magnetic tunnel junctions, *Phys. Rev. B* **79**, p. 224405, doi:10.1103/PhysRevB.79.224405, http://link.aps.org/doi/10.1103/PhysRevB.79.224405.

Bonell, F., Andrieu, S., Bertran, F., Lefevre, P., Ibrahimi, A., Snoeck, E., Tiusan, C.-V., and Montaigne, F. (2009b). MgO-based epitaxial magnetic tunnel junctions using Fe-V electrodes, *IEEE Trans. Magn.* **45**, 10, pp. 3467–3471.

Bonell, F., Andrieu, S., Tiusan, C., Montaigne, F., Snoeck, E., Belhadji, B., Calmels, L., Bertran, F., Le Fvre, P., and Taleb-Ibrahimi, A. (2010). Influence of misfit dislocations on the magnetoresistance of MgO-based epitaxial magnetic tunnel junctions, *Phys. Rev. B* **82**, p. 092405, http://link.aps.org/doi/10.1103/PhysRevB.82.092405.

Bonell, F., Bataille, A. M., Andrieu, S., Tiusan, C., Kierren, B., Lengaigne, G., and Lacour, D. (2008). Influence of interfacial oxygen on single-crystal magnetic tunnel, *Eur. Phys. J. Appl. Phys* **361**, pp. 357–361.

Bonell, F., Hauet, T., Andrieu, S., Bertran, F., Le Fvre, P., Calmels, L., Tejeda, A., Montaigne, F., Warot-Fonrose, B., Belhadji, B., Nicolaou, A., and Taleb-Ibrahimi, A. (2012). Spin-polarized electron tunneling in bcc FeCo/MgO/FeCo(001) magnetic tunnel junctions, *Phys. Rev. Lett.* **108**, p. 176602, http://link.aps.org/doi/10.1103/PhysRevLett.108.176602.

Borovitskaya, E., and Shur, M. (2001). On theory of 1/f noise in semiconductors, *Solid-State Electron.* **45**, 7, pp. 1067–1069, http://www.sciencedirect.com/science/article/pii/S003811010100137X.

Bostwick, A., Ohta, T., Seyller, T., Horn, K., and Rotenberg, E. (2006). Quasiparticle dynamics in graphene, *Nat. Phys.* **3**, doi:10.1038/nphys477.

Bowen, M., Cros, V., Petroff, F., Fert, A., Martinez Boubeta, C., Costa-Krämer, J. L., Anguita, J. V., Cebollada, A., Briones, F., de Teresa, J. M., Morellón, L., Ibarra, M. R., Güell, F., Peiró, F., and Cornet, A. (2001). Large magnetoresistance in Fe/MgO/FeCo(001) epitaxial tunnel junctions on GaAs (001), *Appl. Phys. Lett.* **79**, 11, pp. 1655–1657, http://scitation.aip.org/content/aip/journal/apl/79/11/10.1063/1.1404125.

Brataas, A., Bauer, G. E., and Kelly, P. J. (2006). Non-collinear magnetoelectronics, *Phys. Rep.* **427**, 4, pp. 157–255, doi:http://dx.doi.org/10.1016/j.physrep.2006.01.001, http://www.sciencedirect.com/science/article/pii/S0370157306000238.

Brataas, A., Kent, A. D., and Ohno, H. (2012). Current-induced torques in magnetic materials, *Nat. Mater.* **11**, pp. 372–381.

Braun, S., Salaneck, W. R., and Fahlman, M. (2009). Energy-level alignment at organic/metal and organic/organic interfaces, *Adv. Mater.* **21**, 14-15, pp. 1450–1472, http://dx.doi.org/10.1002/adma.200802893.

Brito, F., Estanislau, J., and Egues, J. (2001). Shot noise in spin-polarized currents, *J. Magnet. Magn. Mater.* **226–230, Part 1**, 0, pp. 457–459, doi:http://dx.doi.org/10.1016/S0304-8853(00)01123-9, http://www.sciencedirect.com/science/article/pii/S0304885300011239, proceedings of the International Conference on Magnetism (ICM 2000).

Brown, R., and Mazey, D. (1828). The Philosophical Magazine, or *Annals of Chemistry, Mathematics, Astronomy, Natural History and General Science* **4**, p. 161.

Brumme, T., Gutierrez, R., and Cuniberti, G. (2012a). Vibrational heating in single-molecule switches: an energy-dependent density-of-states approach, *J. Phys.: Condensed Matter* **24**, 39, p. 394003, http://stacks.iop.org/0953-8984/24/i=39/a=394003.

Brumme, T., Gutierrez, R., and Cuniberti, G. (2012b). Vibrational heating in single-molecule switches: an energy-dependent density-of-states approach, *J. Phys.: Condensed Matter* **24**, 39, p. 394003, http://stacks.iop.org/0953-8984/24/i=39/a=394003.

Brumme, T., Neucheva, O. A., Toher, C., Gutiérrez, R., Weiss, C., Temirov, R., Greuling, A., Kaczmarski, M., Rohlfing, M., Tautz, F. S., and Cuniberti, G. (2011). Dynamical bistability of single-molecule junctions: A combined experimental and theoretical study of PTCDA on Ag(111), *Phys. Rev. B* **84**, p. 115449, doi:10.1103/PhysRevB.84.115449, http://link.aps.org/doi/10.1103/PhysRevB.84.115449.

Bułka, B. R. (2000). Current and power spectrum in a magnetic tunnel device with an atomic-size spacer, *Phys. Rev. B* **62**, pp. 1186–1192, doi:10.1103/PhysRevB.62.1186, http://link.aps.org/doi/10.1103/PhysRevB. 62.1186.

Bułka, B. R., Martinek, J., Michałek, G., and Barnaś, J. (1999). Shot noise in ferromagnetic single-electron tunneling devices, *Phys. Rev. B* **60**, pp. 12246–12255, doi:10.1103/PhysRevB.60.12246, http://link.aps.org/doi/10.1103/PhysRevB.60.12246.

Burtzlaff, A., Weismann, A., Brandbyge, M., and Berndt, R. (2015). Shot noise as a probe of spin-polarized transport through single atoms, *Phys. Rev. Lett.* **114**, p. 016602, doi:10.1103/PhysRevLett.114.016602, http://link.aps.org/doi/10.1103/PhysRevLett.114.016602.

Butler, W. H., Zhang, X.-G., Schulthess, T. C., and MacLaren, J. M. (2001). Spin-dependent tunneling conductance of Fe–MgO–Fe sandwiches, *Phys. Rev. B* **63**, p. 054416, http://link.aps.org/doi/10.1103/PhysRevB.63.054416.

Callegaro, L. (2006). Unified derivation of Johnson and shot noise expressions, *Am. J. Phys.* **74**, p. 438, http://scitation.aip.org/content/aapt/journal/ajp/74/5/10.1119/1.2174034.

Callen, H. B., and Welton, T. A. (1951). Irreversibility and generalized noise, *Phys. Rev.* **83**, pp. 34–40, http://link.aps.org/doi/ 10.1103/PhysRev.83.34.

Camley, R. E. and Barnas, J. (1989). *Phys. Rev. Lett.*, **63**, p. 664. DOI: https://doi.org/10.1103/PhysRevLett.63.664.

Cardoso, S., Leitao, D., Gameiro, L., Cardoso, F., Ferreira, R., Paz, E., and Freitas, P. (2014). Magnetic tunnel junction sensors with ptesla sensitivity, *Microsyst. Technol.* **20**, 4-5, pp. 793–802, doi:10.1007/s00542-013-2035-1, http://dx.doi.org/10.1007/s00542-013-2035-1.

Carpentieri, M., Tomasello, R., Zivieri, R., and Finocchio, G. (2015). Topological, non-topological and instanton droplets driven by spin-transfer torque in materials with perpendicular magnetic anisotropy and Dzyaloshinskii–Moriya interaction, *Sci. Rep.* **5**, p. 16184, doi: 10.1038/srep16184, http://dx.doi.org/10.1038/srep16184.

Carroll, C. E. (1971). *Statistical Physics, Vol. 5: Course of Theoretical Physics*, L.D. Landau and E.M. Lifshitz, 2nd rev. ed., 484 pp., New York, London, Pergamon Press, 1969, vol. 291 (Elsevier Science).

Cascales, J. P. (2015). *Electron Transport and Noise in Magnetic Tunnel Junctions with MgO Barriers*, Ph.D. thesis, Universidad Autónoma de Madrid, Spain, http://www.uam.es/gruposinv/magtran/pages/theses/Tesis_J.P.Cascales_11Feb2015.pdf.

Cascales, J. P., Herranz, D., Aliev, F. G., Szczepański, T., Dugaev, V. K., Barnaś, J., Duluard, A., Hehn, M., and Tiusan, C. (2012). Controlling shot noise in double-barrier magnetic tunnel junctions, *Phys. Rev. Lett.* **109**, p. 066601, http://link.aps.org/doi/10.1103/PhysRevLett.109.066601.

Cascales, J. P., Herranz, D., Ebels, U., Katine, J. A., and Aliev, F. G. (2015a). Detection of spin torque magnetization dynamics through low frequency noise, *Appl. Phys. Lett.* **107**, 5, p. 052401, http://scitation.aip.org/content/aip/journal/apl/107/5/10.1063/1.4927827.

Cascales, J. P., Herranz, D., Sambricio, J. L., Ebels, U., Katine, J. A., and Aliev, F. G. (2013a). Magnetization reversal in sub-100 nm magnetic tunnel junctions with ultrathin MgO barrier biased along the hard axis, *Appl. Phys. Lett.* **102**, 9, 092404, http://scitation.aip.org/content/aip/journal/apl/102/9/10.1063/1.4794537.

Cascales, J. P., Hong, J.-Y., Martinez, I., Lin, M.-T., Szczepański, T., Dugaev, V. K., Barnaś, J., and Aliev, F. G. (2014). Superpoissonian shot noise in organic magnetic tunnel junctions, *Appl. Phys. Lett.* **105**, 23, http://scitation.aip.org/content/aip/journal/apl/105/23/10.1063/1.4903739.

Cascales, J. P., Martin, L., Dulluard, A., Hehn, M., Tiusan, C., Szczepański, T., Dugaev, V. K., Barnaś, J., and Aliev, F. G. (2013b). Shot noise in epitaxial double-barrier magnetic tunnel junctions, *IEEE Trans. Magn.* **49**, 7, pp. 4347–4350, http://ieeexplore.ieee.org/xpl/login.jsp?tp=&arnumber=6559229&url=http%3A%2F%2Fieeexplore.ieee.org%2Fiel7%2F20%2F6558881%2F06559229.pdf%3Farnumber%3D6559229.

Cascales, J. P., Martínez, I., Katmis, F., Chang, C.-Z., Guerrero, R., Moodera, J. S., and Aliev, F. G. (2015b). Band structure of topological insulators from noise measurements in tunnel junctions, *Appl. Phys. Lett.* **107**, 25, 252402, doi:http://dx.doi.org/10.1063/1.4938243, http://scitation.aip.org/content/aip/journal/apl/107/25/10.1063/1.4938243.

Cebollada, A., Martnez, J., Gallego, J., de Miguel, J., Miranda, R., Ferrer, S., Batallán, F., Fillion, G., and Rebouillat, J. (1989). Antiferromagnetic ordering in Co-Cu single-crystal superlattices, *Phys. Rev. B* **39**, pp. 9726–9729, doi:10.1103/PhysRevB.39.9726, http://link.aps.org/ doi/10.1103/PhysRevB.39.9726.

Chanthbouala, A., Matsumoto, R., Grollier, J., Cros, V., Anane, A., Fert, A., Khvalkovskiy, A. V., Zvezdin, K. A., Nishimura, K., Nagamine, Y., Maehara, H., Tsunekawa, K., Fukushima, A., and Yuasa, S. (2011). Vertical-current-induced domain-wall motion in MgO-based magnetic tunnel junctions with low current densities, *Nat. Phys.* **7**, 8, pp. 626–630, doi:10.1063/1.118933, http://link.aip.org/link/?APL/70/2610/1.

Chen, Z., Lin, Y.-M., Rooks, M. J., and Avouris, P. (2007). Graphene nano-ribbon electronics, *Physica E: Low-dimensional Systems and Nanostructures* **40**, 2, pp. 228–232, doi:http://dx.doi.org/10.1016/j.physe.2007.06.020, http://www.sciencedirect.com/science/article/pii/S1386947707001427, international Symposium on Nanometer-Scale Quantum Physics.

Chen, X., Zheng, Q.-R., and Su, G. (2008). Spin transfer and critical current for magnetization reversal in ferromagnet-ferromagnet- ferromagnet double-barrier tunnel junctions, *Phys. Rev. B* **78**, p. 104410, doi:10.1103/PhysRevB.78.104410, http://link.aps.org/doi/10.1103/PhysRevB.78.104410.

Chshiev, M., Stoeffler, D., Vedyayev, A., and Ounadjela, K. (2002). Magnetic diode effect in double-barrier tunnel junctions, *EPL (Europhys. Lett.)* **58**, 2, p. 257, http://stacks.iop.org/0295-5075/58/i=2/a=257.

Chudnovskiy, A. L., Swiebodzinski, J., and Kamenev, A. (2008). Spin-torque shot noise in magnetic tunnel junctions, *Phys. Rev. Lett.* **101**, p. 066601, doi:10.1103/PhysRevLett.101.066601, http://link.aps.org/doi/10.1103/PhysRevLett.101.066601.

Clément, N., Pleutin, S., Seitz, O., Lenfant, S., and Vuillaume, D. (2007). $1/f^{\gamma}$ tunnel current noise through Si-bound alkyl monolayers, *Phys. Rev. B* **76**, p. 205407, doi:10.1103/PhysRevB.76.205407, http://link.aps.org/doi/10.1103/PhysRevB.76.205407.

Cobas, E., Friedman, A. L., vant Erve, O. M. J., Robinson, J. T. and Jonker, B. T. (2012). Graphene as a tunnel barrier: graphene-based magnetic tunnel junctions, *Nano Lett.* **12**, pp. 3000–4. doi: 10.1021/nl3007616.

Colis, S., Gieres, G., Bär, L., and Wecker, J. (2003). Low tunnel magnetoresistance dependence versus bias voltage in double barrier magnetic tunnel junction, *Appl. Phys. Lett.* **83**, 5, pp. 948–950, http://scitation.aip.org/content/aip/journal/apl/83/5/10.1063/1.1597964.

Cottet, A., Belzig, W., and Bruder, C. (2004a). Positive cross-correlations due to dynamical channel blockade in a three-terminal quantum dot, *Phys. Rev. B* **70**, p. 115315, doi:10.1103/PhysRevB.70.115315, http://link.aps.org/doi/10.1103/PhysRevB.70.115315.

Cottet, A., Belzig, W., and Bruder, C. (2004b). Positive cross correlations in a three-terminal quantum dot with ferromagnetic contacts, *Phys. Rev. Lett.* **92**, p. 206801, doi:10.1103/PhysRevLett.92.206801, http://link.aps.org/doi/10.1103/PhysRevLett.92.206801.

Cucchiara, J., Henry, Y., Ravelosona, D., Lacour, D., Fullerton, E. E., Katine, J. A., and Mangin, S. (2009). Telegraph noise due to domain wall motion driven by spin current in perpendicular magnetized nanopillars, *Appl. Phys. Lett.* **94**, 10, 102503, doi:http://dx.doi.org/10.1063/1.3089569, http://scitation.aip.org/content/aip/journal/apl/94/10/10.1063/1.3089569.

Dankert, A., Langouche, L., Kamalakar, M. V., and Dash, S. P. (2014). High-performance molybdenum disulfide field-effect transistors with spin tunnel contacts, *ACS Nano* **8**, 1, pp. 476–482, doi:10.1021/nn404961e, http://dx.doi.org/10.1021/nn404961e, pMID: 24377305.

Danneau, R., Wu, F., Craciun, M. F., Russo, S., Tomi, M. Y., Salmilehto, J., Morpurgo, A. F., and Hakonen, P. J. (2008). Shot noise in ballistic graphene, *Phys. Rev. Lett.* **100**, p. 196802, doi:10.1103/PhysRevLett.100.196802, http://link.aps.org/doi/10.1103/PhysRevLett.100.196802.

Danneau, R., Wu, F., Tomi, M. Y., Oostinga, J. B., Morpurgo, A. F., and Hakonen, P. J. (2010). Shot noise suppression and hopping conduction in graphene nanoribbons, *Phys. Rev. B* **82**, p. 161405, doi:10.1103/PhysRevB.82.161405, http://link.aps.org/doi/10.1103/PhysRevB.82.161405.

Souza, F. M., Jauho, A. P., and Egues, J. C. (2008). Spin-polarized current and shot noise in the presence of spin flip in a quantum dot via nonequilibrium Green's functions, *Phys. Rev. B* **78**, 155303 DOI: https://doi.org/10.1103/PhysRevB.78.155303.

De Teresa, J. M., Barthélémy, A., Fert, A., Contour, J. P., Montaigne, F. and Seneor, P. (1999). Role of metal-oxide interface in determining the spin polarization of magnetic tunnel junctions, *Science* **286**, 5439, pp. 507–509, http://www.sciencemag.org/content/286/5439/507.abstract.

Dediu, V., Hueso, L. E., Bergenti, I., Riminucci, A., Borgatti, F., Graziosi, P., Newby, C., Casoli, F., De Jong, M. P., Taliani, C., and Zhan, Y. (2008). Room-temperature spintronic effects in Alq_3-based hybrid devices, *Phys. Rev. B* **78**, p. 115203, http://link.aps.org/doi/10.1103/ PhysRevB.78.115203.

Dediu, V., Murgia, M., Matacotta, F., Taliani, C., and Barbanera, S. (2002). Room temperature spin polarized injection in organic semiconductor, *Solid State Commun.* **122**, pp. 181–184, http://www.sciencedirect.com/science/article/pii/S003810980200090X.

Devolder, T., Bianchini, L., Kim, J.-V., Crozat, P., Chappert, C., Cornelissen, S., Op de Beeck, M., and Lagae, L. (2009). Auto-oscillation and narrow spectral lines in spin-torque oscillators based on MgO magnetic tunnel junctions, *J. Appl. Phys.* **106**, 10, 103921, http://scitation.aip.org/content/aip/journal/jap/106/10/10.1063/1.3260233.

Diao, Z., Li, Z., Wang, S., Ding, Y., Panchula, A., Chen, E., Wang, L.-C., and Huai, Y. (2007). Spin-transfer torque switching in magnetic tunnel junctions and spin-transfer torque random access memory, *J. Phys. Condensed Matter* **19**, pp. 165209–0, http://iopscience.iop.org/0953-8984/19/16/165209.

Diao, Z., Nowak, E. R., Haughey, K. M., and Coey, J. M. D. (2011). Nanoscale dissipation and magnetoresistive $1/f$ noise in spin valves, *Phys. Rev. B* **84**, p. 094412, doi:10.1103/PhysRevB.84.094412, http://link.aps.org/doi/10.1103/PhysRevB.84.094412.

DiCarlo, L., Williams, J. R., Zhang, Y., McClure, D. T., and Marcus, C. M. (2008). Shot noise in graphene, *Phys. Rev. Lett.* **100**, p. 156801, doi: 10.1103/PhysRevLett.100.156801, http://link.aps.org/doi/10.1103/PhysRevLett.100.156801.

Dimitrov, D. V., Gao, Z., Wang, X., Jung, W., Lou, X., and Heinonen, O. G. (2009). Dielectric breakdown of MgO magnetic tunnel junctions, *Appl. Phys. Lett.* **94**, 12, 123110, http://scitation.aip.org/content/aip/journal/apl/94/12/10.1063/1.3109792.

DiVincenzo, D. P., and Mele, E. J. (1984). Self-consistent effective-mass theory for intralayer screening in graphite intercalation compounds, *Phys. Rev. B* **29**, pp. 1685–1694, doi:10.1103/PhysRevB.29.1685, http://link.aps.org/doi/10.1103/PhysRevB.29.1685.

Donahue, M. J., Porter, D. G., of Standards, N. I., and (U.S.), T. (1999). *OOMMF User's Guide [microform]/M.J. Donahue, D.G. Porter*, version 1.0. edn. (U.S. Dept. of Commerce, Technology Administration, National Institute of Standards and Technology Gaithersburg, MD).

Dong, B., Lei, X. L., and Horing, N. J. M. (2009). Positive current noise cross correlations in capacitively coupled double quantum dots with ferromagnetic leads, *Phys. Rev. B* **80**, p. 153305, doi:10.1103/PhysRevB.80.153305, http://link.aps.org/doi/10.1103/PhysRevB. 80. 153305.

Dong, B., and Lei X.L. (2007). Inelastic cotunneling current and shot noise of an interacting quantum dot with ferromagnetic correlations, *Commun. Theor. Phys.* **48**, pp. 1099–1106.

Droghetti, A., Rungger, I., Cinchetti, M., and Sanvito, S. (2015). Vibron-assisted spin relaxation at a metal/organic interface, *Phys. Rev. B* **91**, p. 224427, doi:10.1103/PhysRevB.91.224427, http://link.aps.org/doi/10.1103/PhysRevB.91.224427.

Du, G. X., Wang, S. G., Ma, Q. L., Wang, Y., Ward, R. C. C., Zhang, X.-G., Wang, C., Kohn, A., and Han, X. F. (2010). Spin-dependent tunneling spectroscopy for interface characterization of epitaxial Fe/MgO/Fe magnetic tunnel junctions, *Phys. Rev. B* **81**, p. 064438, http://link.aps.org/doi/10.1103/PhysRevB.81.064438.

Du Pré, F. (1950). A suggestion regarding the spectral density of flicker noise, *Phys. Rev.* **78**, pp. 615–615, http://journals.aps.org/pr/abstract/10.1103/PhysRev.78.615.

Duan, X., Wang, C., Pan, A., Yu, R., and Duan, X. (2015). Two-dimensional transition metal dichalcogenides as atomically thin semiconductors: opportunities and challenges, *Chem. Soc. Rev.* **44**, pp. 8859–8876, doi:10.1039/C5CS00507H, http://dx.doi.org/10.1039/C5CS00507H.

Durin, G., Falferi, P., Cerdonio, M., Prodi, G. A., and Vitale, S. (1993). Low temperature properties of soft magnetic materials: Magnetic viscosity and 1/f thermal noise, *J. Appl. Phys.* **73**, 10, pp. 5363–5365, http://scitation.aip.org/content/aip/journal/jap/73/10/10.1063/1.353732.

Dutta, P., and Horn, P. M. (1981). Low-frequency fluctuations in solids: 1f noise, *Rev. Mod. Phys.* **53**, pp. 497–516, http://link.aps.org/doi/10.1103/RevModPhys.53.497.

Edelstein, A. S., Fischer, G. A., Pedersen, M., Nowak, E. R., Cheng, S. F., and Nordman, C. A. (2006). Progress toward a thousandfold reduction in 1/f noise in magnetic sensors using an ac microelectromechanical system flux concentrator (invited), *J. Appl. Phys.* **99**, 8, 08.

Egues, J. C., Burkard, G., and Loss, D. (2002). Rashba spin-orbit interaction and shot noise for spin-polarized and entangled electrons, *Phys. Rev. Lett.* **89**, p. 176401, doi:10.1103/PhysRevLett.89.176401, http://link.aps.org/doi/10.1103/PhysRevLett.89.176401.

Eklund, A., Bonetti, S., Sani, S. R., Majid Mohseni, S., Persson, J., Chung, S., Amir Hossein Banuazizi, S., Iacocca, E., Ostling, M., Akerman, J., and Gunnar Malm, B. (2014). Dependence of the colored frequency noise in spin torque oscillators on current and magnetic field, *Appl. Phys. Lett.* **104**, 9, 092405, http://scitation.aip.org/content/aip/journal/apl/104/9/ 10.1063/1.4867257.

Eletskii, A. V., Iskandarova, I. M., Knizhnik, A. A., and Krasikov, D. N. (2011). Graphene: fabrication methods and thermophysical properties, *Physics-Uspekhi* **54**, 3, p. 227, http://stacks.iop.org/1063-7869/54/i=3/a=R01.

Eschrig, M. (2015). Spin-polarized supercurrents for spintronics: A review of current progress, *Rep. Prog. Phys.* **78**, 10, p. 104501, http://stacks.iop.org/0034-4885/78/i=10/a=104501.

Esposito, M., Ochoa, M. A., and Galperin, M. (2015). Efficiency fluctuations in quantum thermoelectric devices, *Phys. Rev. B* **91**, p. 115417, doi:10.1103/PhysRevB.91.115417, http://link.aps.org/doi/10.1103/PhysRevB.91.115417.

Fahy, S., Lindsay, A., Ouerdane, H., and O'Reilly, E. P. (2006). Alloy scattering of *n*-type carriers in GaN_xAs_{1-x}, *Phys. Rev. B* **74**, p. 035203, http://link.aps.org/doi/10.1103/PhysRevB.74.035203.

Fasolino, A., Los, J. H., and Katsnelson, M. I. (2007). Intrinsic ripples in graphene, *Nat. Mater.* **6**, doi:10.1038/nmat2011, http://www.nature.com/nmat/journal/v6/n11/full/nmat2011.html.

Faure-Vincent, J., Tiusan, C., Bellouard, C., Popova, E., Hehn, M., Montaigne, F., and Schuhl, A. (2002). Interlayer magnetic coupling interactions of two ferromagnetic layers by spin polarized tunneling, *Phys. Rev. Lett.* **89**, p. 107206, http://link.aps.org/doi/10.1103/PhysRevLett.89.107206.

Faure-Vincent, J., Tiusan, C., Jouguelet, E., Canet, F., Sajieddine, M., Bellouard, C., Popova, E., Hehn, M., Montaigne, F., and Schuhl, A. (2003). High tunnel magnetoresistance in epitaxial Fe/MgO/Fe tunnel junctions, *Appl. Phys. Lett.* **82**, 25, pp. 4507–4509, http://scitation.aip.org/content/aip/journal/apl/82/25/10.1063/1.1586785.

Fay, A., Danneau, R., Viljas, J. K., Wu, F., Tomi, M. Y., Wengler, J., Wiesner, M., and Hakonen, P. J. (2011). Shot noise and conductivity at high bias in bilayer graphene: Signatures of electron-optical phonon coupling, *Phys. Rev. B* **84**, p. 245427, doi:10.1103/PhysRevB.84.245427, http://link.aps.org/doi/10.1103/PhysRevB.84.245427.

Feng, G., Van Dijken, S., and Coey, J. M. D. (2006). Influence of annealing on the bias voltage dependence of tunneling magnetoresistance in MgO double-barrier magnetic tunnel junctions with CoFeB electrodes, *Appl. Phys. Lett.* **89**, 16, pp. 162501–162501–3.

Feng, G., van Dijken, S., Feng, J. F., Coey, J. M. D., Leo, T., and Smith, D. J. (2009). Annealing of CoFeB/MgO based single and double barrier magnetic tunnel junctions: Tunnel magnetoresistance, bias dependence, and output voltage, *J. Appl. Phys.* **105**, 3, 033916, http://scitation.aip.org/content/aip/journal/jap/105/3/10.1063/1.3068186.

Fert, A., and Jaffrs, H. (2001). Conditions for efficient spin injection from a ferromagnetic metal into a semiconductor, *Phys. Rev. B* **64**, p. 184420, http://link.aps.org/doi/10.1103/PhysRevB.64.184420.

Finocchio, G. (2008). Micromagnetic model of magnetization reversal driven by spin-polarized current in MgO-based magnetic tunnel junctions, *physica status solidi (c)* **5**, 7, pp. 2396–2399, http://dx.doi.org/10.1002/pssc.200777632.

Finocchio, G., Carpentieri, M., Torres, L., Consolo, G., Romeo, A., Lopez-Diaz, L., and Azzerboni, B. (2008). Numerical study of the magnetization reversal driven by spin-polarized current in MgO-based magnetic tunnel junctions, *Phys. B: Condensed Matter* **403**, 2–3, pp. 364–367, http://www.sciencedirect.com/science/article/pii/S092145260700590X, proceedings of the Sixth International Symposium on Hysteresis Modeling and Micromagnetics.

Finocchio, G., Consolo, G., Carpentieri, M., Romeo, A., Azzerboni, B., Torres, L., and Lopez-Diaz, L. (2006). Trends in spin-transfer-driven magnetization dynamics of CoFe/AlO/Py and CoFe/MgO/Py magnetic tunnel junctions, *Appl. Phys. Lett.* **89**, 26, 262509, http://scitation.aip.org/content/aip/journal/apl/89/26/10.1063/1.2425017.

Foros, J., Brataas, A., Bauer, G. E. W., and Tserkovnyak, Y. (2009). Noise and dissipation in magnetoelectronic nanostructures, *Phys. Rev. B* **79**, p. 214407, doi:10.1103/PhysRevB.79.214407, http://link.aps.org/doi/10.1103/PhysRevB.79.214407.

Freitas, P. P., Ferreira, R., Cardoso, S., and Cardoso, F. (2007). Magnetoresistive sensors, *J. Phys.: Condensed Matter* **19**, 16, p. 165221, http://stacks.iop.org/0953-8984/19/i=16/a=165221.

Fuchs, G. D., Emley, N. C., Krivorotov, I. N., Braganca, P. M., Ryan, E. M., Kiselev, S. I., Sankey, J. C., Ralph, D. C., Buhrman, R. A., and Katine, J. A. (2004). Spin-transfer effects in nanoscale magnetic tunnel junctions, *Appl. Phys. Lett.* **85**, 7, pp. 1205–1207, http://scitation.aip.org/content/aip/journal/apl/85/7/10.1063/1.1781769.

Gallagher, W., and Parkin, S. (2006). Development of the magnetic tunnel junction MRAM at IBM: From first junctions to a 16-mb MRAM demonstrator chip, *IBM J. Res. Dev.* **50**, 1, pp. 5–23.

Galperin, M., Nitzan, A., and Ratner, M. A. (2006). Inelastic tunneling effects on noise properties of molecular junctions, *Phys. Rev. B* **74**, p. 075326, http://link.aps.org/doi/10.1103/PhysRevB.74.075326.

Galperin, M., Ratner, M. A., and Nitzan, A. (2007). Molecular transport junctions: vibrational effects, *J. Phys.: Condensed Matter* **19**, 10, p. 103201, http://stacks.iop.org/0953-8984/19/i=10/a=103201.

Garzon, S., Chen, Y., and Webb, R. (2007). Enhanced spin-dependent shot noise in magnetic tunnel barriers, *Phys. E: Low-dimensional Sys., and Nanostruct.* **40**, pp. 133–140, http://arxiv.org/abs/cond-mat/0701458.

Gasiorowicz, S. (2003). *Quantum Physics* (John Wiley & Sons).

George, P. K., Wu, Y., White, R. M., Murdock, E., and Tondra, M. (2002). Shot noise in low-resistance magnetic tunnel junctions, *Appl. Phys. Lett.* **80**, 4, pp. 682–684, http://scitation.aip.org/content/aip/journal/apl/80/4/10.1063/1.1446210.

Gmitra, M., and Fabian, J. (2015). Graphene on transition-metal dichalcogenides: A platform for proximity spin-orbit physics and optospintronics, *Phys. Rev. B* **92**, p. 155403, doi:10.1103/PhysRevB.92.155403, http://link.aps.org/doi/10.1103/PhysRevB.92.155403.

Gokce, A., Nowak, E. R., Yang, S. H., and Parkin, S. S. P. (2006a). 1/f noise in magnetic tunnel junctions with MgO tunnel barriers, *J. Appl. Phys.* **99**, pp. 8–0, http://ieeexplore.ieee.org/xpl/login.jsp?tp=&arnumber=4950539&url=http%3A%2F%2Fieeexplore.ieee.org%2Fiel5%2F4915369%2F4950446%2F04950539.pdf%3Farnumber%3D4950539.

Gokce, A., Nowak, E. R., Yang, S. H., and Parkin, S. S. P. (2006b). 1/f noise in magnetic tunnel junctions with MgO tunnel barriers, *J. Appl. Phys.* **99**, 8, p. 08.

Gokce, A., Stearrett, R., Nowak, E. R., and Nordman, C. (2011). Shot noise suppression in individual and series arrays of magnetic tunnel junctions, *Fluctuation Noise Lett.* **10**, 04, pp. 381–394, doi: 10.1142/S0219477511000648, http://www.worldscientific.com/doi/abs/10.1142/S0219477511000648.

Gokce, A., Stearrett, R., Nowak, E. R., and Nordman, C. (2011). Shot noise suprression in individual and series arrays of magnetic tunnel junctions, *Fluctuation Noise Lett.* **10**, 04, pp. 381–394, http://www.worldscientific.com/doi/abs/10.1142/S0219477511000648.

Goldhaber-Gordon, D. Shtrikman, Hadas Mahalu, D. Abusch-Magder, David Meirav, U. Kastner, M. A. (1999). Kondo effect in a single-electron transistor, *Nature* **391**, 6663.

Greullet, F., Tiusan, C., Montaigne, F., Hehn, M., Halley, D., Bengone, O., Bowen, M., and Weber, W. (2007). Evidence of a symmetry-dependent metallic barrier in fully epitaxial MgO based magnetic tunnel junctions, *Phys. Rev. Lett.* **99**, p. 187202, http://link.aps.org/doi/10.1103/PhysRevLett.99.187202.

Grigorenko, A., and Mapps, D. (2003). Magnetic tunnel junctions with non-collinear anisotropy axes for sensor applications, *J. Phys. D: Appl. Phys.* **36**, 7, p. 791, http://stacks.iop.org/0022-3727/36/i=7/a=304.

Grünberg, P., Schreiber, R., Pang, Y., Brodsky, M. B., and Sowers, H. (1986). Layered magnetic structures: Evidence for antiferromagnetic coupling of Fe layers across Cr interlayers, *Phys. Rev. Lett.* **57**, pp. 2442–2445, doi:10.1103/PhysRevLett.57.2442, http://link.aps.org/doi/10.1103/PhysRevLett.57.2442.

Guan, Y., Sun, J. Z., Jiang, X., Moriya, R., Gao, L., and Parkin, S. S. P. (2009). Thermal-magnetic noise measurement of spin-torque effects on ferromagnetic resonance in MgO-based magnetic tunnel junctions, *Appl. Phys. Lett.* **95**, 8, 082506, http://scitation.aip.org/content/aip/journal/apl/95/8/10.1063/1.3212870.

Guerrero, R. (2007). *Conductancia y ruido en uniones túnel Magnéticas*, Ph.D. thesis, Universidad Autónoma de Madrid, Spain, http://www.uam.es/gruposinv/magtran/publicaciones/TESIS_RG_TOT_JUNE2_07.pdf.

Guerrero, R., Aliev, F. G., Tserkovnyak, Y., Santos, T. S., and Moodera, J. S. (2006b). Shot noise in magnetic tunnel junctions: Evidence for sequential tunneling, *Phys. Rev. Lett.* **97**, p. 266602, http://link.aps.org/doi/10.1103/PhysRevLett.97.266602.

Guerrero, R., Aliev, F. G., Villar, R., Hauch, J., Fraune, M., Güntherodt, G., Rott, K., Brückl, H., and Reiss, G. (2005). Low-frequency noise and tunneling magnetoresistance in Fe(110)/MgO(111)/Fe(110) epitaxial magnetic tunnel junctions, *Appl. Phys. Lett.* **87**, 4, p. 042501, http://scitation.aip.org/content/aip/journal/apl/87/4/10.1063/1.2001128.

Guerrero, R., Aliev, F. G., Villar, R., Hauch, J., Fraune, M., Guntherodt, G., Rott, K., Bruckl, H., and Reiss, G. (2006a). Low-frequency noise and inelastic tunneling spectroscopy in Fe(110)/MgO(111)/Fe(110) epitaxial magnetic tunnel junctions, *J. Magn. Magn. Mater.* **300**, 1, pp. 132–135, http://linkinghub.elsevier.com/retrieve/pii/S0304885305009017.

Guerrero, R., Aliev, F., Villar, R., Ortega-Hertogs, R., Park, W., and Moodera, J. (2002). *J. Phys. D: Appl. Phys.* **35**, p. 1761, http://iopscience.iop.org/0022-3727/35/14/317/.

Guerrero, R., Herranz, D., Aliev, F. G., Greullet, F., Tiusan, C., Hehn, M., and Montaigne, F. (2007). High bias voltage effect on spin-dependent conductivity and shot noise in carbon-doped Fe (001)/MgO (001)/Fe (001) magnetic tunnel junctions, *Appl. Phys. Lett.* **91**, 13, 132504, http://scitation.aip.org/content/aip/journal/apl/91/13/10.1063/1.2793619.

Guerrero, R., Pannetier-Lecoeur, M., Fermon, C., Cardoso, S., Ferreira, R., and Freitas, P. P. (2009). Low frequency noise in arrays of magnetic tunnel junctions connected in series and parallel, *J. Appl. Phys.* **105**, 11, p. 113922, http://link.aip.org/link/JAPIAU/v105/i11/p113922/s1&Agg=doi.

Guerrero, R., Solignac, A., Fermon, C., Pannetier-Lecoeur, M., Lecoeur, P., and Fernández-Pacheco, R. (2012). Low frequency noise in La0.7Sr0.3MnO3 based magnetic tunnel junctions, *Appl. Phys. Lett.* **100**, 14, 142402, doi:http://dx.doi.org/10.1063/1.3698393, http://scitation.aip.org/content/aip/journal/apl/100/14/10.1063/1.3698393.

Guo, F., McKusky, G., and Dahlberg, E. D. (2013). Absence of magnetic state dependent low-frequency noise in spin-valve systems, *Phys. Rev. B* **88**, p. 014409, doi:10.1103/PhysRevB.88.014409, http://link.aps.org/doi/10.1103/PhysRevB.88.014409.

Gupta, M. (1978). Thermal fluctuations in driven nonlinear resistive systems, *Phys. Rev. A* **18**, pp. 2725–2731, http://link.aps.org/doi/10.1103/PhysRevA.18.2725.

Gupta, M. (1982). Thermal noise in nonlinear resistive devices and its circuit representation, *Proc. IEEE* **70**, 8, pp. 788–804.

Harada, T., Ohkubo, I., Lippmaa, M., Sakurai, Y., Matsumoto, Y., Muto, S., Koinuma, H., and Oshima, M. (2012). Spin-filter tunnel junction with matched fermi surfaces, *Phys. Rev. Lett.* **109**, p. 076602, http://link.aps.org/doi/10.1103/PhysRevLett.109.076602.

Hardner, H. T., Weissman, M. B., Salamon, M. B., and Parkin, S. S. P. (1993a). Fluctuation-dissipation relation for giant magnetoresistive $1/f$ noise, *Phys. Rev. B* **48**, pp. 16156–16159, doi:10.1103/PhysRevB. 48. 16156, http:// link.aps.org/doi/10.1103/PhysRevB.48.16156.

Hardner, H. T., Weissman, M. B., Salamon, M. B., and Parkin, S. S. P. (1993b). Fluctuation-dissipation relation for giant magnetoresistive 1/f noise,

Phys. Rev. B **48**, pp. 16156–16159, http://link.aps. org/ doi/ 10.1103/ PhysRevB.48.16156.

Hasan, M. Z., and Kane, C. L. (2010a). Colloquium, *Rev. Mod. Phys.* **82**, pp. 3045–3067, http://link.aps.org/doi/10.1103/RevModPhys. 82.3045.

Haupt, F., Novotný, T. C. V., and Belzig, W. (2009). Phonon-assisted current noise in molecular junctions, *Phys. Rev. Lett.* **103**, p. 136601, doi:10.1103/PhysRevLett.103.136601, http://link.aps.org/ doi/10.1103/PhysRevLett.103.136601.

Hauptmann J. R., P. J., and E., L. P. (2008). Electric-field-controlled spin reversal in a quantum dot with ferromagnetic contacts, *Nat. Phys.* **4**, pp. 373–376.

Heikkilä, T. T., and Nagaev, K. E. (2013). Nonuniversal shot noise in quasiequilibrium spin valves, *Phys. Rev. B* **87**, p. 235411, doi:10. 1103/PhysRevB.87.235411, http://link.aps.org/doi/10.1103/Phys RevB.87.235411.

Helmer, A., Cornelissen, S., Devolder, T., Kim, J.-V., van Roy, W., Lagae, L. and Chappert, C. (2010). Quantized spin-wave modes in magnetic tunnel junction nanopillars, *Phys. Rev. B* **81**, p. 094416, http:// link.aps.org/doi/10.1103/PhysRevB.81.094416.

Herranz, D. (2012). *Electron Transport and Noise in Magnetic Tunnel Junctions with MgO Barriers*, Ph.D. thesis, Universidad Autónoma de Madrid, Spain, http://www.uam.es/gruposinv/magtran/Tesis_David %20Herranz_4April_2012.pdf.

Herranz, D., Aliev, F. G., Tiusan, C., Hehn, M., Dugaev, V. K., and Barnaś, J. (2010a). Tunneling in double barrier junctions with "hot spots", *Phys. Rev. Lett.* **105**, p. 047207, http://link.aps.org/doi/ 10. 1103/ PhysRevLett.105.047207.

Herranz, D., Aliev, F. G., Tiusan, C., Hehn, M., Dugaev, V. K., and Barnas, J. (2010b). Tunneling in double barrier junctions with hot spots, *Phys. Rev. Lett.* **105**, p. 047207, doi:10.1103/PhysRevLett.105.047207, http://link.aps.org/doi/10.1103/PhysRevLett.105.047207.

Herranz, D., Bonell, F., Gomez-Ibarlucea, A., Andrieu, S., Montaigne, F., Villar, R., Tiusan, C., and Aliev, F. G. (2010c). Strongly suppressed 1/f noise and enhanced magnetoresistance in epitaxial Fe-V/MgO/Fe magnetic tunnel junctions, *Appl. Phys. Lett.* **96**, 20, 202501, http://link.aip.org/link/?APL/96/202501/1.

Herranz, D., Gomez-Ibarlucea, A., Schäfers, M., Lara, A., Reiss, G., and Aliev, F. G. (2011). Low frequency noise due to magnetic inhomogeneities in submicron FeCoB/MgO/FeCoB magnetic tunnel junctions, *Appl. Phys. Lett.* **99**, 6, 062511, http://scitation.aip.org/content/aip/journal/apl/99/6/10.1063/1.3615798.

Herranz, D., Guerrero, R., Cascales, J. P., Aliev, F. G., Hehn, M., and Tiusan, C. (2012). Strong reduction of 1/f noise by carbon doping in epitaxial Fe/MgO (100) 12 ml/Fe magnetic tunnel junctions with barrier defects, *Acta Phys. Pol. A* **121**, 5-6, http://www.uam.es/gruposinv/magtran/publicaciones/2012/A131_Herranz_PM11_2012.pdf.

Herranz, D., Guerrero, R., Villar, R., Aliev, F. G., Swaving, A. C., Duine, R. A., van Haesendonck, C., and Vavra, I. (2009). Anomalous low-frequency noise in synthetic antiferromagnets: Possible evidence of current-induced domain-wall motion, *Phys. Rev. B* **79**, p. 134423, http://link.aps.org/doi/10.1103/PhysRevB.79.134423.

Hershfield, S., Davies, J. H., Hyldgaard, P., Stanton, C. J., and Wilkins, J. W. (1993). Zero-frequency current noise for the double-tunnel-junction Coulomb blockade, *Phys. Rev. B* **47**, pp. 1967–1979, http://link.aps.org/doi/10.1103/PhysRevB.47.1967.

Hong, J.-Y., Ou Yang, K.-H., Wang, B.-Y., Li, K.-S., Shiu, H.-W., Chen, C.-H., Chan, Y.-L., Wei, D.-H., Chang, F.-H., Lin, H.-J., Chiang, W.-C., and Lin, M.-T. (2014). Interfacial spectroscopic characterization of organic/ferromagnet hetero-junction of 3,4,9,10-perylene-teracarboxylic dianhydride-based organic spin valves, *Appl. Phys. Lett.* **104**, 8, 083301, http://scitation.aip.org/content/aip/journal/apl/104/8/10.1063/1.4866164.

Hooge, F. N., Kleinpenning, T. G. M., and Vandamme, L. K. J. (1981a). Experimental studies on 1/f noise, *Rep. Prog. Phys.* **44**, pp. 479–532, http://iopscience.iop.org/0034-4885/44/5/001/.

Hooge, F. N., Kleinpenning, T. G. M., and Vandamme, L. K. J. (1981b). Experimental studies on l/f noise, *Rep. Prog. Phys* **44**, December 1980.

Houssameddine, D. (2009). *Magnetization Dynamics in Spin Torque Microwave Nano-Oscillators*, Thesis, Université Joseph-Fourier- Grenoble I, https://tel.archives-ouvertes.fr/tel-00466637.

Houssameddine, D., Ebels, U., Delaët, B., Rodmacq, B., Firastrau, I., Ponthenier, F., Brunet, M., Thirion, C., Michel, J.-P., Prejbeanu-Buda, L. et al. (2007). Spin-torque oscillator using a perpendicular polarizer and a planar free layer, *Nat. Mater.* **6**, 6, pp. 447–453.

Houssameddine, D., Ebels, U., Dieny, B., Garello, K., Michel, J.-P., Delaet, B., Viala, B., Cyrille, M.-C., Katine, J. A., and Mauri, D. (2009).

Temporal coherence of MgO based magnetic tunnel junction spin torque oscillators, *Phys. Rev. Lett.* **102**, p. 257202, http://link.aps.org/doi/10.1103/PhysRevLett.102.257202.

Houssameddine, D., Florez, S. H., Katine, J. A., Michel, J.-P., Ebels, U., Mauri, D., Ozatay, O., Delaet, B., Viala, B., Folks, L., Terris, B. D., and Cyrille, M.-C. (2008). Spin transfer induced coherent microwave emission with large power from nanoscale MgO tunnel junctions, *Appl. Phys. Lett.* **93**, 2, 022505, http://scitation.aip.org/content/aip/journal/apl/93/2/10.1063/1.2956418.

Huang, B., Liu, M., Su, N., Wu, J., Duan, W., Gu, B.-l., and Liu, F. (2009). Quantum manifestations of graphene edge stress and edge instability: A first-principles study, *Phys. Rev. Lett.* **102**, p. 166404, doi: 10.1103/PhysRevLett.102.166404, http://link.aps.org/doi/10.1103/PhysRevLett.102.166404.

Hui-Qiang, G., Wei-Yue, T., Liang, L., Jian, W., Da-Lai, L., Jia-Feng, F., and Xiu-Feng, H. (2015). Low frequency noise in asymmetric double barrier magnetic tunnel junctions with a top thin MgO layer, *Chin. Phys. B* **24**, 7, p. 078504, http://stacks.iop.org/1674-1056/24/i=7/a=078504.

Hung, K., Ko, P.-K., Hu, C., and Cheng, Y. C. (1990). Random telegraph noise of deep-submicrometer mosfets, *Electron Device Lett., IEEE* **11**, 2, pp. 90–92.

Iannaccone, G., Lombardi, G., Macucci, M., and Pellegrini, B. (1998). Enhanced shot noise in resonant tunneling: Theory and experiment, *Phys. Rev. Lett.* **80**, pp. 1054–1057, http://link.aps.org/doi/10.1103/PhysRevLett.80.1054.

Ikeda, S., Hayakawa, J., Ashizawa, Y., Lee, Y. M., Miura, K., Hasegawa, H., Tsunoda, M., Matsukura, F., and Ohno, H. (2008). Tunnel magnetoresistance of 604CoFeB/MgO/CoFeB pseudo-spin-valves annealed at high temperature, *Appl. Phys. Lett.* **93**, 8, 082508, http://scitation.aip.org/content/aip/journal/apl/93/8/10.1063/1.2976435.

Ikegawa, S., Asao, Y., Saito, Y., Takahashi, S., Kai, T., Tsuchida, K., and Yoda, H. (2003). A fully integrated 1 kb magnetoresistive random access memory with a double magnetic tunnel junction, *Jpn. J. Appl. Phys.* **42**, 7A, p. L745, http://stacks.iop.org/1347-4065/42/i=7A/a=L745.

Ingvarsson, S., Xiao, G., Wanner, R. A., Trouilloud, P., Lu, Y., Gallagher, W. J., Marley, A., Roche, K. P., and Parkin, S. S. P. (1999). Electronic noise in magnetic tunnel junctions, *J. Appl. Phys.* **85**, 8, pp. 5270–5272, http://scitation.aip.org/content/aip/journal/jap/85/8/10.1063/1.369851.

Ingvarsson, S., Xiao, G., Parkin, S. S., Gallagher, W. J., Grinstein, G., and Koch, R. H. (2000). Low-frequency magnetic noise in micron-scale magnetic tunnel junctions. *Phys. Rev. Lett.* **85**, 15, pp. 3289–92, http://www.ncbi.nlm.nih.gov/pubmed/11019323.

Iovan, A., Andersson, S., Naidyuk, Y. G., Vedyaev, A., Dieny, B., and Korenivski, V. (2008). Spin diode based on Fe/MgO double tunnel junction, *Nano Lett.* **8**, 3, pp. 805–809, http://pubs.acs.org/doi/abs/10.1021/nl072676z.

Jaklevic, R. C., and Lambe, J. (1966). Molecular vibration spectra by electron tunneling, *Phys. Rev. Lett.* **17**, pp. 1139–1140, http://link.aps.org/doi/10.1103/PhysRevLett.17.1139.

Jander, A., Nordman, C. A., Pohm, A. V., and Anderson, J. M. (2003). Chopping techniques for low-frequency nanotesla spin-dependent tunneling field sensors, *J. Appl. Phys.* **93**, 10.

Jayaraman, R., and Sodini, C. (1989). A 1/f noise technique to extract the oxide trap density near the conduction band edge of silicon, *IEEE Trans. Electron Device* **36**, 9, pp. 1773–1782.

Jiang, L., Nowak, E., Scott, P., Johnson, J., Slaughter, J., Sun, J., and Dave, R. (2004a). Low-frequency magnetic and resistance noise in magnetic tunnel junctions, *Phys. Rev. B* **69**, 5, pp. 1–9, http://link.aps.org/doi/10.1103/PhysRevB.69.054407.

Jiang, L., Skovholt, J. F., Nowak, E. R., and Slaughter, J. M. (2004b). Low-frequency magnetic and resistance noise in magnetoresistive tunnel junctions, *Proc. SPIE 5469*, Fluctuations and Noise in Materials, doi:10.1117/12.544659.

Johnson, J. B. (1928). Thermal agitation of electricity in conductors, *Phys. Rev.* **32**, pp. 97–109, http://link.aps.org/doi/10.1103/PhysRev.32.97.

Johnson, D. D., Pinski, F. J., and Staunton, J. B. (1987). The Slater-Pauling curve: First principles calculations of the moments of $Fe_{1-c}Ni_c$ and V_cFe_{1-c}, *J. Appl. Phys.* **61**, 8, pp. 3715–3717, doi:10.1063/1.338670, http://link.aip.org/link/?JAP/61/3715/1.

Jonscher, A. K. (2004). *Electronic Processes in Non-Crystalline Materials: By N. F. Mott and E. A. Davis, Clarendon Press: Oxford University Press, Oxford, 1971 and 437 pp.* (Wiley, New York), http://onlinelibrary.wiley.com/doi/10.1002/crat.19720070420/abstract.

Jozwiak, C., Park, C.-H., Gotlieb, K., Hwang, C., Lee, D.-H., Louie, S. G., Denlinger, J. D., Rotundu, C. R., Birgeneau, R. J., Hussain, Z., and

Lanzara, A. (2013). Photoelectron spin-flipping and texture manipulation in a topological insulator, *Nat. Phys.* **9**, pp. 1745–2473, doi:10.1038/nphys2572, http://dx.doi.org/10.1038/nphys2572.

Jr., W. E., Pong, P., Unguris, J., McMichael, R., Nowak, E., Edelstein, A., Burnette, J., and Fischer, G. (2009). Critical challenges for picotesla magnetic-tunnel-junction sensors, *Sensors Actuators A: Physical* **155**, 2, pp. 217–225, doi:http://dx.doi.org/10.1016/j.sna.2009.08.016, http://www.sciencedirect.com/science/article/pii/S0924424709003781.

Julliere, M. (1975). Tunneling between ferromagnetic films, *Phys. Lett. A* **54**, 3, pp. 225–226.

Kamalakar, M. V., Madhushankar, B. N., Dankert, A., and Dash, S. P. (2015). Low Schottky barrier black phosphorus field-effect devices with ferromagnetic tunnel contacts, *Small*, doi:10.1002/smll.201402900, http://dx.doi.org/10.1002/smll.201402900.

Kamra, A., and Belzig, W. (2016). Super-poissonian shot noise of squeezed-magnon mediated spin transport, *Phys. Rev. Lett.* **116**, p. 146601, doi: 10.1103/PhysRevLett.116.146601, http://link.aps.org/doi/10.1103/PhysRevLett.116.146601.

Kamra, A., Witek, F. P., Meyer, S., Huebl, H., Geprägs, S., Gross, R., Bauer, G. E. W., and Goennenwein, S. T. B. (2014). Spin Hall noise, *Phys. Rev. B* **90**, p. 214419, doi:10.1103/PhysRevB.90.214419, http://link.aps.org/doi/10.1103/PhysRevB.90.214419.

Karpan, V. M., Giovannetti, G., Khomyakov, P. A., Talanana, M., Starikov, A. A., Zwierzycki, M., van den Brink, J., Brocks, G., and Kelly, P. J. (2007). Graphite and graphene as perfect spin filters, *Phys. Rev. Lett.* **99**, p. 176602, doi:10.1103/PhysRevLett.99.176602, http://link.aps.org/doi/10.1103/PhysRevLett.99.176602.

Katine, J. A., Albert, F. J., Buhrman, R. A., Myers, E. B., and Ralph, D. C. (2000). Current-driven magnetization reversal and spin-wave excitations in Co/Cu/Co pillars, *Phys. Rev. Lett.* **84**, pp. 3149–3152, http://link.aps.org/doi/10.1103/PhysRevLett.84.3149.

Kaun, C.-C., and Seideman, T. (2005). Current-driven oscillations and time-dependent transport in nanojunctions, *Phys. Rev. Lett.* **94**, p. 226801, doi:10.1103/PhysRevLett.94.226801, http://link.aps.org/doi/10.1103/PhysRevLett.94.226801.

Kaverzin, A. A., Mayorov, A. S., Shytov, A., and Horsell, D. W. (2012). Impurities as a source of $1/f$ noise in graphene, *Phys. Rev. B* **85**, p. 075435, doi:10.1103/PhysRevB.85.075435, http://link.aps.org/doi/10.1103/PhysRevB.85.075435.

Keevers, T. L., Danos, A., Schmidt, T. W., and McCamey, D. R. (2013). An agnostic approach, *Nat. Nanotech.* **8**, p. 886, http://www.nature.com/nnano/journal/v8/n12/full/nnano.2013.261.html.

Khajetoorians, A. A., Baxevanis, B., Hübner, C., Schlenk, T., Krause, S., Wehling, T. O., Lounis, S., Lichtenstein, A., Pfannkuche, D., Wiebe, J., and Wiesendanger, R. (2013). Current-driven spin dynamics of artificially constructed quantum magnets, *Science* **339**, 6115, pp. 55–59, doi:10.1126/science.1228519, http://science.sciencemag.org/content/339/6115/55.

Kießlich, G., Schöll, E., Brandes, T., Hohls, F., and Haug, R. J. (2007). Noise enhancement due to quantum coherence in coupled quantum dots, *Phys. Rev. Lett.* **99**, p. 206602, http://link.aps.org/doi/10.1103/PhysRevLett.99.206602.

Kim, J.-V., Mistral, Q., Chappert, C., Tiberkevich, V. S., and Slavin, A. N. (2008a). Line shape distortion in a nonlinear auto-oscillator near generation threshold: Application to spin-torque nano-oscillators, *Phys. Rev. Lett.* **100**, p. 167201, http://link.aps.org/doi/10.1103/PhysRevLett.100.167201.

Kim, Y., Song, H., Kim, D., Lee, T., and Jeong, H. (2010a). Noise characteristics of charge tunneling via localized states in metal-molecule-metal junctions, *ACS Nano* **4**, 8, pp. 4426–4430, doi:10.1021/nn100255b, http://dx.doi.org/10.1021/nn100255b, pMID: 20677808.

Kim, Y., Song, H., Kim, D., Lee, T., and Jeong, H. (2010b). Noise characteristics of charge tunneling via localized states in metal-molecule-metal junctions, *ACS Nano* **4**, 8, pp. 4426–4430, doi:10.1021/nn100255b, http://dx.doi.org/10.1021/nn100255b, pMID: 20677808.

Kim, J.-V., Tiberkevich, V., and Slavin, A. N. (2008b). Generation linewidth of an auto-oscillator with a nonlinear frequency shift: Spin-torque nano-oscillator, *Phys. Rev. Lett.* **100**, p. 017207, http://link.aps.org/doi/10.1103/PhysRevLett.100.017207.

Kim, J.-V., Tiberkevich, V., and Slavin, A. N. (2008c). Generation linewidth of an auto-oscillator with a nonlinear frequency shift: Spin-torque nano-oscillator, *Phys. Rev. Lett.* **100**, p. 017207, http://link.aps.org/doi/10.1103/PhysRevLett.100.017207.

Kirschenbaum, L., Rogers, C., Russek, S., and Sanders, S. (1995). Telegraph noise in silver-permalloy giant magnetoresistance test structures, *IEEE Trans. Magn.* **31**, 6, pp. 3943–3945.

Kiselev, S. I., Sankey, J. C., Krivorotov, I. N., Emley, N. C., Schoelkopf, R. J., Buhrman, R. A., and Ralph, D. C. (2003). Microwave oscillations of a nanomagnet driven by a spin-polarized current, *Nature* **425**, pp. 380–383, http://www.nature.com/nature/journal/v425/n6956/abs/nature01967.html.

Kittel, C. (2004). *Introduction to Solid State Physics, 8th edition* (Wiley, New York), http://www.wiley.com/WileyCDA/WileyTitle/productCd-EHEP000803.html.

Kittel, C. (1951). Ferromagnetic resonance, *Journal de Physique et le Radium* **12**(3), pp. 291–302.

Kogan, S. (1996). *Electronic Noise and Fluctuations in Solids* (Cambridge University Press), http://dx.doi.org/10.1017/CBO9780511551666.

Krause, S., Berbil-Bautista, L., Herzog, G., Bode, M., and Wiesendanger, R. (2007). Current-induced magnetization switching with a spin-polarized scanning tunneling microscope, *Science* **317**, 5844, p. 1537, doi:10.1126/science.1145336, http://science.sciencemag.org/content/ 317/5844/1537.

Kresse, G., and Furthmüller, J. (1996). Efficiency of ab-initio total energy calculations for metals and semiconductors using a plane-wave basis set, *Comput. Mater. Sci.* **6**, 1, pp. 15–50, http://www.sciencedirect.com/science/article/pii/0927025696000080.

Kresse, G., and Furthmüller, J. (1996). Efficient iterative schemes for ab initio total-energy calculations using a plane-wave basis set, *Phys. Rev. B* **54**, pp. 11169–11186, http://link.aps.org/doi/10.1103/PhysRevB.54.11169.

Kresse, G., and Hafner, J. (1993). Ab initio, *Phys. Rev. B* **47**, pp. 558–561, http://link.aps.org/doi/10.1103/PhysRevB.47.558.

Kresse, G., and Joubert, D. (1999). From ultrasoft pseudopotentials to the projector augmented-wave method, *Phys. Rev. B* **59**, pp. 1758–1775, http://link.aps.org/doi/10.1103/PhysRevB.59.1758.

Kroemer, H. (2001). Nobel lecture: Quasielectric fields and band offsets: teaching electrons new tricks, *Rev. Mod. Phys.* **73**, pp. 783–793, http://link.aps.org/doi/10.1103/RevModPhys.73.783.

Kubota, H., Fukushima, A., Yakushiji, K., Nagahama, T., Yuasa, S., Ando, K., Maehara, H., Nagamine, Y., Tsunekawa, K., and Djayaprawira, D. D. (2007). Quantitative measurement of voltage dependence of spin-transfer torque in MgO-based magnetic tunnel junctions, *Nat. Phys.* **4**, pp. 37–41, http://www.nature.com/nphys/journal/v4/n1/full/nphys784.html.

Kumar, M., Avriller, R., Yeyati, A. L., and van Ruitenbeek, J. M. (2012). Detection of vibration-mode scattering in electronic shot noise,

Phys. Rev. Lett. **108**, p. 146602, http://link.aps.org/doi/10.1103/PhysRevLett.108.146602.

Kung, B., Rossler, C., Beck, M., Faist, J., Ihn, T., and Ensslin, K. (2012). Quantum dot occupation and electron dwell time in the cotunneling regime, *New J. Phys.* **14**, 8, p. 083003, http://stacks.iop.org/1367-2630/14/i=8/a=083003.

Kuznetsov, V. V., Mendez, E. E., Zuo, X., Snider, G. L., and Croke, E. T. (2000). Partially suppressed shot noise in hopping conduction: Observation in sige quantum wells, *Phys. Rev. Lett.* **85**, pp. 397–400, http://link.aps.org/doi/10.1103/PhysRevLett.85.397.

Lamacraft, A. (2004). Shot noise of spin-polarized electrons, *Phys. Rev. B* **69**, p. 081301, doi:10.1103/PhysRevB.69.081301, http://link.aps.org/doi/10.1103/PhysRevB.69.081301.

Lambert, N., Aguado, R., and Brandes, T. (2007). Nonequilibrium entanglement and noise in coupled qubits, *Phys. Rev. B* **75**, p. 045340, http://link.aps.org/doi/10.1103/PhysRevB.75.045340.

Landauer, R. (1993). Solid-state shot noise, *Phys. Rev. B* **47**, pp. 16427–16432, http://link.aps.org/doi/10.1103/PhysRevB.47.16427.

Lee, H. S., Min, S.-W., Chang, Y.-G., Park, M. K., Nam, T., Kim, H., Kim, J. H., Ryu, S., and Im, S. (2012). MoS2 nanosheet phototransistors with thickness-modulated optical energy gap, *Nano Lett.* **12**, 7, pp. 3695–3700, doi:10.1021/nl301485q, http://dx.doi.org/10.1021/nl301485q, pMID: 22681413.

Lei, Z., Li, G., Egelhoff, W., Lai, P., and Pong, P. (2011). Review of noise sources in magnetic tunnel junction sensors, *IEEE Trans. Magn.* **47**, 3, pp. 602–612, doi:10.1109/TMAG.2010.2100814.

Li, K.-S., Chang, Y.-M., Agilan, S., Hong, J.-Y., Tai, J.-C., Chiang, W.-C., Fukutani, K., Dowben, P. A., and Lin, M.-T. (2011). Organic spin valves with inelastic tunneling characteristics, *Phys. Rev. B* **83**, p. 172404, http://link.aps.org/doi/10.1103/PhysRevB.83.172404.

Li, D., Feng, J., Yu, G., Wei, H., Han, X., and Coey, J. (2013). MgO-based double barrier magnetic tunnel junctions with synthetic antiferromagnetic free layer, *IEEE Trans. Magn.* **49**, 10, pp. 5204–5207.

Likharev, K. (1999). Single-electron devices and their applications, *Proc. IEEE* **87**, 4, pp. 606–632, doi:10.1109/5.752518.

Lin, Y.-M., and Avouris, P. (2008). Strong suppression of electrical noise in bilayer graphene nanodevices, *Nano Lett.* **8**, 8, pp. 2119–2125, doi:10.1021/nl080241l, http://dx.doi.org/10.1021/nl080241l.

Lindebaum, S., Urban, D., and König, J. (2009). Spin-induced charge correlations in transport through interacting quantum dots with ferromagnetic leads, *Phys. Rev. B* **79**, p. 245303, doi:10.1103/PhysRevB. 79. 245303, http://link.aps.org/doi/10.1103/PhysRevB.79.245303.

Liu, H.-W., Chang, H.-J., Li, G.-P., and Bachman, M. (2009). Performance improvement of organic thin-film transistors by solution-processed crystallization of pentacene at room temperature, *Electron Device Lett., IEEE* **30**, 4, pp. 346–348.

Liu, L., Niu, J., Guo, H., Wei, J., Li, D. L., Feng, J. F., Han, X. F., Coey, J. M. D., and Zhang, X.-G. (2016). Spin-flip noise due to nonequilibrium spin accumulation, *Phys. Rev. B* **93**, p. 180401, doi:10.1103/PhysRevB. 93. 180401, http://link.aps.org/doi/10.1103/PhysRevB.93.180401.

Liu, G., Rumyantsev, S., Shur, M. S., and Balandin, A. A. (2013). Origin of 1/f noise in graphene multilayers: Surface vs. volume, *Appl. Phys. Lett.* **102**, 9, p. 093111, doi:http://dx.doi.org/10.1063/1.4794843, http://scitation.aip.org/content/aip/journal/apl/102/9/10.1063/1.4794843.

Liu, K., Xia, K., and Bauer, G. E. W. (2012). Shot noise in magnetic tunnel junctions from first principles, *Phys. Rev. B* **86**, p. 020408, http://link.aps.org/doi/10.1103/PhysRevB.86.020408.

López, R., and Sánchez, D. (2003). Nonequilibrium spintronic transport through an artificial Kondo impurity: Conductance, magnetoresistance, and shot noise, *Phys. Rev. Lett.* **90**, p. 116602, doi:10. 1103/PhysRevLett.90.116602, http://link.aps.org/doi/10.1103/PhysRevLett.90.116602.

Lu, Y., Li, X. W., Gong, G. Q., Xiao, G., Gupta, A., Lecoeur, P., Sun, J. Z., Wang, Y. Y., and Dravid, V. P. (1996). Large magnetotunneling effect at low magnetic fields in micrometer-scale epitaxial $La_{0.67}Sr_{0.33}MnO_3$ tunnel junctions, *Phys. Rev. B* **54**, pp. R8357–R8360, doi:10.1103/PhysRevB. 54.R8357, http://link.aps.org/doi/10.1103/PhysRevB.54.R8357.

Lu, Y., Yang, H.-X., Tiusan, C., Hehn, M., Chshiev, M., Duluard, A., Kierren, B., Lengaigne, G., Lacour, D., Bellouard, C., and Montaigne, F. (2012). Spin-orbit coupling effect by minority interface resonance states in single-crystal magnetic tunnel junctions, *Phys. Rev. B* **86**, p. 184420, http://link.aps.org/doi/10.1103/PhysRevB.86.184420.

M. Wilczyński, J. B., R. Świrkowicz (2009). Enhancing spin-transfer torque through the proximity of quantum well states, *Acta Phys. Pol. A* **115**, p. 269.

MacEachern, S. (1997). *Kendall's Advanced Theory of Statistics. Vol. 2B: Bayesian Inference* Anthony O'Hagan, Edward Arnold, London, 1994. pp. 330. ISBN: 0-340-52922-9, Vol. 16 (John Wiley and Sons).

Machlup, S. (1954). Noise in semiconductors: Spectrum of a two-parameter random signal, *J. Appl. Phys.* **25**, 3.

Maekawa, S., Valenzuela, S. O., Saitoh, E., and Kimura T. (2012). *Spin Current* (Oxford University Press), http://eu.wiley.com/WileyCDA/WileyTitle/productCd-EHEP000803.html.

Malard, L., Pimenta, M., Dresselhaus, G., and Dresselhaus, M. (2009). Raman spectroscopy in graphene, *Phys. Rep.* **473**, 5–6, pp. 51–87, doi:http://dx.doi.org/10.1016/j.physrep.2009.02.003, http://www.sciencedirect.com/science/article/pii/S0370157309000520.

Manchon, A., Koo, H. C. Nitta, J., Frolov S. M., and Duine R. A. (2015). New perspectives for Rashba spin-orbit coupling, *Nat. Mater.* **14**, pp. 871–882, doi:10.1038/nmat4360, http://dx.doi.org/10.1038/nmat4360.

Mann, B., and Kuhn, H. (1971). Tunneling through fatty acid salt monolayers, *J. Appl. Phys.* **42**, 11, pp. 4398–4405, doi:http://dx.doi.org/10.1063/1.1659785, http://scitation.aip.org/content/aip/journal/jap/42/11/10.1063/1.1659785.

Martinek, J., Barnas, J., Fert, A., Maekawa, S., and Schön, G. (2003). Transport in magnetic nanostructures in the presence of Coulomb interaction (invited), *J. Appl. Phys.* **93**, 10, pp. 8265–8270, doi:http://dx.doi.org/10.1063/1.1557831, http://scitation.aip.org/content/aip/journal/jap/93/10/10.1063/1.1557831.

Martinez, I., Cascales, J., Hong, J.-Y., Lin, M.-T., Prezioso, M., Riminucci, A., Dediu, V. A., and Aliev, F. G. (2016). Electron transport and noise spectroscopy in organic magnetic tunnel junctions with PTCDA and alq3 barriers, *Proc. SPIE* **9931**, pp. 99313P–1, doi:10.1117/12.2237721.

Martinez, I., Ribeiro, M., Andres, P., Hueso, L. E., Casanova, F., and Aliev, F. G. (2017). *Physical Review Applied*, **7**, 034034. DOI:https://doi.org/10.1103/PhysRevApplied.7.034034

Mathon, J., and Umerski, A. (2001). Theory of tunneling magnetoresistance of an epitaxial Fe/MgO/Fe (001) junction, *Phys. Rev. B* **63**, p. 220403, http://link.aps.org/doi/10.1103/PhysRevB.63.220403.

Mazumdar, D., Liu, X., Schrag, B. D., Carter, M., Shen, W., and Xiao, G. (2007). Low frequency noise in highly sensitive magnetic tunnel junctions with (001) MgO tunnel barrier, *Appl. Phys. Lett.* **91**, 3, 033507, doi:http://dx.doi.org/10.1063/1.2754352, http://scitation.aip.org/content/aip/journal/apl/91/3/10.1063/1.2754352.

McClure, J. W. (1956). Diamagnetism of graphite, *Phys. Rev.* **104**, pp. 666–671, doi:10.1103/PhysRev.104.666, http://link.aps.org/doi/10.1103/PhysRev.104.666.

Meair, J., Stano, P., and Jacquod, P. (2011). Measuring spin accumulations with current noise, *Phys. Rev. B* **84**, p. 073302, doi:10.1103/PhysRevB.84.073302, http://link.aps.org/doi/10.1103/PhysRevB.84.073302.

Meng, H., and Wang, J.-P. (2005). Spin transfer effect in magnetic tunnel junction with a nano-current-channel layer in free layer, *IEEE Trans. Magn.* **41**, 10, pp. 2612–2614.

Meric, I., Dean, C. R., Young, A. F., Baklitskaya, N., Tremblay, N. J., Nuckolls, C., Kim, P., and Shepard, K. L. (2011). Channel length scaling in graphene field-effect transistors studied with pulsed current-voltage measurements, *Nano Lett.* **11**, 3, pp. 1093–1097, doi:10.1021/nl103993z, http://dx.doi.org/10.1021/nl103993z.

Mermin, N. D. (1968). Crystalline order in two dimensions, *Phys. Rev.* **176**, pp. 250–254, doi:10.1103/PhysRev.176.250, http://link.aps.org/doi/10.1103/PhysRev.176.250.

Mirzanian, S., Shokri, A., Agah, K. M., and Elahi, S. (2015). Angular dependent study on spin transport in magnetic semiconductor heterostructures with Dresselhaus spin–orbit interaction, *Physica E: Low-Dimensional Syst. Nanostruct.* **73**, pp. 220–225, doi:http://dx.doi.org/10.1016/j.physe.2015.05.034, http://www.sciencedirect.com/science/article/pii/S1386947715300709.

Mishchenko, E. G. (2003). Shot noise in a diffusive ferromagnetic-paramagnetic-ferromagnetic spin valve, *Phys. Rev. B* **68**, p. 100409, doi:10.1103/PhysRevB.68.100409, http://link.aps.org/doi/10.1103/PhysRevB.68.100409.

Mistral, Q., Kim, J.-V., Devolder, T., Crozat, P., Chappert, C., Katine, J., Carey, M., and Ito, K. (2006). Current-driven microwave oscillations in current perpendicular-to-plane spin-valve nanopillars, *Appl. Phys. Lett.* **88**, 19, 192507, http://scitation.aip.org/content/aip/journal/apl/88/19/10.1063/1.2201897.

Miyazaki, T., and Tezuka, N. (1995). Giant magnetic tunneling effect in Fe/Al2O3/Fe junction, *J. Magnet. Magn. Mater.* **139**, pp. 0–0, http://www.sciencedirect.com/science/article/pii/0304885395900012.

Montaigne, F., Tiusan, C., and Hehn, M. (2010). Angular dependence of tunnel magnetoresistance in magnetic tunnel junctions and specific aspects in spin-filtering devices, *J. Appl. Phys.* **108**, 6, 063912, http://scitation.aip.org/content/aip/journal/jap/108/6/10.1063/1.3466778.

Moodera, J. S., Kinder, L. R., Wong, T. M., and Meservey, R. (1995). Large magnetoresistance at room temperature in ferromagnetic thin film tunnel junctions, *Phys. Rev. Lett.* **74**, pp. 3273–3276, http://link.aps.org/doi/10.1103/PhysRevLett.74.3273.

Moodera, J. S., Nassar, J., and Mathon, G. (1999a). Spin-tunneling in ferromagnetic junctions, *Annu. Rev. Mater. Sci.* **29**, pp. 381–432, http://www.sciencedirect.com/science/article/pii/S0304885399005156.

Moodera, J. S., Nowak, J., Kinder, L. R., Tedrow, P. M., van de Veerdonk, R. J. M., Smits, B. A., van Kampen, M., Swagten, H. J. M., and de Jonge, W. J. M. (1999b). Quantum well states in spin-dependent tunnel structures, *Phys. Rev. Lett.* **83**, pp. 3029–3032, http://link.aps.org/doi/10.1103/PhysRevLett.83.3029.

Mott, N. (1969). Conduction in non-crystalline materials: III. localized states in a pseudogap and near extremities of conduction and valence bands, *Philos. Mag.* **19**, 160, pp. 835–852.

Myers, E. B., Ralph, D. C., Katine, J. A., Louie, R. N., and Buhrman, R. A. (1999). Current-induced switching of domains in magnetic multilayer devices, *Science* **285**, 5429, pp. 867–870, http://www. sciencemag.org/ content/285/5429/867.abstract.

Na, J., Joo, M.-K., Shin, M., Huh, J., Kim, J.-S., Piao, M., Jin, J.-E., Jang, H.-K., Choi, H. J., Shim, J. H., and Kim, G.-T. (2014). Low-frequency noise in multilayer MoS2 field-effect transistors: the effect of high-k passivation, *Nanoscale* **6**, pp. 433–441, doi:10.1039/C3NR04218A, http://dx.doi.org/10.1039/C3NR04218A.

Nam Do, V., Dollfus, P., and Lien Nguyen, V. (2007). Zero-frequency current noise for the double-tunnel-junction Coulomb blockade, *Comput. Electron* **6**, p. 125.

Nazarov, Y., and Struben, J. (1996). Universal excess noise in resonant tunneling via strongly localized states, *Phys. Rev. B* **53**, pp. 15466–15468, http://link.aps.org/doi/10.1103/PhysRevB.53.15466.

Néel, N., Kröger, J., and Berndt, R. (2011). Two-level conductance fluctuations of a single-molecule junction, *Nano Lett.* **11**, 9, pp. 3593–3596, doi:10.1021/nl201327c, http://dx.doi.org/10.1021/nl201327c.

Ngo, T. D., Chang, J.-W., Lee, K., Han, S., Lee, J. S., Kim, Y. H., Jung, M.-H., Doh, Y.-J., Choi, M.-S., Song, J., and Kim, J. (2014). Polarity-tunable magnetic tunnel junctions based on ferromagnetism at oxide heterointerfaces, *Nat. Commun.* **6**, p. 8035, doi:10.1038/ncomms9035, http://www.nature.com/articles/ncomms9035.

Nikolić, K., and MacKinnon, A. (1994). Conductance and conductance fluctuations of narrow disordered quantum wires, *Phys. Rev. B* **50**, pp. 11008–11017, http://link.aps.org/doi/10.1103/PhysRevB.50.11008.

Nitzan, A., and Ratner, M. A. (2003). Electron transport in molecular wire junctions, *Science* **300**, 5624, pp. 1384–1389.

Novoselov, K. S. (2005). Two-dimensional atomic crystals, *Proc. Natl. Acad. Sci.* **102**, doi:10.1073/pnas.0502848102, http://www.pnas.org/content/102/30/10451.abstract.

Novoselov, K. S., Geim, A. K., Morozov, S. V., Jiang, D., Katsnelson, M. I., Grigorieva, I. V., Dubonos, S. V., and Firsov, A. A. (2005). Two-dimensional gas of massless Dirac fermions in graphene, *Nature* **438**, doi:10.1038/nature04233, http://www.nature.com/nature/journal/v438/n7065/abs/nature04233.html.

Novoselov, K. S., Geim, A. K., Morozov, S. V., Jiang, D., Zhang, Y., Dubonos, S. V., Grigorieva, I. V., and Firsov, A. A. (2004). Electric field effect in atomically thin carbon films, *Science* **306**, 5696, pp. 666–669, doi:10.1126/science.1102896, http://www.sciencemag.org/content/306/5696/666.abstract.

Novotný, T., Donarini, A., Flindt, C., and Jauho, A.-P. (2004). Shot noise of a quantum shuttle, *Phys. Rev. Lett.* **92**, p. 248302, doi:10.1103/PhysRevLett.92.248302, http://link.aps.org/doi/10.1103/PhysRevLett.92.248302.

Nowak, E. R., Merithew, R. D., Weissman, M. B., Bloom, I., and Parkin, S. S. P. (1998). Noise properties of ferromagnetic tunnel junctions, *J. Appl. Phys.* **84**, 11, pp. 6195–6201, http://scitation.aip.org/content/aip/journal/jap/84/11/10.1063/1.368936.

Nowak, E. R., Weissman, M. B., and Parkin, S. S. P. (1999). Electrical noise in hysteretic ferromagnet/insulator/ferromagnet tunnel junctions, *Appl. Phys. Lett.* **74**, 4, pp. 600–602, http://scitation.aip.org/content/aip/journal/apl/74/4/10.1063/1.123158.

Nozaki, T., Tezuka, N., and Inomata, K. (2006a). Quantum oscillation of the tunneling conductance in fully epitaxial double barrier magnetic tunnel junctions, *Phys. Rev. Lett.* **96**, p. 027208, http://link.aps.org/doi/10.1103/PhysRevLett.96.027208.

Nozaki, T., Tezuka, N., and Inomata, K. (2006b). Quantum oscillation of the tunneling conductance in fully epitaxial double barrier magnetic tunnel junctions, *Phys. Rev. Lett.* **96**, p. 027208, doi:

10.1103/PhysRevLett.96.027208, http://link.aps.org/doi/10.1103/PhysRevLett.96.027208.

Nuñez, A. S., Duine, R., Haney, P., and MacDonald, A. (2006). Theory of spin torques and giant magnetoresistance in antiferromagnetic metals, *Phys. Rev. B* **73**, p. 214426.

Nyquist, H. (1928). Thermal agitation of electric charge in conductors, *Phys. Rev.* **32**, pp. 110–113, http://link.aps.org/doi/10.1103/PhysRev.32.110.

Oestreich, M., Römer, M., Haug, R. J., and Hägele, D. (2005). Spin noise spectroscopy in GaAs, *Phys. Rev. Lett.* **95**, p. 216603, doi:10.1103/PhysRevLett.95.216603, http://link.aps.org/doi/10.1103/PhysRevLett.95.216603.

Okazaki, Y., Sasaki, S., and Muraki, K. (2013). Shot noise spectroscopy on a semiconductor quantum dot in the elastic and inelastic cotunneling regimes, *Phys. Rev. B* **87**, p. 041302, http://link.aps.org/doi/10.1103/PhysRevB.87.041302.

Onac, E., Balestro, F., Trauzettel, B., Lodewijk, C. F. J., and Kouwenhoven, L. P. (2006). Shot-noise detection in a carbon nanotube quantum dot, *Phys. Rev. Lett.* **96**, p. 026803, http://link.aps.org/doi/10.1103/PhysRevLett.96.026803.

Oshima, T., and Kawaguchi, Y. (2016). Spin Hall effect in a spinor dipolar bose-einstein condensate, *Phys. Rev. A* **93**, p. 053605, doi:10.1103/PhysRevA.93.053605, http://link.aps.org/doi/10.1103/PhysRevA.93.053605.

Ozbay, A., Gokce, A., Flanagan, T., Stearrett, R. A., Nowak, E. R., and Nordman, C. (2009). Low frequency magnetoresistive noise in spin-valve structures, *Appl. Phys. Lett.* **94**, 20, 202506, doi:http://dx.doi.org/10.1063/1.3139067, http://scitation.aip.org/content/aip/journal/apl/94/20/10.1063/1.3139067.

Ozbay, A., Nowak, E. R., Edelstein, A. S., Fischer, G. A., Nordman, C. A., and Cheng, S. F. (2006). Magnetic-field dependence of the noise in a magnetoresistive sensor having mems flux concentrators, *IEEE Trans. Magn.* **42**, 10, pp. 3306–3308, doi:10.1109/TMAG.2006.879752.

Pan, F., Gao, S., Chen, C., Song, C., and Zeng, F. (2014). Recent progress in resistive random access memories: Materials, switching mechanisms, and performance, *Mater. Sci. Eng. R: Reports* **83**, pp. 1–59, doi:http://dx.doi.org/10.1016/j.mser.2014.06.002, http://www.sciencedirect.com/science/article/pii/S0927796X14000692.

Pannetier, M., Fermon, C., Le Goff, G., Simola, J., and Kerr, E. (2004). Femtotesla magnetic field measurement with magnetoresistive sensors, *Science* **304**, 5677, pp. 1648–1650, doi:10.1126/science.1096841, http://www.sciencemag.org/content/304/5677/1648.abstract.

Paperno, E., and Kaplan, B.-Z. (1995). Suppression of barkhausen noise in magnetoresistive sensors employing ac bias, *IEEE Trans. Magn.* **31**, 6, pp. 3161–3163.

Parkin, S. S. P., Hayashi, M., and Thomas, L. (2008). Magnetic Domain-Wall Racetrack Memory, *Science* **320**, 5873, pp. 190–194, doi:10.1126/science.1145799, http://www.sciencemag.org/content/320/5873/190.abstract.

Parkin, S. S. P., Kaiser, C., Panchula, A., Rice, P. M., Hughes, B., Samant, M. and Yang, S.-H. (2004). Giant tunnelling magnetoresistance at room temperature with MgO (100) tunnel barriers, *Nat. Mater.* **3**, pp. 862–867, http://www.nature.com/nmat/journal/v3/n12/full/nmat1256.html.

Pasupathy, A. N., Bialczak, R. C., Martinek, J., Grose, J. E., Donev, L. A. K., McEuen, P. L., and Ralph, D. C. (2004). The Kondo effect in the presence of ferromagnetism, *Science* **306**, 5693, pp. 86–89, doi:10.1126/science.1102068, http://www.sciencemag.org/content/306/5693/86.abstract.

Peres, N. M. R. (2010). *Colloquium*: The transport properties of graphene: An introduction, *Rev. Mod. Phys.* **82**, pp. 2673–2700, doi:10.1103/RevModPhys.82.2673, http://link.aps.org/doi/10.1103/RevModPhys.82.2673.

Perkins, F. K., Friedman, A. L., Cobas, E., Campbell, P. M., Jernigan, G. G. and Jonker, B. T. (2013). Chemical vapor sensing with monolayer MoS2, *Nano Lett.* **13**, 2, pp. 668–673, doi:10.1021/nl3043079, http://dx.doi.org/10.1021/nl3043079, pMID: 23339527.

Pershin, Y. V., and Di Ventra, M. (2008). Spin memristive systems: Spin memory effects in semiconductor spintronics, *Phys. Rev. B* **78**, p. 113309, doi:10.1103/PhysRevB.78.113309, http://link.aps.org/doi/10.1103/PhysRevB.78.113309.

Petit, S., Baraduc, C., Thirion, C., Ebels, U., Liu, Y., Li, M., Wang, P., and Dieny, B. (2007). Spin-torque influence on the high-frequency magnetization fluctuations in magnetic tunnel junctions, *Phys. Rev. Lett.* **98**, p. 077203, http://link.aps.org/doi/10.1103/PhysRevLett.98.077203.

Pioro-Ladrière, M., Ciorga, M., Lapointe, J., Zawadzki, P., Korkusiński, M., Hawrylak, P., and Sachrajda, A. S. (2003). Spin-blockade spectroscopy of a two-level artificial molecule, *Phys. Rev. Lett.* **91**, p. 026803, doi:10.1103/PhysRevLett.91.026803, http://link.aps.org/doi/10.1103/PhysRevLett.91.026803.

Popova, E., Faure-Vincent, J., Tiusan, C., Bellouard, C., Fischer, H., Hehn, M., Montaigne, F., Alnot, M., Andrieu, S., Schuhl, a., Snoeck, E., and da Costa, V. (2002). Epitaxial MgO layer for low-resistance and coupling-free magnetic tunnel junctions, *Appl. Phys. Lett.* **81**, 6, p. 1035, doi:10.1063/1.1498153, http://link.aip.org/link/APPLAB/v81/i6/p1035/s1&Agg=doi.

Prezioso, M., Riminucci, A., Bergenti, I., Graziosi, P., Brunel, D., and Dediu, V. A. (2011). Electrically programmable magnetoresistance in multifunctional organic-based spin valve devices, *Adv. Mater.* **23**, pp. 1371–1375, http://onlinelibrary.wiley.com/doi/10.1002/adma.201003974/abstract.

Prodi, G. A., Vitale, S., Cerdonio, M., and Falferi, P. (1989). Thermal magnetization noise as a function of frequency in amorphous ferromagnets, *J. Appl. Phys.* **66**, 12, pp. 5984–5987, http://scitation.aip.org/content/aip/journal/jap/66/12/10.1063/1.343626.

Puglisi, F. M., Pavan, P., Padovani, A., Larcher, L., and Bersuker, G. (2013). RT noise characterization of HfOx RRAM in high resistive state, *Solid-State Electron.* **84**, 0, pp. 160–166, doi:http://dx.doi.org/10.1016/j.sse.2013.02.023, http://www.sciencedirect.com/science/article/pii/S0038110113000841, selected papers from the ESSDERC 2012 Conference.

Quinlan, F., Fortier, T. M., Jiang, H., Hati, A., Nelson, C., Fu, Y., Campbell, J. C., and Diddams, S. A. (2013). Exploiting shot noise correlations in the photodetection of ultrashort optical pulse trains, *Nat. Photonics*, pp. 290–293, http://www.nature.com/nphoton/journal/v7/n4/full/nphoton.2013.33.html.

Quinsat, M., Gusakova, D., Sierra, J. F., Michel, J. P., Houssameddine, D., Delaet, B., Cyrille, M.-C., Ebels, U., Dieny, B., Buda-Prejbeanu, L. D., Katine, J. A., Mauri, D., Zeltser, A., Prigent, M., Nallatamby, J.-C., and Sommet, R. (2010). Amplitude and phase noise of magnetic tunnel junction oscillators, *Appl. Phys. Lett.* **97**, 18, 182507, http://scitation.aip.org/content/aip/journal/apl/97/18/10.1063/1.3506901.

Meservey, R., and Tedrow, P. (1994). Spin-polarized electron tunneling, *Phys. Rep.* **238**, pp. 173–243, doi:10.1016/0370-1573(94)90105-8.

Ralls, K. S., and Buhrman, R. A. (1988). Defect interactions and noise in metallic nanoconstrictions, *Phys. Rev. Lett.* **60**, pp. 2434–2437, http://link.aps.org/doi/10.1103/PhysRevLett.60.2434.

Ralph, D., and Stiles, M. (2008). Spin transfer torques, *J. Magn. Magn. Mater.* **320**, pp. 1190–1216, http://www.sciencedirect.com/science/article/pii/S0304885307010116.

Ramachandran, G. K., Hopson, T. J., Rawlett, A. M., Nagahara, L. A., Primak, A. and Lindsay, S. M. (2003). A bond-fluctuation mechanism for stochastic switching in wired molecules, *Science* **300**, 5624, pp. 1413–1416, doi:10.1126/science.1083825.

Raman, K. V., Watson, S. M., Shim, J. H., Borchers, J. A., Chang, J., and Moodera, J. S. (2009). Effect of molecular ordering on spin and charge injection in rubrene, *Phys. Rev. B* **80**, p. 195212, http://link.aps.org/doi/10.1103/PhysRevB.80.195212.

Rebei, A. (2011). Suppression of noise in magnetic layers by a periodic spin torque, *Phys. Rev. E* **83**, p. 061148, doi:10.1103/PhysRevE.83.061148, http://link.aps.org/doi/10.1103/PhysRevE.83.061148.

Rebei, A., Berger, L., Chantrell, R., and Covington, M. (2005). 1/f-type noise in a biased current perpendicular to the plane spin valve: A numerical study, *J. Appl. Phys.* **97**, 10, 10.

Reig, C., Cubells-Beltrán, M.-D., and Ramírez Muñoz, D. (2009). Magnetic field sensors based on giant magnetoresistance (GMR) technology: Applications in electrical current sensing, *Sensors* **9**, 10, p. 7919, doi:10.3390/s91007919, http://www.mdpi.com/1424-8220/9/10/7919.

Renteria, J., Samnakay, R., Rumyantsev, S. L., Jiang, C., Goli, P., Shur, M. S. and Balandin, A. A. (2014). Low-frequency 1/f noise in MoS2 transistors: Relative contributions of the channel and contacts, *Appl. Phys. Lett.* **104**, 15, 153104, doi:http://dx.doi.org/10.1063/1.4871374, http://scitation.aip.org/content/aip/journal/apl/104/15/10.1063/1.4871374.

Rogge, M. C., Fühner, C., Keyser, U. F., and Haug, R. J. (2004). Spin blockade in capacitively coupled quantum dots, *Appl. Phys. Lett.* **85**, 4.

Rogge, M. C., Fühner, C., and Haug, R. J. (2006). Multiple transitions of the spin configuration in quantum dots, *Phys. Rev. Lett.* **97**, p. 176801, doi: 10.1103/PhysRevLett.97.176801, http://link.aps.org/doi/10.1103/PhysRevLett.97.176801.

Romeike, C., Wegewijs, M. R., and Schoeller, H. (2006). Spin quantum tunneling in single molecular magnets: Fingerprints in transport spectroscopy of current and noise, *Phys. Rev. Lett.* **96**, p. 196805, doi: 10.1103/PhysRevLett.96.196805, http://link.aps.org/doi/10.1103/PhysRevLett.96.196805.

Rumyantsev, S., Liu, G., Stillman, W., Shur, M., and Balandin, A. (2010). Electrical and noise characteristics of graphene field-effect transistors: ambient effects, noise sources and physical mechanisms, *J. Phys.: Condensed Matter* **22**, 39, p. 395302.

Rungger, I., Mryasov, O., and Sanvito, S. (2009). Resonant electronic states and *i-v* curves of Fe/MgO/Fe (100) tunnel junctions, *Phys. Rev. B* **79**, p. 094414, http://link.aps.org/doi/10.1103/PhysRevB.79.094414.

Safonov, V. L., and Bertram, H. N. (2005). Fluctuation-dissipation considerations and damping models for ferromagnetic materials, *Phys. Rev. B* **71**, p. 224402, doi:10.1103/PhysRevB.71.224402, http://link.aps.org/doi/10.1103/PhysRevB.71.224402.

Safonov, S. S., Savchenko, A. K., Bagrets, D. A., Jouravlev, O. N., Nazarov, Y. V., Linfield, E. H., and Ritchie, D. A. (2003). Enhanced shot noise in resonant tunneling via interacting localized states, *Phys. Rev. Lett.* **91**, p. 136801, http://link.aps.org/doi/10.1103/PhysRevLett.91.136801.

Sangwan, V. K., Arnold, H. N., Jariwala, D., Marks, T. J., Lauhon, L. J., and Hersam, M. C. (2013). Low-frequency electronic noise in single-layer MoS2 transistors, *Nano Lett.* **13**, 9, pp. 4351–4355, doi:10.1021/nl402150r, http://dx.doi.org/10.1021/nl402150r, pMID: 23944940.

Sankey, J. C., Braganca, P. M., Garcia, A. G. F., Krivorotov, I. N., Buhrman, R. A., and Ralph, D. C. (2006). Spin-transfer-driven ferromagnetic resonance of individual nanomagnets, *Phys. Rev. Lett.* **96**, p. 227601, http://link.aps.org/doi/10.1103/PhysRevLett.96.227601.

Sankey, J. C., Cui, Y.-T., Sun, J. Z., Slonczewski, J. C., Buhrman, R. A., and Ralph, D. C. (2007). Measurement of the spin-transfer-torque vector in magnetic tunnel junctions, *Nat. Phys.* **4**, pp. 67–71, http://www.nature.com/nphys/journal/v4/n1/full/nphys783.html.

Sanvito, S. (2010). Molecular spintronics: The rise of spinterface science, *Nat. Phys.* **6**, p. 562–564, doi:10.1038/nphys1714, http://www.nature.com/nphys/journal/v6/n8/abs/nphys1714.html.

Sattari, F. (2015). Spin-dependent shot noise in magnetic graphene superlattice, *Superlattices Microstruct.* **86**, pp. 29–35, doi:http://dx.doi.org/10.1016/j.spmi.2015.07.028, http://www.sciencedirect.com/science/article/pii/S0749603615301038.

Schaffert, J., Cottin, M. C., Sonntag, A., Karacuban, H., Bobisch, C. A., Lorente, N., Gauyacq, J.-P., and Möller, R. (2012). Imaging the dynamics of individually adsorbed molecules, *Nat. Mater.* **12**, doi:10.1038/nmat3527.

Scheck, C., Cheng, L., Barsukov, I., Frait, Z., and Bailey, W. E. (2007). Low relaxation rate in epitaxial vanadium-doped ultrathin iron films, *Phys. Rev. Lett.* **98**, 11, p. 117601, doi:10.1103/PhysRevLett.98.117601.

Schlenk, T., Bianchi, M., Koleini, M., Eich, A., Pietzsch, O., Wehling, T. O., Frauenheim, T., Balatsky, A., Mi, J.-L., Iversen, B. B., Wiebe,

J., Khajetoorians, A. A., Hofmann, P., and Wiesendanger, R. (2013). Controllable magnetic doping of the surface state of a topological insulator, *Phys. Rev. Lett.* **110**, p. 126804, http://link.aps.org/doi/10.1103/PhysRevLett.110.126804.

Schmidt, G., Ferrand, D., Molenkamp, L. W., Filip, A. T., and van Wees, B. J. (2000). Fundamental obstacle for electrical spin injection from a ferromagnetic metal into a diffusive semiconductor, *Phys. Rev. B* **62**, pp. R4790–R4793, http://link.aps.org/doi/10.1103/PhysRevB.62.R4790.

Schmidt, H., Giustiniano, F., and Eda, G. (2015). Electronic transport properties of transition metal dichalcogenide field-effect devices: surface and interface effects, *Chem. Soc. Rev.* **44**, pp. 7715–7736, doi:10.1039/C5CS00275C, http://dx.doi.org/10.1039/C5CS00275C.

Schoelkopf, R. J., Burke, P. J., Kozhevnikov, A. A., Prober, D. E., and Rooks, M. J. (1997). Frequency dependence of shot noise in a diffusive mesoscopic conductor, *Phys. Rev. Lett.* **78**, pp. 3370–3373, http://link.aps.org/doi/10.1103/PhysRevLett.78.3370.

Schomerus, H. (2007). Effective contact model for transport through weakly-doped graphene, *Phys. Rev. B* **76**, p. 045433, doi:10.1103/PhysRevB.76.045433, http://link.aps.org/doi/10.1103/PhysRevB.76.045433.

Schoonus, J. J. H. M., Lumens, P. G. E., Wagemans, W., Kohlhepp, J. T., Bobbert, P. A., Swagten, H. J. M., and Koopmans, B. (2009). Magnetoresistance in hybrid organic spin valves at the onset of multiple-step tunneling, *Phys. Rev. Lett.* **103**, p. 146601, http://link.aps.org/doi/10.1103/PhysRevLett.103.146601.

Schwierz, F. (2010). Graphene transistors, *Nat. Nanotechnol.* **5**, 7, pp. 487–496, doi:10.1038/nnano.2010.89, http://www.nature.com/nnano/journal/v5/n7/abs/nnano.2010.89.html.

Scola, J., Polovy, H., Fermon, C., Pannetier-Lecoeur, M., Feng, G., Fahy, K. and Coey, J. M. D. (2007a). Noise in MgO barrier magnetic tunnel junctions with CoFeB electrodes: Influence of annealing temperature, *Appl. Phys. Lett.* **90**, 25, 252501, http://scitation.aip.org/content/aip/journal/apl/90/25/10.1063/1.2749433.

Scola, J., Polovy, H., Fermon, C., Pannetier-Lecœur, M., Feng, G., Fahy, K. and Coey, J. M. D. (2007b). Noise in MgO barrier magnetic tunnel junctions with CoFeB electrodes: Influence of annealing temperature, *Appl. Phys. Lett.* **90**, 25, 252501, doi:http://dx.doi.org/10.1063/1.2749433, http://scitation.aip.org/content/aip/journal/apl/90/25/10.1063/1.2749433.

Secker, D., Wagner, S., Ballmann, S., Härtle, R., Thoss, M., and Weber, H. B. (2011). Resonant vibrations, peak broadening, and noise in single molecule contacts: The nature of the first conductance peak, *Phys. Rev. Lett.* **106**, p. 136807, doi:10.1103/PhysRevLett.106.136807, http://link.aps.org/doi/10.1103/PhysRevLett.106.136807.

Sekiguchi, K., Arakawa, T., Yamauchi, Y., Chida, K., Yamada, M., Takahashi, H., Chiba, D., Kobayashi, K., and Ono, T. (2010). Observation of full shot noise in CoFeB/MgO/CoFeB-based magnetic tunneling junctions, *Appl. Phys. Lett.* **96**, 25, 252504, http://scitation.aip.org/content/aip/journal/apl/96/25/10.1063/1.3456548.

Semenoff, G. W. (1984). Condensed-matter simulation of a three-dimensional anomaly, *Phys. Rev. Lett.* **53**, pp. 2449–2452, doi:10.1103/PhysRevLett.53.2449, http://link.aps.org/doi/10.1103/PhysRevLett.53.2449.

Semenov, Y. G., Kim, K. W., and Zavada, J. M. (2007). Spin field effect transistor with a graphene channel, *Appl. Phys. Lett.* **91**, 15, 153105, doi:http://dx.doi.org/10.1063/1.2798596, http://scitation.aip.org/content/aip/journal/apl/91/15/10.1063/1.2798596.

Shao, Q., Liu, G., Teweldebrhan, D., Balandin, A. A., Rumyantsev, S., Shur, M. S., and Yan, D. (2009). Flicker noise in bilayer graphene transistors, *Electron Device Lett., IEEE* **30**, 3, pp. 288–290.

Sharma, R., Durrenfeld, P., Iacocca, E., Heinonen, O. G., Akerman, J., and Muduli, P. K. (2014). Mode-hopping mechanism generating colored noise in a magnetic tunnel junction based spin torque oscillator, *Appl. Phys. Lett.* **105**, 13, p. 132404.

Shen, X. Y., Dong, B., Lei, X. L., and Horing, N. J. M. (2007). Vibration-mediated resonant tunneling and shot noise through a molecular quantum dot, *Phys. Rev. B* **76**, p. 115308, http://link.aps.org/doi/10.1103/PhysRevB.76.115308.

Shi, S., Liu, F., Smith, D. L., and Ruden, P. P. (2015). Effects of disorder on spin injection and extraction for organic semiconductor spin-valves, *J. Appl. Phys.* **117**, 8, 085501, doi:http://dx.doi.org/10.1063/1.4913281, http://scitation.aip.org/content/aip/journal/jap/117/8/10.1063/1.4913281.

Shim, J. H., Raman, K. V., Park, Y. J., Santos, T. S., Miao, G. X., Satpati, B. and Moodera, J. S. (2008). Large spin diffusion length in an amorphous organic semiconductor, *Phys. Rev. Lett.* **100**, p. 226603, http://link.aps.org/doi/10.1103/PhysRevLett.100.226603.

Sicot, M., Andrieu, S., Tiusan, C., Montaigne, F., and Bertran, F. (2006). On the quality of molecular-beam epitaxy grown Fe/MgO and Co/MgO(001) interfaces, *J. Appl. Phys.* **99**, 8, 08D301, http://link.aip.org/link/?JAP/99/08D301/1.

Sierra, J.F. et al. (2009). *Appl. Phys. Lett.* **94**, 012506, DOI:10.1063/1.3054642)

Simmons, J. G. (1963). Generalized formula for the electric tunnel effect between similar electrodes separated by a thin insulating film, *J. Appl. Phys.* **34**, 6, pp. 1793–1803.

Simoen, E., and Claeys, C. (1999). On the flicker noise in submicron silicon mosfets, *Solid-State Electron.* **43**, pp. 865–882.

Slonczewski, J. (1996). Current-driven excitation of magnetic multilayers, *J. Magnet. Magn. Mater.* **159**, pp. L1–L7, http://www. sciencedirect.com/science/article/pii/0304885396000625.

Sperl, A., Kröger, J., and Berndt, R. (2010). Direct observation of conductance fluctuations of a single-atom tunneling contact, *Phys. Rev. B* **81**, p. 035406, doi:10.1103/PhysRevB.81.035406, http://link.aps.org/doi/10.1103/PhysRevB.81.035406.

Spietz, L., Lehnert, K. W., Siddiqi, I., and Schoelkopf, R. J. (2003). Primary electronic thermometry using the shot noise of a tunnel junction, *Science* **300**, 5627, pp. 1929–1932, doi:10.1126/science.1084647, http://www.sciencemag.org/content/300/5627/1929.abstract.

Stearrett, R., Wang, W. G., Shah, L. R., Gokce, A., Xiao, J. Q., and Nowak, E. R. (2010a). Evolution of barrier-resistance noise in CoFeB/MgO/CoFeB tunnel junctions during annealing, *J. Appl. Phys.* **107**, 6, 064502, http://scitation.aip.org/content/aip/journal/jap/107/6/10.1063/1.3327440.

Stearrett, R., Wang, W. G., Shah, L. R., Xiao, J. Q., and Nowak, E. R. (2010b). Magnetic noise evolution in CoFeB/MgO/CoFeB tunnel junctions during annealing, *Appl. Phys. Lett.* **97**, 24, 243502, http://scitation.aip.org/content/aip/journal/apl/97/24/10.1063/1.3526721.

Stewart, D. A. (2010). New type of magnetic tunnel junction based on spin filtering through a reduced symmetry oxide: FeCo—Mg3B2O6—FeCo, *Nano Lett.* **10**, 1, pp. 263–267, http://dx.doi.org/10.1021/nl9034362.

Strukov, D. B., Snider, G. S., Stewart, D. R., and Williams, R. S. (2008). The missing memristor found, *Nature* **453**, doi:10.1038/nature06932.

Stutzke, N. A., Russek, S. E., Pappas, D. P., and Tondra, M. (2005). Low-frequency noise measurements on commercial magnetoresistive magnetic field sensors, *J. Appl. Phys.* **97**, 10Q107, doi:http://dx.doi.org/10.1063/1.1861375.

Sun, J. (1999). Current-driven magnetic switching in manganite trilayer junctions, *J. Magnet. Magn. Mater.* **202**, 1, pp. 157–162, http://www.sciencedirect.com/science/article/pii/S0304885399002899.

Szczepański, T., Dugaev, V. K., Barnaś, J., Cascales, J. P., and Aliev, F. G. (2013). Shot noise in magnetic double-barrier tunnel junctions, *Phys. Rev. B* **87**, p. 155406, http://link.aps.org/doi/10.1103/PhysRevB.87.155406.

Szczepański, T., Dugaev, V. K., Barnaś, J., Martinez, I., Cascales, J. P., Hong, J.-Y., Lin, M.-T., and Aliev, F. G. (2016). Shot noise in magnetic tunneling structures with two-level quantum dots, *Phys. Rev. B* **94**, p. 235429, doi:10.1103/PhysRevB.94.235429, http://link.aps.org/doi/10.1103/PhysRevB.94.235429.

Szulczewski, G., Sanvito, S., and Coey, M. (2009). A spin of their own, *Nat. Mater.* **8**, pp. 693–695, http://www.nature.com/nmat/journal/v8/n9/abs/nmat2518.html.

Tanaka, T., Arakawa, T., Maeda, M., Kobayashi, K., Nishihara, Y., Ono, T., Nozaki, T., Fukushima, A., and Yuasa, S. (2014). Leak current estimated from the shot noise in magnetic tunneling junctions, *Appl. Phys. Lett.* **105**, 4, 042405, https://doi.org/10.1063/1.4891556

Tao, B. S., Yang, H. X., Zuo, Y. L., Devaux, X., Lengaigne, G., Hehn, M., Lacour, D., Andrieu, S., Chshiev, M., Hauet, T., Montaigne, F., Mangin, S., Han, X. F., and Lu, Y. (2015). Long-range phase coherence in double-barrier magnetic tunnel junctions with a large thick metallic quantum well, *Phys. Rev. Lett.* **115**, p. 157204, doi:10.1103/PhysRevLett. 115. 157204, http://link.aps.org/doi/10.1103/PhysRevLett.115.157204.

Teixeira, J. M., Ventura, J., Araujo, J. P., Sousa, J. B., Wisniowski, P., Cardoso, S., and Freitas, P. P. (2011). Resonant tunneling through electronic trapping states in thin MgO magnetic junctions, *Phys. Rev. Lett.* **106**, p. 196601, http://link.aps.org/doi/10.1103/PhysRevLett.106.196601.

Thadani, K. V., Finocchio, G., Li, Z.-P., Ozatay, O., Sankey, J. C., Krivorotov, I. N., Cui, Y.-T., Buhrman, R. A., and Ralph, D. C. (2008). Strong linewidth variation for spin-torque nano-oscillators as a function of in-plane magnetic field angle, *Phys. Rev. B* **78**, p. 024409, http://link.aps.org/doi/10.1103/PhysRevB.78.024409.

Theodonis, I., Kalitsov, A., and Kioussis, N. (2007). Enhancing spin-transfer torque through the proximity of quantum well states, *Phys.*

Rev. B **76**, p. 224406, http://link.aps.org/doi/10.1103/PhysRevB.76.224406.

Thomson, W. (1857). On the electrodynamic qualities of metals: effects of magnetization on the electric conductivity of nickel and iron, *Proc. Roy. Soc. London* **8**, pp. 546–550. doi: 10.1098/rspl.1856.0144.

Tiberkevich, V., Slavin, A., and Kim, J.-V. (2007). Microwave power generated by a spin-torque oscillator in the presence of noise, *Appl. Phys. Lett.* **91**, 19, 192506, http://scitation.aip.org/content/aip/journal/apl/91/19/10.1063/1.2812546.

Tiusan, C., Greullet, F., Hehn, M., Montaigne, F., Andrieu, S., and Schuhl, A. (2007). Spin tunnelling phenomena in single-crystal magnetic tunnel junction systems, *J. Phys.: Condensed Matter* **19**, 16, p. 165201, http://stacks.iop.org/0953-8984/19/i=16/a=165201.

Tiusan, C., Greullet, F., Sicot, M., Hehn, M., Bellouard, C., Montaigne, F., Andrieu, S., and Schuhl, A. (2006). Engineering of spin filtering in double epitaxial tunnel junctions, *J. Appl. Phys.* **99**, 08A903, doi: http://dx.doi.org/10.1063/1.2166592.

Tserkovnyak, Y., and Brataas, A. (2001). Shot noise in ferromagnet–normal metal systems, *Phys. Rev. B* **64**, p. 214402, http://link.aps.org/doi/10.1103/PhysRevB.64.214402.

Tserkovnyak, Y., and Brataas, A. (2002). Current and spin torque in double tunnel barrier ferromagnet-superconductor-ferromagnet systems, *Phys. Rev. B* **65**, p. 094517, http://link.aps.org/doi/10.1103/PhysRevB.65.094517.

Tserkovnyak, Y., Brataas, A., Bauer, G. E. W., and Halperin, B. I. (2005). Nonlocal magnetization dynamics in ferromagnetic heterostructures, *Rev. Mod. Phys.* **77**, pp. 1375–1421, doi:10.1103/RevModPhys.77.1375, http://link.aps.org/doi/10.1103/RevModPhys.77.1375.

Tsoi, M., Jansen, A. G. M., Bass, J., Chiang, W.-C., Seck, M., Tsoi, V., and Wyder, P. (1998). Excitation of a magnetic multilayer by an electric current, *Phys. Rev. Lett.* **80**, pp. 4281–4284, http://link.aps.org/doi/10.1103/PhysRevLett.80.4281.

Tsutsui, M., Taniguchi, M., and Kawai, T. (2010). Single-molecule identification via electric current noise, *Nat. Commun.* **1**, p. 138, doi:doi:10.1038/ncomms1141.

Tsymbal, E., and Zutic, I. (2011). *Handbook of Spin Transport and Magnetism* (Taylor & Francis), ISBN 9781439803776.

Tulapurkar, A. A., Suzuki, Y., Fukushima, A., Kubota, H., Maehara, H., Tsunekawa, K., Djayaprawira, D. D., Watanabe, N., and Yuasa, S.

(2005). Spin-torque diode effect in magnetic tunnel junctions, *Nature* **438**, pp. 339–342, http://www.nature.com/nature/journal/v438/n7066/full/nature04207.html.

Tworzydło, J., Trauzettel, B., Titov, M., Rycerz, A., and Beenakker, C. W. J. (2006). Sub-poissonian shot noise in graphene, *Phys. Rev. Lett.* **96**, p. 246802, doi:10.1103/PhysRevLett.96.246802, http://link.aps.org/doi/10.1103/PhysRevLett.96.246802.

Ubbelohde, N., Fricke, C., Hohls, F., and Haug, R. J. (2013). Spin-dependent shot noise enhancement in a quantum dot, *Phys. Rev. B* **88**, p. 041304, doi:10.1103/PhysRevB.88.041304, http://link.aps.org/doi/10.1103/PhysRevB.88.041304.

Ulstrup, S., Frederiksen, T., and Brandbyge, M. (2012). Nonequilibrium electron-vibration coupling and conductance fluctuations in a C_{60} junction, *Phys. Rev. B* **86**, p. 245417, doi:10.1103/PhysRevB.86.245417, http://link.aps.org/doi/10.1103/PhysRevB.86.245417.

Useinov, A. N., Kosel, J., Useinov, N. K., and Tagirov, L. R. (2011). Resonant tunnel magnetoresistance in double-barrier planar magnetic tunnel junctions, *Phys. Rev. B* **84**, p. 085424, http://link.aps.org/doi/10.1103/PhysRevB.84.085424.

Useinov, A., Ye, L.-X., Useinov, N., Wu, T.-H., and Lai, C.-H. (2015). Anomalous tunnel magnetoresistance and spin transfer torque in magnetic tunnel junctions with embedded nanoparticles, *Sci. Rep.* **5**, doi:10.1038/srep18026, http://dx.doi.org/10.1038/srep18026.

Valadeiro, J., Amaral, J. P., Leitao, D., Ferreira, R., Freitas Cardoso, S., and Freitas, P. (2015). Strategies for ptesla field detection using magnetoresistive sensors with a soft pinned sensing layer, *J. Appl. Phys.* **51**, 1, 4400204, doi:10.1109/TMAG.2014.2352115, 10.1109/TMAG.2014.2352115.

Vandamme, L. K. J., Cölle, M., De Leeuw, D. M., and Verbakel, F. (2011). Low-frequency noise to characterise resistive switching of metal oxide on polymer memories, *Fluctuation Noise Lett.* **10**, 04, pp. 497–514, doi:10.1142/S0219477511000739, http://www.worldscientific.com/doi/abs/10.1142/S0219477511000739.

Vázquez, H., Oszwaldowski, R., Pou, P., Ortega, J., Pérez, R., Flores, F. and Kahn, A. (2004). Dipole formation at metal/PTCDA interfaces: Role of the charge neutrality level, *EPL (Europhys. Lett.)* **65**, 6, p. 802, http://stacks.iop.org/0295-5075/65/i=6/a=802.

Vedyayev, A., Ryzhanova, N., Dieny, B., and Strelkov, N. (2006a). Resonant spin-torque in double barrier magnetic tunnel junctions, *Phys.*

Lett. A **355**, 3, pp. 243–246, http://www.sciencedirect.com/science/article/pii/S0375960106002271.

Vedyayev, A., Ryzhanova, N., Dieny, B., and Strelkov, N. (2006b). Resonant spin-torque in double barrier magnetic tunnel junctions, *Phys. Lett. A* **355**, 3, pp. 243–246, http://www.sciencedirect.com/science/article/pii/S0375960106002271.

Velev, J. P., Belashchenko, K. D., Jaswal, S. S., and Tsymbal, E. Y. (2007). Effect of oxygen vacancies on spin-dependent tunneling in Fe/MgO/Fe magnetic tunnel junctions, *App. Phys. Lett.* **90**, 7, p. 072502, doi: DOI:10.1063/1.2643027, http://dx.doi.org/doi/10.1063/1.2643027.

Vinzelberg, H., Schumann, J., Elefant, D., Gangineni, R., Thomas, J., and Buchner, B. (2008). Low temperature tunneling magnetoresistance on (La,Sr)MnO3Co junctions with organic spacer layers, *J. Appl. Phys.* **103**, 9, 093720, http://scitation.aip.org/content/aip/journal/jap/103/9/10.1063/1.2924435.

Vitale, S., Tommasini, R., Cerdonio, M., Bonaldi, M., Cavalleri, A., and Durin, G. (1992). Magnetic viscosity, thermal relaxation, and thermal equilibrium noise in Co-based amorphous alloys at millikelvin temperatures, *J. Appl. Phys.* **72**, 10, pp. 4820–4825, http://scitation.aip.org/content/aip/journal/jap/72/10/10.1063/1.352045.

Waldron, D., Timoshevskii, V., Hu, Y., Xia, K., and Guo, H. (2006). First principles modeling of tunnel magnetoresistance of Fe/MgO/Fe trilayers, *Phys. Rev. Lett.* **97**, 22, p. 226802, doi:10.1103/PhysRevLett.97.226802.

Wallace, P. R. (1947). The band theory of graphite, *Phys. Rev.* **71**, pp. 622–634, doi:10.1103/PhysRev.71.622, http://link.aps.org/doi/10.1103/PhysRev.71.622.

Wang, S. X., and Guanxiong, L. (2008). Advances in giant magnetoresistance biosensors with magnetic nanoparticle tags: Review and outlook, *IEEE Trans. Magn.* **44**, pp. 1687–1702, doi:10.1109/TMAG.2008.920962.

Wang, W. G., Jordan-sweet, J., Miao, G. X., Ni, C., Rumaiz, A. K., Shah, L. R., Fan, X., Parsons, P., Stearrett, R., Nowak, E. R., Moodera, J. S., and Xiao, J. Q. (2009). In-situ characterization of rapid crystallization of amorphous CoFeB electrodes in CoFeB/MgO/CoFeB junctions during thermal annealing, *Appl. Phys. Lett.* **95**, 24, 242501, http://scitation.aip.org/content/aip/journal/apl/95/24/10.1063/1.3273397.

Wang, Z., Ki, D., Chen, H., Berger, H., MacDonald, A. H., and Morpurgo, A. F. (2015). Strong interface-induced spin–orbit interaction in graphene on Ws2, *Nat. Commun.* **6**, p. 8339, doi:10.1038/ncomms9339, http://dx.doi.org/10.1038/ncomms9339.

Wang, T. X., Li, Y., Lee, K. J., Cho, J. U., Kim, D. K., Noh, S. J., and Kim, Y. K. (2011b). Influence of interface state in Fe/MgO/Fe magnetic tunnel junction system: C modified interfaces-a first principle study, *J. Appl. Phys.* **109**, 8, 083714, doi:10.1063/1.3575337, http://link.aip.org/link/?JAP/109/083714/1.

Wang, Y., Lu, Z.-Y., Zhang, X.-G., and Han, X. F. (2006). First-principles theory of quantum well resonance in double barrier magnetic tunnel junctions, *Phys. Rev. Lett.* **97**, p. 087210, http://link.aps.org/doi/10.1103/PhysRevLett.97.087210.

Wang, F., Macià, F., Wohlgenannt, M., Kent, A. D., and Flatté, M. E. (2012a). Magnetic fringe-field control of electronic transport in an organic film, *Phys. Rev. X* **2**, p. 021013, http://link.aps.org/doi/10.1103/PhysRevX.2.021013.

Wang, D., Nordman, C., Daughton, J., Qian, Z., and Fink, J. (2004a). 70junctions with CoFeB as free and reference layers, *IEEE Trans. Magn.* **40**, 4, pp. 2269–2271.

Wang, D., Nordman, C., Daughton, J., Qian, Z., and Fink, J. (2004b). 70junctions with CoFeB as free and reference layers, *IEEE Trans. Magn.* **40**, 4, pp. 2269–2271.

Wang, Y., and Perdew, J. P. (1991). Correlation hole of the spin-polarized electron gas, with exact small-wave-vector and high-density scaling, *Phys. Rev. B* **44**, pp. 13298–13307, http://link.aps.org/doi/10.1103/PhysRevB.44.13298.

Wang, R.-Q., Sheng, L., Hu, L.-B., Wang, B., and Xing, D. Y. (2011a). Coexistence of super-poissonian mechanisms in quantum dots with ferromagnetic leads, *Phys. Rev. B* **84**, p. 115304, doi:10.1103/PhysRevB.84.115304, http://link.aps.org/doi/10.1103/PhysRevB.84.115304.

Wang, H., Yu, L., Lee, Y.-H., Shi, Y., Hsu, A., Chin, M. L., Li, L.-J., Dubey, M., Kong, J., and Palacios, T. (2012b). Integrated circuits based on bilayer MoS2 transistors, *Nano Lett.* **12**, 9, pp. 4674–4680, doi:10.1021/nl302015v, http://dx.doi.org/10.1021/nl302015v, pMID: 22862813.

Watanabe, M., Okabayashi, J., Toyao, H., Yamaguchi, T., and Yoshino, J. (2008). Current-driven magnetization reversal at extremely low threshold current density in (Ga,Mn)As-based double-barrier magnetic tunnel junctions, *Appl. Phys. Lett.* **92**, 8, p. 082506, http://dx.doi.org/doi/10.1063/1.2841703.

Werner, R., Petrov, A. Y., Miño, L. A., Kleiner, R., Koelle, D., and Davidson, B. A. (2011). Improved tunneling magnetoresistance at low temperature in manganite junctions grown by molecular beam epitaxy, *Appl. Phys. Lett.* **98**, 16, 162505, doi:http://dx.doi.org/10.1063/1.3581885, http://scitation.aip.org/content/aip/journal/apl/98/16/10.1063/1.3581885.

Weymann, I. (2008). Effects of different geometries on the conductance, shot noise, and tunnel magnetoresistance of double quantum dots, *Phys. Rev. B* **78**, p. 045310, doi:10.1103/PhysRevB.78.045310, http://link.aps.org/doi/10.1103/PhysRevB.78.045310.

Wilczynski, M., and Barnaś, J. (2000). Tunnel magnetoresistance in ferromagnetic double-barrier planar junctions: coherent tunneling regime, *J. Magnet. Magn. Mater.* **221**, pp. 373–381, http://www.sciencedirect.com/science/article/pii/S030488530000514X.

Wisniowski, P., Dabek, M., Skowronski, W., Stobiecki, T., Cardoso, S., and Freitas, P. P. (2014). Reduction of low frequency magnetic noise by voltage-induced magnetic anisotropy modulation in tunneling magnetoresistance sensors, *Appl. Phys. Lett.* **105**, 8, 082404, doi:http://dx.doi.org/10.1063/1.4894172, http://scitation.aip.org/content/aip/journal/apl/105/8/10.1063/1.4894172.

Wolf, D., and Holler, E. (1967). *J. Appl. Phys.* **38**, 189 http://aip.scitation.org/doi/10.1063/1.1708950.

Wortmann, D., Ishida, H., and Blügel, S. (2005). Embedded green-function formulation of tunneling conductance: Bardeen versus Landauer approaches, *Phys. Rev. B* **72**, p. 235113, http://link.aps.org/doi/10.1103/PhysRevB.72.235113.

Xiao, J., Zangwill, A., and Stiles, M. D. (2005). Macrospin models of spin transfer dynamics, *Phys. Rev. B* **72**, p. 014446, http://link.aps.org/doi/10.1103/ PhysRevB.72.014446.

Xie, L., and Cui, X. (2016). Manipulating spin-polarized photocurrents in 2d transition metal dichalcogenides, *Proc. Natl. Acad. Sci.* **113**, 14, pp. 3746–3750, doi:10.1073/pnas.1523012113, http://www.pnas.org/content/113/14/3746.abstract.

Xu, M., and Mai, Z. (1999). Angular dependence of giant magnetoresistance in spin-valve sandwiched structures, *Phys. Rev. B* **60**, pp. 9224–9227, doi:10.1103/PhysRevB.60.9224, http://link.aps.org/doi/10.1103/PhysRevB.60.9224.

Xu, G., Torres, C. M., Song, E. B., Tang, J., Bai, J., Duan, X., Zhang, Y., and Wang, K. L. (2010a). Enhanced conductance fluctuation by quantum confinement effect in graphene nanoribbons, *Nano Lett.* **10**, 11, pp. 4590–4594, http://dx.doi.org/10.1021/nl1025979.

Xu, G., Torres, C. M., Song, E. B., Tang, J., Bai, J., Duan, X., Zhang, Y., and Wang, K. L. (2010b). Enhanced conductance fluctuation by quantum confinement effect in graphene nanoribbons, *Nano Lett.* **10**, 11, pp. 4590–4594, doi:10.1021/nl1025979, http://dx.doi.org/10.1021/nl1025979.

Xue, H.-B., Liang, J.-Q., and Liu, W.-M. (2015). Negative differential conductance and super-poissonian shot noise in single-molecule magnet junctions, *Sci. Rep.* **5**, p. 8730, doi:10.1038/srep08730, http://dx.doi.org/10.1038/srep08730.

Yamaguchi, A., Nasu, S., Tanigawa, H., Ono, T., Miyake, K., Mibu, K., and Shinjo, T. (2005). Effect of joule heating in current-driven domain wall motion, *Appl. Phys. Lett.* **86**, 1, 012511, doi:10.1063/1.1847714, http:// link.aip.org/ link/?APL/86/012511/1.

Yamauchi, Y., Sekiguchi, K., Chida, K., Arakawa, T., Nakamura, S., Kobayashi, K., Ono, T., Fujii, T., and Sakano, R. (2011a). Evolution of the Kondo effect in a quantum dot probed by shot noise, *Phys. Rev. Lett.* **106**, p. 176601, http://link.aps.org/doi/10.1103/PhysRevLett.106.176601.

Yamauchi, Y., Sekiguchi, K., Chida, K., Arakawa, T., Nakamura, S., Kobayashi, K., Ono, T., Fujii, T., and Sakano, R. (2011b). Evolution of the Kondo effect in a quantum dot probed by shot noise, *Phys. Rev. Lett.* **106**, p. 176601, http://link.aps.org/doi/10.1103/PhysRevLett.106.176601.

Yan, W., Txoperena, O., Llopis, R., Dery, H., Hueso, L. E., Casanova, F. (2016). A two-dimensional spin field-effect switch, *Nat. Commun.* **7**, p. 13372, doi:10.1038/ncomms13372, http://dx.doi.org/10.1038/ncomms13372.

Yang, X., Liu, G., Rostami, M., Balandin, A. A., and Mohanram, K. (2011). Graphene ambipolar multiplier phase detector, *Electron Device Lett., IEEE* **32**, 10, pp. 1328–1330.

Yennie, D. R. (1987). Integral quantum Hall effect for nonspecialists, *Rev. Mod. Phys.* **59**, pp. 781–824, doi:10.1103/RevModPhys.59.781, http://link.aps.org/doi/10.1103/RevModPhys.59.781.

Yogev, S., Matsubara, R., Nakamura, M., Zschieschang, U., Klauk, H., and Rosenwaks, Y. (2013). Fermi level pinning by gap states in organic semiconductors, *Phys. Rev. Lett.* **110**, p. 036803, http://link.aps.org/doi/10.1103/PhysRevLett.110.036803.

Yu, G. Q., Diao, Z., Feng, J. F., Kurt, H., Han, X. F., and Coey, J. M. D. (2011). 1/f noise in MgO double-barrier magnetic tunnel junctions, *Appl. Phys. Lett.* **98**, 11, 112504, http://scitation.aip.org/content/aip/journal/apl/98/11/10.1063/1.3562951.

Yu, G., Upadhyaya, P., Fan, Y., Alzate, J. G., Jiang, W., Wong, K. L., Takei, S., Bender, S. A., Chang, L.-T., Jiang, Y., Lang, M., Tang, J., Wang, Y., Tserkovnyak, Y., Amiri, P. K., and Wang, K. L. (2014). Switching of perpendicular magnetization by spin-orbit torques in the absence of external magnetic fields, *Nat. Nanotechnol.* **9**, 548–554, doi:10.1038/nnano.2014.94.

Yuasa, S., and Djayaprawira, D. D. (2007). Giant tunnel magnetoresistance in magnetic tunnel junctions with a crystalline MgO (0 0 1) barrier, *J. Phys. D: Appl. Phys.* **40**, 21, p. R337, http://stacks.iop.org/0022-3727/40/i=21/a=R01.

Yuasa, S., Katayama, T., Nagahama, T., Fukushima, A., Kubota, H., Suzuki, Y. and Ando, K. (2005). Giant tunneling magnetoresistance in fully epitaxial body-centered-cubic Co/MgO/Fe magnetic tunnel junctions, *Appl. Phys. Lett.* **87**, 22, p. 222508, doi:DOI:10.1063/1.2138355, http://dx.doi.org/doi/10.1063/1.2138355.

Yuasa, S., Nagahama, T., Fukushima, A., Suzuki, Y., and Ando, K. (2004). Giant room-temperature magnetoresistance in single-crystal Fe/MgO/Fe magnetic tunnel junctions, *Nat. Mater.* **3**, pp. 868–871, http://www.nature.com/nmat/journal/v3/n12/full/nmat1257.html.

Zahid Hossain, M., Rumyantsev, S., Shur, M. S., and Balandin, A. A. (2013). Reduction of 1/f noise in graphene after electron-beam irradiation, *Appl. Phys. Lett.* **102**, 15, p. 153512, doi:http://dx.doi.org/10.1063/1.4802759, http://scitation.aip.org/content/aip/journal/apl/102/15/10.1063/1.4802759.

Zeng, Z., Finocchio, G., and Jiang, H. (2013). Spin transfer nano-oscillators, *Nanoscale* **5**, pp. 2219–2231, http://pubs.rsc.org/en/content/articlelanding/2012/nr/c2nr33407k#!divAbstract.

Zermatten, P.-J., Gaudin, G., Maris, G., Miron, M., Schuhl, A., Tiusan, C., Greullet, F., and Hehn, M. (2008). Experimental evidence of interface resonance states in single-crystal magnetic tunnel junctions, *Phys. Rev. B* **78**, p. 033301, http://link.aps.org/doi/10.1103/PhysRevB.78.033301.

Zhan, F., Denisov, S., and Hänggi, P. (2011). Electronic heat transport across a molecular wire: Power spectrum of heat fluctuations, *Phys. Rev. B* **84**, p. 195117, doi:10.1103/PhysRevB.84.195117, http://link.aps.org/doi/10.1103/PhysRevB.84.195117.

Zhang, Y., Tan, Y.-W., Stormer, H. L., and Kim, P. (2005). Experimental observation of the quantum Hall effect and Berry's phase in graphene, *Nature* **438**, doi:10.1038/nature04235.

Zhong, B., Chen, Y., Garzon, S., Crawford, T. M., and Webb, R. A. (2011). Low frequency noise characteristics of submicron magnetic tunnel junctions, *J. Appl. Phys.* **109**, 7, 07.

Zhou, Y., Roesler, A., and Zhu, J.-G. (2002). Experimental observations of thermally excited ferromagnetic resonance and mag-noise spectra in spin valve heads, *J. Appl. Phys.* **91**, 10, pp. 7276–7278, http://scitation.aip.org/content/aip/journal/jap/91/10/10.1063/1.1452676.

Zhu, J.-G. (2002). Thermal magnetic noise and spectra in spin valve heads, *J. Appl. Phys.* **91**, 10, 7273.

Zhu, J.-G. (2005). Pinholes and spin transfer effect in magnetic tunnel junction heads, *J. Appl. Phys.* **97**, 10, p. 10.

Zhu, J.-G., and Zhu, X. (2004). Spin transfer induced noise in CPP read heads, *IEEE Trans. Magn.* **40**, 1, pp. 182–188.

Zibouche, N., Kuc, A., Musfeldt, J., and Heine, T. (2014). Transition-metal dichalcogenides for spintronic applications, *Annalen der Physik* **526**, 9–10, pp. 395–401, doi:10.1002/andp.201400137, http://dx.doi.org/10.1002/andp.201400137.

Ziel, A. V. D. (1950). On the noise spectra of semi-conductor noise and of flicker effect, *Physica* **16**, pp. 359–372.

Žutić, I., Fabian, J., and Das Sarma, S. (2004). Spintronics: Fundamentals and applications, *Rev. Mod. Phys.* **76**, pp. 323–410, http://link.aps.org/doi/10.1103/RevModPhys.76.323.

Index

For Product Safety Concerns and Information please contact our EU representative GPSR@taylorandfrancis.com
Taylor & Francis Verlag GmbH, Kaufingerstraße 24, 80331 München, Germany

www.ingramcontent.com/pod-product-compliance
Ingram Content Group UK Ltd.
Pitfield, Milton Keynes, MK11 3LW, UK
UKHW022159190926
475702UK00003B/110
* 9 7 8 9 8 1 4 7 7 4 3 4 5 *